Applications of Mössbauer Spectroscopy

Volume II

Contributors

D. P. E. Dickson
James A. Dumesic
Paul A. Flinn
Ulrich Gonser
G. P. Huffman
C. E. Johnson
G. Lang
Steen Mørup
Moshe Ron
K. Spartalian
Henrik Topsøe

APPLICATIONS OF MÖSSBAUER SPECTROSCOPY

Volume II

Edited by

Richard L. Cohen

Bell Telephone Laboratories, Incorporated
Murray Hill, New Jersey

 1980

ACADEMIC PRESS

A Subsidiary of Harcourt Brace Jovanovich, Publishers

New York   London   Toronto   Sydney   San Francisco

6461-2284 ✓

CHEMISTRY

ACADEMIC PRESS, INC.
111 Fifth Avenue, New York, New York 10003

United Kingdom Edition published by
ACADEMIC PRESS, INC. (LONDON) LTD.
24/28 Oval Road, London NW1 7DX

Library of Congress Cataloging in Publication Data
Main entry under title:

Applications of Mössbauer spectroscopy.

Includes bibliographies and indexes.
1. Mössbauer spectroscopy. I. Cohen, Richard
Lewis, Date.
QC491.A66   537.5'352   75–26349
ISBN 0–12–178401–0 (v. 1)
ISBN 0–12–178402–9 (v. 2)

PRINTED IN THE UNITED STATES OF AMERICA

80 81 82 83   9 8 7 6 5 4 3 2 1

Contents

COLLOID AND INTERFACE CHEMISTRY

1 Magnetic Microcrystals
Steen Mørup, James A. Dumesic, and Henrik Topsøe

2 Catalysis and Surface Science
Henrik Topsøe, James A. Dumesic, and Steen Mørup

3 Mössbauer Studies of Surface-Treated Steels
G. P. Huffman

BIOLOGICAL STUDIES

METALLURGY

List of Contributors

Numbers in parentheses indicate the pages on which the authors' contributions begin.

D. P. E. Dickson (209), Department of Physics, University of Liverpool, Liverpool, England

James A. Dumesic (1, 55), Department of Chemical Engineering, University of Wisconsin, Madison, Wisconsin 53706

Paul A. Flinn (393), Intel Corporation, Santa Clara, California 95051

Ulrich Gonser (281), Fachbereich Angewandte Physik, Universität des Saarlaandes, 6600 Saarbrücken, West Germany

G. P. Huffman (189), United States Steel Corporation, Research Laboratory, Monroeville, Pennsylvania 15146

C. E. Johnson (209), Department of Physics, University of Liverpool, Liverpool, England

G. Lang (249), Department of Physics, The Pennsylvania State University, University Park, Pennsylvania 16802

Steen Mørup (1, 55), Laboratory of Applied Physics II, Technical University of Denmark, DK-2800 Lyngby, Denmark

Moshe Ron (281, 329), Department of Materials Engineering, Technion, Haifa, Israel

K. Spartalian[†] (249), Department of Physics, The Pennsylvania State University, University Park, Pennsylvania 16802

Henrik Topsøe (1, 55), Haldor Topsøe Research Laboratories, DK-2800 Lyngby, Denmark

[†] Present address: Department of Physics, The University of Vermont, Burlington, Vermont 05405.

Preface

In the four years since the publication of the first volume, Mössbauer spectroscopy has been increasingly used for applied research. The range of problems being studied, and the detail achieved in these studies, have increased substantially. The reviews in this book reflect these changes.

We continue to emphasize the results obtained from the Mössbauer spectroscopy investigations, rather than the investigatory techniques and detailed interpretations of the data. Our goal is to make the results understandable and useful to scientists and engineers who are not Mössbauer spectroscopists.

To a great extent, this volume completes the surveys of the major areas of applied studies that use Mössbauer spectroscopy. Inevitably, many of the reviews bring out still-unsolved problems and propose further work in the same areas. Even as this book is going to press, however, it is clear that over the next few years there will be considerable expansion of Mössbauer spectroscopy work in new fields—studies of coal and electrochemistry, for example, directly related to energy-utilization problems. It is clear that the range of topics studied using Mössbauer spectroscopy, and the fraction of the research effort devoted to applied topics, will continue to increase.

Contents of Volume I

COLLOID AND
INTERFACE CHEMISTRY

1

Magnetic Microcrystals

Steen Mørup

Laboratory of Applied Physics II
Technical University of Denmark
Lyngby, Denmark

James A. Dumesic

Department of Chemical Engineering
University of Wisconsin
Madison, Wisconsin

Henrik Topsøe

Haldor Topsøe Research Laboratories
Lyngby, Denmark

I. Introduction

Mössbauer spectroscopic studies of composite materials have in many cases revealed the presence of magnetic microcrystals. This is due to

1

the existence of magnetic fluctuation effects and the great sensitivity of Mössbauer spectroscopy to these effects. An understanding of these phenomena as well as their influence on Mössbauer spectra is in many cases essential in applications of Mössbauer spectroscopy in materials science.

In addition to its sensitivity to relaxation effects, Mössbauer spectroscopy has several other advantages in studies of microcrystalline samples, and it has therefore become an important technique for the study of magnetic microcrystals. The applicability of Mössbauer spectroscopy to *in situ* studies is particularly important because small particles have a large fraction of their atoms located at the surface and consequently their properties may depend on the environment of the surface atoms.

In studies of composite materials such as lavas, clays, ceramics, alloys, etc., the interpretation of the results is facilitated by the fact that only one element (isotope) contributes to the spectrum, and the information obtained is not an average of all atoms, but rather consists of separable contributions from Mössbauer ions in different surroundings.

^{57}Fe has been the most commonly used Mössbauer isotope. This is due to the ease with which it can be employed, as well as to the large number of important magnetic materials containing iron. Magnetic compounds not containing a Mössbauer isotope may be studied by introducing, for example, ^{57}Fe or ^{57}Co as a substitutional impurity in a sufficiently small concentration so that the properties of the material are not altered.

In this chapter we shall give a review of Mössbauer studies of magnetic microcrystals, emphasizing the recent developments. Earlier reviews have been given by Collins et al. (1967), Schroeer (1970), and Dumesic and Topsøe (1977). The last authors lay a special emphasis on discussing the applications in catalysis.

Section II deals with the theory of magnetic anisotropy and magnetic fluctuations (i.e., superparamagnetic relaxation and collective magnetic excitations) in microcrystals and the influence of the fluctuation effects on Mössbauer spectra. Moreover, the influence of the demagnetizing field on the magnetic splitting in Mössbauer spectra of single-domain particles is discussed. This section forms a basis for the discussion of the applications. This "theory" section may be omitted when first reading this chapter. However, to fully utilize the potential of the technique this section is essential.

In Section III various observations of superparamagnetism are discussed with examples from different fields such as ceramics, geology, lunar mineralogy, corrosion, catalysis, biology, and alloys.

Experimental investigations of collective magnetic excitations are given in Section IV. It is shown how information on particle size distributions can be obtained from the particle size dependence of the observed hyperfine fields.

Section V provides examples illustrating how magnetic moments of particles can be estimated from magnetic field dependence of Mössbauer spectra.

The possibility of obtaining quantitative information about the magnetic anisotropies is discussed in Section VI. Since the anisotropy energy constant is related to properties such as the structure, shape, stress state, and surface structure of the particles, such information is most useful for the characterization of microcrystals.

The magnetic order in microcrystals has in several cases been found to be different from that in macroscopic crystals. This is the subject of Sections VII and VIII, with the last section concentrating on surface phenomena.

Finally, in Section IX, some examples of observations of micromagnetic behavior in macrocrystals are presented. Such situations may arise, for example, in ferrites and alloys, where small magnetic clusters are formed as a result of magnetic dilution.

II. Theory

A. Magnetic Anisotropy

1. General Remarks

For small particles of a magnetically ordered material, a critical size exists below which a particle consists of a *single domain* in a zero applied field. This is due to the fact that in a particle smaller than the critical domain size, the decrease in magnetic energy obtained by splitting the single domain into smaller domains would be less than the energy increase created by the formation of *domain walls*. For typical magnetic materials this critical domain size is of the order of several tens of nanometers. In this chapter we shall discuss mainly the magnetic properties of particles smaller than 10 nm, and these particles will therefore be assumed to be single-domain particles.

Within each single-domain particle, the *exchange interaction* leads to a coupling between spins of neighboring ions, which may result in ferromagnetism, ferrimagnetism, or antiferromagnetism. In all three cases, the relative orientation of the atomic spins is specified, but the entire spin system may be rotated in space without changing the exchange energy, if the rotation is done coherently (that is, the rotation does not change the relative orientation between spins). However, other interactions lead to a dependence of the total magnetic energy on the direction of the exhange-coupled spin system. The so-called *easy directions*, defined as the low energy directions of the spin system, are separated by *anisotropy energy barriers.* The origin and the magnitude of these anisotropic magnetic interactions are briefly discussed in this section. More comprehensive reviews of these interactions can be found elsewhere [e.g., Bozorth (1951), Morrish (1965), Herpin (1968), Jacobs and Bean (1963), Kneller (1969), and Cullity (1972)].

2. Magnetocrystalline Anisotropy

Although the *spin–orbit coupling* is normally weak compared to the exchange interaction, it tends to tie the electron spin to the orbital electronic state. The latter is, in turn, strongly coupled to the crystal structure (lattice) of the solid. This is the origin of *magnetocrystalline anisotropy*.

Because the magnetocrystalline anisotropy is specified by the crystal lattice, symmetry considerations can be used to deduce the form of the expression representing this anisotropy. Thus for cubic crystals the magnetocrystalline energy per unit volume can be written as

$$E_{xtal}^{cubic} = K_0 + K_1(\alpha_1^2\alpha_2^2 + \alpha_2^2\alpha_3^2 + \alpha_3^2\alpha_1^2) + K_2(\alpha_1^2\alpha_2^2\alpha_3^2) + \cdots, \quad (1)$$

where K_0, K_1, and K_2 are the magnetocrystalline anisotropy energy constants, which depend on the material and the temperature, and α_i are the cosines of the angles between the spin direction and the three crystal axes. For example, if K_2 is small, the (100) direction is an easy direction for positive K_1 (this is the case for metallic iron), while the (111) direction is easy for negative K_1. In general, the easy directions and the anisotropy energy barriers separating them can be easily calculated when the values of K_1 and K_2 are known.

For hexagonal crystals (such as cobalt), the magnetocrystalline energy is specified by (i) the angle θ between the spin direction and the c axis and by (ii) the azimuthal angle Φ of the spin projection on the basal plane. If the anisotropy in the latter direction is negligible compared to that of the former, the symmetry is approximately *uniaxial*, and the magnetocrystalline energy per unit volume can be written as

$$E_{xtal}^{hex} = K_0' + K_1'\sin^2\theta + K_2'\sin^4\theta + \cdots. \quad (2)$$

Again, from the values of K_1' and K_2', the easy directions can be determined and the anisotropy energy barriers calculated.

3. Shape Anisotropy

The origin of shape anisotropy is the *magnetostatic energy*. In a ferro- or ferrimagnetic particle, opposite magnetic poles are present at opposite ends of the spin direction. The magnetic field lines that connect these poles then create a *demagnetizing field* inside the system, and this is responsible for the magnetostatic energy. When the spin system rotates, the two poles sweep over the surface of the system. Therefore, the shape of the system determines the magnitude of the magnetostatic energy as a function of the orientation of the spin direction. The magnetic anisotropy thus produced is called *shape anisotropy*.

A shape of particular interest for the calculation of shape anisotropy is the prolate ellipsoid, with semimajor axis c and semiminor axis a, for which the angle-dependent part of the magnetostatic energy per unit volume is given by

$$E_{\text{shape}}^{\text{ellipsoid}} = \tfrac{1}{2} M_s^2 (N_a - N_c) \sin^2 \theta, \tag{3}$$

where M_s is the magnetization (magnetic moment per unit volume), N_a and N_c are the *demagnetizing coefficients* along the a and c axes, respectively, and θ is the angle between the spin direction (magnetization) and the c axis. In general, (for $c > a$) the semimajor axis is the easy magnetic direction. As the ellipsoid approaches a sphere, $N_a - N_c$ approaches zero, and as the ellipsoid approaches a long cylinder (c much longer than a), $N_a - N_c$ approaches a value of 2π. Even for relatively modest values of c/a, the value of $N_a - N_c$ becomes significant. For example, the value of $N_a - N_c$ is 3.01, when c/a equals 2, and when c/a equals 5, the value of $N_a - N_c$ is 5.23.

4. Stress Anisotropy

Magnetostriction is the normalized change in length (strain) in a given direction that a sample experiences when it is magnetized in that direction. Because strain is also related mechanically to any stress that may be acting on the sample, the spin direction is indirectly coupled to the stress state of the system. This is the origin of *stress anisotropy*. It is again the *spin–orbit coupling* that is primarily responsible for magnetostriction. *Magnetostriction constants* λ_{hkl} are defined for various crystal directions. For a cubic crystal with $\lambda_{100} \simeq \lambda_{111} = \lambda$, the magnetoelastic energy per unit volume may be written

$$E_{\text{elastic}}^{\text{isotropic}} = -\tfrac{3}{2} \lambda \sigma \cos^2 \theta, \tag{4}$$

where σ is the stress (force per unit area) and θ is the angle between the spin and stress directions. If λ is positive (metallic iron for example), then the easy magnetic direction will be along a direction of tensile stress (positive σ) or perpendicular to a direction of compressive stress (negative σ). The reverse case holds when λ is negative (metallic nickel, for example). Stress anisotropy induced by surface tension will be discussed in Section II, A, 7.

5. Interaction Anisotropy

If the magnetic system is ferromagnetic or ferrimagnetic, then the system as a whole has an associated net magnetic moment μ. Furthermore, if two such systems are in close proximity, then the net magnetic moments of the two systems interact. This interaction leads to an additional anisotropic magnetic energy (*interaction anisotropy*), with the easy direction determined by the relative positions of the two coupled systems.

As a simple model for interaction anisotropy, imagine two identical spherical magnetic systems at a relative separation (center to center) of distance r. Let the net magnetic moment μ of the first sphere form an angle θ with the vector $\mathbf{r}$, connecting the centers of the two spheres. The magnetostatic coupling between the two spheres then tends to align the magnetic moment of the second sphere at an angle of approximately $-\theta$ with respect to $\mathbf{r}$. The resulting interaction leads to a θ dependence of the total magnetostatic energy. This can be written as

$$E_{\text{interaction}} = - (\mu^2/r^3)(1 + \cos^2 \theta). \tag{5}$$

(Note that this is an expression for the total magnetostatic interaction energy, in contrast to the previous expressions that give the energy per unit volume of a single particle.) An easy magnetic direction for interaction anisotropy is according to Eq. (5) formed along $\mathbf{r}$.

For an assembly of interacting magnetic particles, the total interaction energy is given by

$$E_{\text{interaction}} = \sum_{i>j} \left[\frac{\boldsymbol{\mu}_i \cdot \boldsymbol{\mu}_j}{r_{ij}^3} - \frac{3(\boldsymbol{\mu}_i \cdot \mathbf{r}_{ij})(\boldsymbol{\mu}_j \cdot \mathbf{r}_{ij})}{r_{ij}^5} \right], \tag{6}$$

where $\mathbf{r}_{ij}$ is the vector connecting the two particles with dipole moments $\boldsymbol{\mu}_i$ and $\boldsymbol{\mu}_j$. In this case the situation is very complicated, and the energy will depend on the geometric arrangement of the particles.

6. Exchange Anisotropy

Exchange anisotropy, like interaction anisotropy, is produced by the coupling between two magnetic systems. While interaction anisotropy is due to magnetic dipole interaction, exchange anisotropy arises from an exchange interaction at the interface between two systems. The exchange interaction is short-ranged, so the two systems must be in close contact for this anisotropy to be possible. For this reason, systems for which exchange anisotropy is of particular importance are epitaxial oxide coatings on metals and inhomogeneous alloy phases (Jacobs and Bean, 1963; Meiklejohn, 1962). In the first case the exchange anisotropy is between two different phases that are in close crystallographic contact at an interface, while in the second case the exchange interaction is between regions of different local composition within the single inhomogeneous phase.

If one of the magnetic systems has a weaker magnetic anisotropy than the other, the exchange interaction between the systems introduces a unidirectional anisotropy in the former. This is because the exchange interaction between the two systems tends to couple the spins in the first system to the spins in the second. The anisotropy energy per unit volume of the first system is then given by an expression of the form

$$E_{\text{exchange}} = -K_{\text{ex}} \cos \theta, \tag{7}$$

where K_{ex} is defined as the exchange anisotropy constant per unit volume of the first system and θ is the angle between the spin direction and the single easy direction.

Due to the interfacial nature of exchange anisotropy, the atomic structure of the interface can be of critical importance to the ultimate coupling. For example, in the case of a ferromagnet interacting with an antiferromagnet (e.g., an oxide layer on a metal), the exchange between phases depends on the number of uncompensated spins at the surface of the antiferromagnet. Consider Co–CoO, in particular. The (100) plane of antiferromagnetic CoO has an equal number of spin-up and spin-down magnetic moments. This plane would not be expected to couple strongly to ferromagnetic metallic cobalt via exchange interactions. However, the (111) plane of CoO is the ferromagnetic layer plane of the material (i.e., the magnetic moments are aligned in parallel for this plane), and a strong exchange coupling with metallic cobalt at this type of interface is expected. These expectations have been confirmed experimentally (Jacobs and Bean, 1963).

7. Surface Anisotropy

The lower symmetry of a surface atom compared to that of an atom within the solid leads to another form of magnetic anisotropy. The relative magnitude of this anisotropy compared, for example, to magnetocrystalline and shape anisotropies, increases as the size of the system decreases due to the corresponding increase in the surface to volume ratio. These surface anisotropy effects are expected to be significant when a characteristic dimension of the system is less than approximately 10 nm (Néel, 1954).

A phenomenological theory of *surface anisotropy*, in which the magnetic energy for the system is written as a summation of pairwise interactions, was proposed by Néel (1953, 1954). This pairwise interaction between two atoms is a function of the distance between the atoms r and the angle θ, which the magnetic moments μ form with a vector connecting the two atoms. The pairwise interaction energy $W(r, \theta)$ is expanded in a series of

Legendre polynomials as

$$W(r, \theta) = g_1(r)P_2(\cos\theta) + g_2(r)P_4(\cos\theta) + \cdots, \qquad (8)$$

where $g_i(r)$ are the coefficients of the expansion and $P_{2i}(\cos\theta)$ are Legendre polynomials of an even degree. In the first term of the expansion, $P_2(\cos\theta)$ is equal to $\cos^2\theta - \frac{1}{3}$, and this is the form of a dipole–dipole interaction for parallel moments. (It should be noted that the dipole–dipole coupling for interaction anisotropy was proportional to $1 + \cos^2\theta$ since this interaction was between nonparallel moments.) Following Néel, there are two primary contributions to this term. One is the magnetostatic dipole–dipole interaction between the magnetic moments on each atom, and the other is the first term (dipolar) in the representation of the short-range interactions that lead to such phenomena as magnetocrystalline anisotropy and magnetostriction effects. This second dipolar term is short ranged, and to a first approximation one needs only to consider nearest neighbor interactions.

For a cubic crystal lattice, the summation of $g_1(r)P_2(\cos\theta)$ over all atom pairs gives zero. For a distorted lattice, however, this term does not vanish. Therefore, the coefficient $g_1(r)$ of this term can be estimated from magnetostriction data, and its value so determined was found by Néel to be significantly greater than that due to the magnetostatic dipole–dipole interaction. In addition, at a surface, the $g_1(r)P_2(\cos\theta)$ term does not average to zero when $W(r, \theta)$ is summed over all atom pairs. This is the origin of Néel's anisotropy. Since short-range interactions are the primary contribution to the $g_1(r)P_2(\cos\theta)$ term, the sum over atom pairs is calculated for nearest neighbors, and this leads to a dependence of the surface anisotropy on the atomic structure of the surface. For example, a (100) surface of a body-centered-cubic crystal (e.g., iron) has zero surface anisotropy, while the (100) surface of a face-centered cubic-crystal (e.g., nickel) does generate surface anisotropy. An order of magnitude estimate for surface anisotropy is 10^{-3} J m^{-2}. In addition, the overall shape of the system must be considered in evaluating the contribution from the surface to the total magnetic anisotropy. For example, a face-centered-cubic system having a cubic shape with (100) planes exposed at the surface has zero total surface anisotropy because the intrinsic surface anisotropies of each of the six (100) planes cancel. Although details of Néel's phenomenological theory of surface anisotropy have been criticized (Pincus, 1960; Corciovei et al., 1972), the basic origin of surface anisotropy follows from rather general symmetry arguments.

Surface anisotropy may well be dependent on the presence of adsorbed species on the surface. This adsorption-dependent interaction may facilitate detection of the presence of surface anisotropy in magnetic systems of small size.

A special type of anisotropy originating from the presence of the surface was proposed by Janssen (1970) who considered the surface tension induced stress anisotropy. The anisotropy energy will be given by Eq. (4) with σ now being proportional to the surface tension divided by the characteristic dimension of the system. Thus this type of anisotropy will also be most important for small particles. Furthermore, it may depend on the structure of the surface and on adsorption of gases.

8. Induced Magnetic Anisotropy

Diffusion in solids is slow at ordinary temperatures, and, therefore, the magnetic anisotropy may show marked *hysteresis*. That is, the anisotropy is sensitive to previous treatments of the system, and this sensitivity is termed *induced magnetic anisotropy*.

One such treatment that affects the anisotropy is *magnetic annealing*, defined as a heat treatment in the presence of an external magnetic field. This effect is particularly important for magnetic materials containing more than one element (e.g., alloys, mixed ferrites). The heat treatment serves to increase the solid state diffusion rate, and the external magnetic field defines atomic configurations with low energy. For example, in a substitutional alloy, pairs of like atoms may tend to align with the applied field, and for an interstitial alloy certain interstices increase in size at the expense of others depending on the direction of the magnetic field (due to magnetostriction). In this way, magnetic annealing tends to create new easy magnetic directions; on cooling the system and removing the external field, an induced magnetic anisotropy remains.

Similarly, magnetic anisotropy can be induced by heat treatment with the system under stress and plastic deformation. In the former case, the stress at high temperatures induces movement of atoms to low energy positions (e.g., in an interstitial alloy, the interstices that grow with application of the stress will be preferentially occupied), and this atomic distribution is frozen in place by rapid cooling. The anisotropic atomic distribution then leads to the induced magnetic anisotropy. With plastic deformation, atomic movement is achieved at low temperatures. Once again, the anisotropic atomic configuration (e.g., slip planes, residual stresses) leads to induced magnetic anisotropy.

9. Magnitudes of Anisotropy Energy Constants

The overall observed anisotropy energy constant may, in general, have contributions from several effects. Furthermore, the relative magnitudes of the different anisotropies, may well change as the particle size is decreased. In Fig. 1 some typical values of anisotropy energy constants are summarized. For clarity, a brief description is necessary.

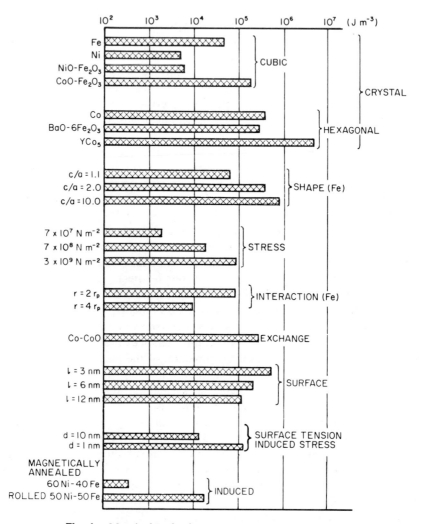

Fig. 1. Magnitudes of anisotropy energy constants (see text).

The entries under magnetocrystalline anisotropy are values for $|K_1|$ or $|K_1'|$, depending on whether the material is cubic or hexagonal, respectively. For these materials $|K_1| > |K_2|$ and $|K_1'| > |K_2'|$.

For shape anisotropy, $\frac{1}{2} M_s^2 (N_a - N_c)$ determines the magnitude of the anisotropy, and this value is plotted for different c/a ratios. The magnetization used in the calculation was 1724 G, a value typical of metallic iron.

The magnitude for stress anisotropy is determined by $\frac{3}{2} \lambda \sigma$. Using a value of 2×10^{-5} for λ (typical of many materials), the entries in the figure for various stresses σ follow immediately.

The total energy barrier for interaction anisotropy is μ^2/r^3. But $\mu = M_sV$, where V is the volume of each spherical particle. Thus the energy barrier per volume of magnetic material is $(\mu^2/r^3)/(2V)$, which is equal to $M_s^2V/2r^3$. If the spheres touch, then $r = 2r_p$, where r_p is the radius of each sphere. In the figure, the interaction anisotropy per unit volume ($M_s^2V/2r^3$) is given for two values of r, using the magnetization for metallic iron (1724 G).

The value for exchange anisotropy was obtained experimentally for the system Co–CoO. The measured anisotropy is for the metallic cobalt phase with particle size (diameter) equal to 20 nm (Meiklejohn and Bean, 1957).

The estimate for surface anisotropy comes from Néel's value of 10^{-3} J m^{-2}. The conversion of this value to a volume normalized anisotropy constant involves the surface to volume ratio S/V and the shape of the system. For the figure, the system was chosen to be a film of thickness l, for which $S/V \simeq 2/l$. Therefore, the volume normalized anisotropy constant is 2×10^{-3} J m$^{-2}/l$, where l is the unit of m.

The surface tension induced stress anisotropy was calculated for Ni, and σ was taken to be the surface tension (~ 2.5 N m^{-1}) divided by the particle size.

The two examples of induced magnetic anisotropy are for the substitutional alloy Ni–Fe (Chikazumi and Oomura, 1955; Rathenau and Snoek, 1941). Both of these cases have been interpreted as orientation (directional order) of pairs of like atoms during the magnetic annealing or during the cold rolling (plastic deformation).

B. Fluctuations of the Magnetization Direction in Microcrystals

In the absence of an applied magnetic field, the magnetization direction of a large magnetically ordered crystal is along an easy direction. However, since the anisotropy energy decreases when the particle size decreases, the thermal energy may become comparable to the anisotropy energy of a small particle ($\lesssim 10$ nm), even at temperatures below room temperature. The magnetization is then no longer fixed to an easy direction, but fluctuates in a random way. For simplicity these fluctuations can be divided into two categories. Small fluctuations around an easy direction are called *collective magnetic excitations*, whereas one talks about *superparamagnetic relaxation* when the magnetization direction fluctuates among the various easy directions. When the superparamagnetic relaxation time is short compared with the time scale of the experimental method used for the study of the magnetic properties, the sample resembles a paramagnet.

The *blocking temperature* T_B is defined as the temperature below which the particles behave as magnetically ordered crystals. The blocking temperature is therefore not uniquely defined, but is related to the time scale of the experimental technique used for the studies of the magnetic properties of the particle.

Superparamagnetic relaxation can take place via different modes of spin excitations. The most important of these are (i) spin rotation in unison (coherent rotation), (ii) buckling, and (iii) curling. A concise discussion of these processes has been given by Kneller (1969). The modes are schematically pictured in Fig. 2, but a brief description may be appropriate. The process of buckling involves a periodic fluctuation of the spin direction along the easy direction. In any plane perpendicular to the easy direction, the spins are approximately aligned, but the overall direction of the aligned spins varies along the easy direction. The curling mode can be described as a twist of the spins around an easy direction. The spin direction now depends on the radial distance from an axis parallel to the easy direction, but it is independent of the position along the easy direction.

For simplicity, consider spin rotation in a long cylindrical system with the easy direction being the cylinder axis. For the first mode of rotation (rotation in unison), the exchange energy remains constant, while the magnetostatic energy gives rise to magnetic anisotropy. The two other modes of spin rotation are incoherent processes (the atomic spins do not all remain internally aligned) and, therefore, the exchange energy does not remain constant during rotation. Yet these modes of rotation exist because the magnetostatic energy is less than that of rotation in unison. In the buckling mode alternating north and south poles are formed along the surface of the cylinder, and the curling mode takes place without the formation of magnetic poles on the cylinder surfaces. The order of decreasing magnetostatic energy for the three modes of rotation is rotation in unison, buckling, followed by curling.

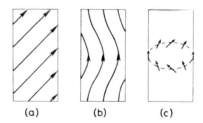

(a)   (b)   (c)

Fig. 2.   Modes of spin rotation: (a) rotation in unison, (b) buckling, (c) curling (Kneller, 1969).

The key to the qualitative understanding of the size dependence of the rotation process is that the magnetic exchange forces are short ranged, while the magnetostatic forces are long ranged. Thus for systems larger than a certain critical size, the spins will rotate incoherently (usually by curling rather than buckling), while for systems smaller than the critical size, the spins will remain uniformly aligned during the rotation. The actual value of the critical size depends on the shape of the system and on the inclusion of other magnetic interaction terms besides exchange and magnetostatic. For metallic iron, the critical cylinder diameter below which spin rotation takes place coherently is approximately 15 nm. In general, this estimate is typical of magnetic systems with other shapes (Cullity, 1972).

Most of the systems in which superparamagnetic relaxation has been studied by use of Mössbauer spectroscopy are less than around 15 nm. In these cases the rotation of the spins can normally be considered to take place coherently. Therefore, the theoretical model for rotation in unison, which is briefly outlined in what follows, can be used in most cases of interest in connection with Mössbauer spectroscopy.

It was seen in Section II, A that the magnetic anisotropy can often be considered uniaxial with the total magnetic anisotropy energy given by the expression

$$E(\theta) = KV \sin^2 \theta, \qquad (9)$$

where K is the *anisotropy energy constant*, V the volume of the particle, and θ the angle between the magnetization direction and the easy direction of magnetization. According to the above equation, two energy minima at $\theta = 0$ and $\theta = \pi$ are separated by an energy barrier equal to KV.

In the presence of an external magnetic field H along the symmetry axis, the energy of a ferromagnetic particle is given by

$$E(\theta) = KV \sin^2 \theta - HM_s V \cos \theta. \qquad (10)$$

For $H < 2K/M_s$ there are two minima at $\theta = 0$ and $\theta = \pi$, but for larger applied fields the minimum at $\theta = \pi$ disappears. Figure 3 shows $E(\theta)$ for various values of HM_s/K.

The probability that the magnetization vector of a small particle forms an angle between θ and $\theta + d\theta$ with an easy direction is given by

$$f(\theta) \, d\theta = \frac{\exp[-E(\theta)/kT] \sin \theta \, d\theta}{\int_0^\pi \exp[-E(\theta)/kT] \sin \theta \, d\theta}. \qquad (11)$$

Here k is Boltzmann's constant and T the temperature. Equation (11) is easily generalized to forms of the magnetic energy other than the form

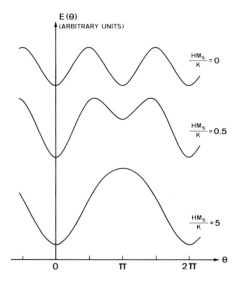

Fig. 3. Magnetic energy as given by Eq. (10) for various values of HM_s/K.

given by Eq. (10) (e.g., cubic anisotropy or other directions of the applied magnetic field).

The magnetic fluctuations may be described as a "random walk" of the magnetization vector in all directions with the probability of finding the magnetization in a given direction given by Eq. (11). When KV and/or HM_sV are very large compared to kT, $f(\theta) \approx 0$ except at the energy minima. The magnetization can then be considered fixed in one of these directions. This is the situation in large magnetically ordered crystals. Smaller values of KV/kT and/or HM_sV/kT lead to a broadening of $f(\theta)$ near the minima, i.e., the magnetization vector may fluctuate around the directions corresponding to the minima (collective magnetic excitations). Finally, for $KV/kT \lesssim 1$ and $H < 2K/M_s$ the magnetization has a significant probability of surmounting the energy barrier separating the minima (superparamagnetic relaxation).

The superparamagnetic relaxation frequency (τ^{-1}) is defined as the rate at which the magnetization vector passes the energy barrier. The first calculation of the superparamagnetic relaxation time τ of a ferromagnetic microcrystal was made by Néel (1949) who (for $H = 0$) found an expression of the form

$$\tau = \tau_0 \exp(KV/kT), \tag{12}$$

where τ_0 is of the order of 10^{-9}–10^{-11} s.

The theory of superparamagnetic relaxation was later refined by Brown (1959, 1963a, 1963b). He considered a small ferromagnetic crystal with the energy given by Eq. (10). The relaxation time was given in terms of the smallest nonvanishing eigenvalue of a Fokker–Planck type differential equation. In the presence of an applied field, the transition probabilities for transitions in the two directions are different because of the difference in the probability for the magnetization to be in each of the two energy wells at $\theta = 0$ and $\theta = \pi$. In the high energy barrier approximation $(KV/kT \gg 1)$, the two relaxation times $\tau_{1 \to 2}$ and $\tau_{2 \to 1}$ for relaxation between the lower minimum 1 and the higher minimum 2 are approximately given by

$$\left.\begin{array}{c} \tau_{1 \to 2} \\ \tau_{2 \to 1} \end{array}\right\} = \frac{M_s \pi^{\frac{1}{2}}}{K \gamma_0} \alpha^{-\frac{1}{2}} (1 - h^2)^{-1} (1 \pm h)^{-1} \exp\{\alpha(1 \pm h)^2\}, \quad (13)$$

where $h = HM_s/2K$, $\alpha = KV/kT$, and γ_0 is the gyromagnetic ratio. The probabilities of finding the magnetization vector in minima 1 and 2 are given by $p_1 = \tau_{2 \to 1}/(\tau_{1 \to 2} + \tau_{2 \to 1})$ and $p_2 = \tau_{1 \to 2}/(\tau_{1 \to 2} + \tau_{2 \to 1})$, respectively. For $H = 0$ Eq. (13) reduces to

$$\left.\begin{array}{c} \tau_{1 \to 2} \\ \tau_{2 \to 1} \end{array}\right\} = \frac{M_s \pi^{\frac{1}{2}}}{K \gamma_0} \alpha^{-\frac{1}{2}} \exp\{\alpha\}. \quad (14)$$

Often the relaxation time is given by a single parameter τ defined by

$$\tau^{-1} = (\tau_{1 \to 2})^{-1} + (\tau_{2 \to 1})^{-1}. \quad (15)$$

Aharoni (1964, 1969) has calculated the relaxation times numerically. He found that the expressions (13) and (14) can be used without introducing significant errors for $\alpha > 1.0$.

More recently the superparamagnetic relaxation time of small particles with cubic anisotropy has also been calculated numerically (Aharoni, 1973; Aharoni and Eisenstein, 1975; Eisenstein and Aharoni, 1977a, b, c).

The relaxation time for magnetization reversal by curling in spherical particles has also been calculated (Afanas'ev et al., 1970, 1972; Eisenstein and Aharoni, 1976a, b). It has been found that curling can take place only in particles larger than a critical size. Below this size rotation in unison can take place. In a region just above the critical size, the superparamagnetic relaxation time decreases with increasing particle size due to the onset of spin reversal by curling.

As mentioned earlier, the magnetic energy given by Eq. (10) has only one minimum for $H \geqslant 2K/M_s$; hence we cannot talk about superparamagnetic relaxation as it is defined here. However, since the magnetization vector is generally not fixed in space, the fluctuation phenomena may still influence the magnetic properties considerably.

For example, the magnetization of a ferromagnetic microcrystal, averaged over a time that is long compared to the characteristic time of the fluctuations, can be derived by using Eq. (11) and is given by

$$\langle M \rangle = M_s \frac{\int_0^\pi \exp[-E(\theta)/kT]\cos\theta\sin\theta\,d\theta}{\int_0^\pi \exp[-E(\theta)/kT]\sin\theta\,d\theta}. \tag{16}$$

When the effect of the anisotropy is negligible in contrast to the influence of the applied field, Eq. (16) reduces to

$$\langle M \rangle = M_s L\left(\frac{\mu H}{kT}\right), \tag{17}$$

where $\mu = M_s V$ is the magnetic moment of the particle and

$$L\left(\frac{\mu H}{kT}\right) = \coth\left(\frac{\mu H}{kT}\right) - \frac{kT}{\mu H} \tag{18}$$

is the classic Langevin function. In the limit of low and high fields $L(\mu H/kT)$ is given by simple expressions

$$L\left(\frac{\mu H}{kT}\right) = \frac{\mu H}{3kT}, \qquad \frac{\mu H}{kT} \ll 1; \tag{19}$$

$$L\left(\frac{\mu H}{kT}\right) = 1 - \frac{kT}{\mu H}, \qquad \frac{\mu H}{kT} \gg 1. \tag{20}$$

If the magnetic anisotropy is not negligible, Eq. (17) should be replaced by a more general expression that has been derived by West (1961). Since μ is often very large (e.g., of the order of $10^4\mu_B$), magnetic saturation is generally achieved at relatively small applied magnetic fields even at room temperature.

In studies of magnetic microcrystals it is important to realize that the result of a measurement depends on the time scale of the experimental technique (the observation time) compared to the time scale of the fluctuation phenomena.

Let us consider, for example, the case where no applied field is present. According to Eq. (16), the average magnetization is zero. However, if the superparamagnetic relaxation time is long compared to the observation time (i.e., the magnetization vector remains near one of the easy directions during the observation time), a finite value of the magnetization is measured. If the correlation time of collective magnetic excitations is short compared to observation time, the measured magnetization is given by

$$M(V, T) = M_s(T)\langle\cos\theta\rangle_\mathrm{T}, \tag{21}$$

where $\langle\cos\theta\rangle_\mathrm{T}$ is the thermal average of $\cos\theta$ near one of the minima.

Setting $kT/KV = \beta$, we find (Mørup and Topsøe, 1976)

$$\langle\cos\theta\rangle_T = \frac{\int_0^{\pi/2}\exp\left[-(1/\beta)\sin^2\theta\right]\cos\theta\sin\theta\,d\theta}{\int_0^{\pi/2}\exp\left[-(1/\beta)\sin^2\theta\right]\sin\theta\,d\theta}$$

$$= \frac{\beta^{\frac{1}{2}}}{2}\frac{\exp(1/\beta)-1}{\int_0^{\beta-\frac{1}{2}}\exp(x^2)\,dx}. \tag{22}$$

In the low temperature limit ($KV/kT \gg 1$) it is found that

$$\langle\cos\theta\rangle_T \simeq 1 - kT/2KV. \tag{23}$$

The general expression for the average magnetization of a particle, with the magnetization bound to an energy minimum at $\theta = \theta_0$ with uniaxial symmetry, is in the low temperature approximation given by

$$\langle\cos(\theta - \theta_0)\rangle_T = 1 - kT/\left[\partial^2 E/\partial\theta^2\right]_{\theta=\theta_0}. \tag{24}$$

C. Mössbauer Spectra in the Presence of Fluctuations of the Magnetization

In Section II, B we discussed the fluctuation effects and their influence on the measured magnetization. In this section we shall see how information on superparamagnetic relaxation and collective magnetic excitations can be obtained using Mössbauer spectroscopy.

The time scale for observation of magnetically split Mössbauer spectra is approximately given by the Larmor precession time τ_L of the nuclear magnetic moment. Roughly speaking, we find a magnetically split Mössbauer spectrum (with six lines in the case of ^{57}Fe Mössbauer spectroscopy) when $\tau \gg \tau_L$, and a paramagnetic spectrum (with one or two lines) when $\tau \ll \tau_L$. In the intermediate range ($\tau \approx \tau_L$), complex spectra with broadened lines can be observed. In Mössbauer studies of magnetic microcrystals the blocking temperature T_B is often defined as the temperature at which τ equals the nuclear Larmor precession time of the nucleus in its excited state.

When considering superparamagnetic relaxation in a particle with uniaxial symmetry, a more rigorous definition of T_B can be made. For $H = 0$ we may assume as a first approximation that the hyperfine field in such a particle can take on the two values $+H_{hf}$ and $-H_{hf}$. Then it can be shown (Blume and Tjon, 1968) that a given pair of lines (e.g., lines n and $7 - n$ of a ^{57}Fe spectrum) collapse at a relaxation time τ_c given by

$$\tau_c = \tau_{1\to2} = \tau_{2\to1} = \hbar/\Delta E. \tag{25}$$

Here ΔE is the Zeeman shift of each of the two lines and $\tau_{1 \to 2}$ and $\tau_{2 \to 1}$ are defined in Section II, B.

In ^{57}Fe Mössbauer spectroscopy it is convenient to define T_B as the temperature where the magnetically split $\pm \frac{3}{2} \to \pm \frac{1}{2}$ lines (lines 1 and 6) disappear. According to Eq. (25), these lines collapse for $\tau_c = \hbar / |(\frac{3}{2} g_1 - \frac{1}{2} g_0)\mu_N H_{hf}|$, where g_1 and g_0 are the g factors of the nucleus in the excited and ground state, respectively, and μ_N is the nuclear magneton. For $H_{hf} = 550$ kG (which is a typical value of ferric compounds), we find $\tau_c = 1.5 \times 10^{-9}$ s. Theoretical Mössbauer spectra at various relaxation times that have been calculated by Wickmann (1966) are shown in Fig. 4.

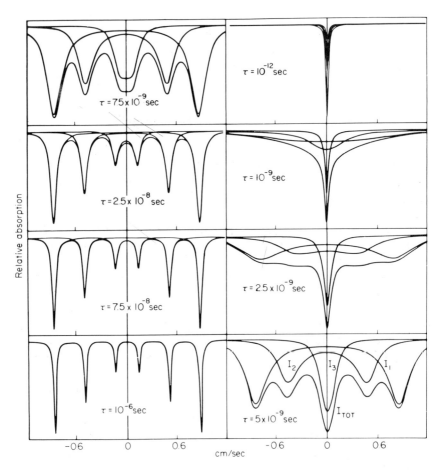

Fig. 4.   Theoretical Mössbauer spectra for various relaxation times. The hyperfine field is assumed to take on the two values $+550$ and -550 kG (Wickman, 1966).

From these spectra we see that the magnetic splitting is resolved for $\tau \gtrsim 2.5 \times 10^{-9}$ s, and this value may be used for an empirical definition of T_B.

A particle with cubic symmetry has several easy directions, and this complicates the calculation of theoretical spectra. However, the dependence of the spectral shape on τ is qualitatively similar to that found in the case of uniaxial symmetry (Dattagupta and Blume, 1974; Manykin and Onishchenko, 1976; Afanas'ev and Onishchenko, 1976).

In practice, a sample of small particles always contains a distribution of particle sizes, and the experimental spectra will then consist of a sum of spectra with different relaxation times. According to Eqs. (13) and (14), τ is very sensitive to the volume, and the distribution in relaxation times will therefore be very broad. This implies that in many cases only a small fraction of the particles have relaxation times in the critical region $(10^{-9}-10^{-8}$ s), and the spectra can therefore be described as consisting of a magnetically split component and a paramagnetic component.

Theoretical spectra in the presence of a particle size distribution have been calculated by Belozerskii and Pavlyukhin (1977).

So far we have discussed the influence of superparamagnetic relaxation on the Mössbauer spectra. As will be shown in what follows, the small fluctuations of the magnetization around an energy minimum (collective magnetic excitations) may also influence the Mössbauer spectra even far below the blocking temperature (Mørup and Topsøe, 1976).

Since the correlation time of collective magnetic excitations is much shorter than the observation time (the nuclear Larmor precession time), the magnetic splitting of the Mössbauer spectrum is proportional to the average value of the hyperfine field, i.e.,

$$H_{hf}(V, T) = H_{hf}(V = \infty, T)\langle\cos\theta\rangle_T, \qquad (26)$$

where $H_{hf}(V = \infty, T)$ is the hyperfine field in a large crystal at the same temperature (i.e., in the absence of collective magnetic excitations), and $\langle\cos\theta\rangle_T$ is given for example, by, Eqs. (22), (23), or (24).

Thus the magnetic splitting in a Mössbauer spectrum of a small magnetic particle is generally smaller than that found in a macroscopic crystal. If a sample contains a broad particle size distribution, the magnetic splitting of the spectra of particles with different volumes will be different. This gives rise to a broadening of the Mössbauer lines from which the distribution in KV may be estimated.

In many cases additional information can be obtained by application of magnetic fields. Mössbauer spectra are particularly sensitive to magnetic fields above the blocking temperature. The magnetic splitting in this case is proportional to the sum of the external field H and the induced average hyperfine field. The latter is proportional to $\langle M \rangle$, given by Eq. (16). When the anisotropy is negligible, we find

$$H_{hf}(V, T, H) = H + H_{hf}(V = \infty, T)L(\mu H/kT). \qquad (27)$$

In ^{57}Fe Mössbauer spectroscopy the direction of the induced hyperfine field is generally opposite to the applied magnetic field. In practice, H is often much smaller than the second term in Eq. (27).

The relative intensities of the Mössbauer lines depend on the direction of the magnetic field at the nucleus relative to the gamma-ray direction. If the magnetic field at the nucleus is parallel or perpendicular to the gamma-ray direction, the relative areas of the six lines of a ^{57}Fe spectrum are given by $3:0:1:1:0:3$ and $3:4:1:1:4:3$, respectively.

D. The Demagnetizing Field in Small Particles

In a ferromagnetic particle the arrangement of the atomic magnetic dipoles may give rise to uncompensated magnetic poles at the surface, and these create the so-called demagnetizing field. This field adds to the magnetic hyperfine field at the Mössbauer nucleus, and it therefore contributes to the magnetic splitting of the Mössbauer spectrum.

In multidomain particles the magnetization is split into domains of homogeneous magnetization, arranged in a way that minimizes the magnetic energy. Both at domain boundaries and at the surface of such multidomain particles, the domain magnetization is arranged in such a way that the demagnetizing field is negligible.

The contribution of the demagnetizing field to the magnetic splitting of the nuclear states may, however, have to be considered for single-domain systems. To such systems belong magnetic microcrystals when their dimensions are smaller than a critical value (≈ 20 nm) and large particles that are magnetized to saturation by application of an external magnetic field. For single-domain systems the uncompensated poles at the particle surface then give rise to a demagnetizing field $\mathbf{H_D} = -N\mathbf{M_S}$ in the interior of the particle, where N is the demagnetizing coefficient, which depends on the shape of the particle and the magnetization direction (see Section II, A, 3). For example, for a spherical particle $N = 4\pi/3$, whereas for a foil, magnetized parallel or perpendicular to the foil plane, $N = 0$ and $N = 4\pi$, respectively. Thus for a magnetized particle the magnetic splitting of a Mössbauer spectrum depends on the particle shape and the direction of magnetization (Knudsen and Mørup, 1980). The demagnetizing field is the origin of shape anisotropy discussed in Section II, A, and if the magnetization direction is mainly determined by the shape anisotropy, the demagnetizing field is minimized. However, in cases where the magnetization direction is determined by other contributions to the magnetic anisotropy energy constant (see Section II, A), greater contributions from the demagnetizing field may be expected (Knudsen and Mørup, 1980).

In metallic iron the demagnetizing field increases the magnetic splitting of the Mössbauer spectrum because H_D and the hyperfine field are both antiparallel to the magnetization. For example, the magnetic field acting at the nuclei in a spherical single-domain particle of a α-Fe should be larger than that of a multidomain particle by about 7 kG (Von Eynatten and Bömmel, 1977; Knudsen and Mørup, 1980).

It should be emphasized that in studies in which the magnetic hyperfine splitting of small ferromagnetic particles is compared to that of macroscopic crystals, it may be necessary to perform appropriate corrections for the different demagnetizing fields in single-domain particles and multidomain particles. Such corrections may be necessary, for example, in studies of collective magnetic excitations in ferromagnetic microcrystals.

III. Mössbauer Studies of Superparamagnetic Relaxation

In Section II, C we discussed the potential of Mössbauer spectroscopy to give information on magnetic fluctuations and thereby on magnetic anisotropy, and particle size. In this section we shall give a short review of Mössbauer studies in which superparamagnetic behavior has been observed. The examples will be taken from a number of different fields and will therefore illustrate the wide range of materials that contain superparamagnetic particles. In many studies only qualitative information has been obtained. In later sections we shall see that quantitative information about both particle size and anisotropy energy can be derived.

The first observation of superparamagnetic relaxation using Mössbauer spectroscopy was made by Nakamura *et al.* (1964) who studied small particles of α-Fe$_2$O$_3$ with an average diameter of approximately 5 nm. These authors observed a magnetically split spectrum at 120 K, but at 300 K (which is well below the Néel temperature of α-Fe$_2$O$_3$) the magnetic splitting was absent due to superparamagnetic relaxation. Subsequently, several other Mössbauer studies of the magnetic properties of small particles of α-Fe$_2$O$_3$ have been made. Constabaris *et al.* (1965) observed superparamagnetic iron oxide particles dispersed on various high surface area supports. Detailed investigations of various samples of small particles of α-Fe$_2$O$_3$ supported on a high area silica were made by Kündig *et al.* (1966, 1967a, b). They obtained Mössbauer spectra at various temperatures of samples with different average particle sizes. Some of the spectra for a sample with an average particle diameter of 13.5 nm are shown in Fig. 5. It

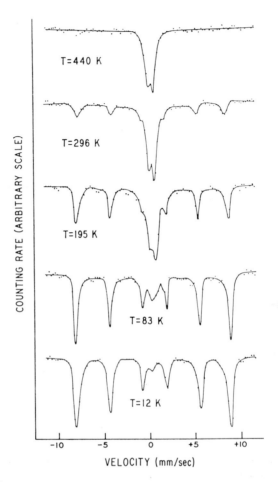

Fig. 5. Mössbauer spectra of microcrystals of α-Fe$_2$O$_3$ with an average diameter of 13.5 nm at various temperatures (Kündig *et al.*, 1966).

is shown that between 83 and 296 K the spectra contain both a Zeeman split component and a paramagnetic component. At a given temperature the area ratio of these two components reflects the relative number of particles that are below and above the blocking temperature. From the temperature dependence of the area ratio and by using a theoretical expression for the temperature dependence of the relaxation time, the anisotropy constant K was determined. The result was $K = (8.2 \pm 2.4) \times 10^3 \ \mathrm{J\,m^{-3}}$. Moreover, by assuming that K is independent of the particle size, they were able to determine the particle-size distribution from the

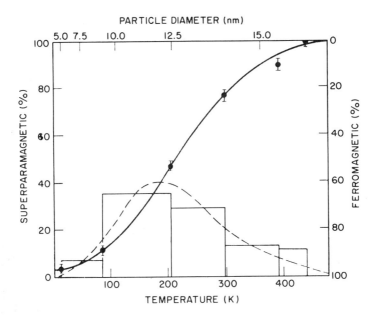

Fig. 6. Percentage of superparamagnetic fraction in the spectra shown in Fig. 5. The estimated particle-size distribution is also shown (Kündig *et al.*, 1966).

spectra shown in Fig. 5. The result is shown in Fig. 6. The value of K was also determined from the dependence of the spectra on an average volume at a constant temperature. The result was in good agreement with that found from the temperature dependence of the spectra. Van der Kraan (1972) made a similar analysis of Mössbauer spectra of 7-nm particles of $\alpha\text{-Fe}_2\text{O}_3$ and found an anisotropy energy constant of $K = 6.0 \times 10^4 \, \text{J m}^{-3}$.

Hobson and Gager (1970) studied various samples of $\alpha\text{-Fe}_2\text{O}_3$ particles supported on silica and $\gamma\text{-Al}_2\text{O}_3$. From the superparamagnetic behavior they were able to draw conclusions regarding changes in the particle-size distributions after different heat treatments of the samples.

Small particles of iron oxide ($\alpha\text{-Fe}_2\text{O}_3$) have been found in atmospheric aerosols. Kopcewicz and Kopcewicz (1976) compared the iron-containing particles from the atmosphere in Wroclaw (a town with industrial air pollution) and Kasprowy Wierck (a summit in the Tatra mountains). They found from superparamagnetic relaxation effects that the particles from the latter location had a radius of about 5.5 nm, whereas the particles collected in Wroclaw contained a considerable amount of larger particles. They suggested that the smaller particles originate from meteors and the larger ones from industrial air pollution.

In geological samples the presence of superparamagnetic particles has been discovered in several cases. Gangas *et al.* (1973) used Mössbauer spectroscopy to characterize soil from Attica. They were able to identify superparamagnetic particles of α-Fe$_2$O$_3$ and determined the particle-size distribution from Mössbauer spectra at various temperatures. Govaert *et al.* (1976) used Mössbauer spectroscopy to characterize some naturally occurring samples of α-FeOOH. Spectra of samples of different origin (Fig. 7) showed different relaxation behavior, and this reflects a difference in particle-size distributions.

In several other studies of microcrystalline α-FeOOH and α-Fe$_2$O$_3$ in soil, Mössbauer spectroscopy has been used to obtain particle-size information (Kodama *et al.*, 1977; Singh *et al.*, 1977; Hill *et al.*, 1978). Ferric oxyhydroxyde gels have been studied extensively (van der Giessen, 1967; Mathalone *et al.*, 1970; Coey and Readman, 1973a; Hanzel and Sevsek, 1976; Povitskii *et al.*, 1976; Belozerskii *et al.*, 1976; Saraswat *et al.*, 1977). Superparamagnetic behavior has been found both in the FeOOH and α-Fe$_2$O$_3$ formed on thermal decomposition.

Johnson and Glasby (1969) observed superparamagnetic behavior in iron-manganese nodules from the Indian Ocean. At 4.2 K the hyperfine splitting was consistent with that expected for a mixture of α − and γ − FeOOH. From the temperature dependence of the spectra they estimated a mean particle size of 9 nm. Helgason *et al.* (1976) observed superparamagnetic particles in a lava sample from Iceland. Superparamagnetic particles have also been found in lunar samples. Housley *et al.* (1971) observed superparamagnetic metallic iron in samples from Mare Tranquillitatis. They concluded that iron particles with a diameter smaller than about 4 nm were present in the samples. Superparamagnetic behavior has also been observed in lunar olivine (Huffman and Dunmyre, 1975).

Mössbauer studies of the magnetic properties of small iron oxide particles in ceramics have yielded some interesting results. In particular, much information on ancient pottery has been obtained, and this subject has been reviewed by Kostikas *et al.* (1974, 1976). It is of special interest that the blocking temperature estimated from Mössbauer spectra seems to depend on the age of the pottery (Danon *et al.*, 1976; Gangas *et al.*, 1976). The experiments have shown that the temperature at which 50% of the iron oxides are superparamagnetic decreases with increasing age. This phenomenon has been explained in terms of a decrease in the size of the magnetic particles due to a weathering mechanism. It has also been found that a refiring of the pottery at a typical firing temperature ($\sim$ 1300 K) causes an increase in the temperature at which 50% of the oxide is superparamagnetic.

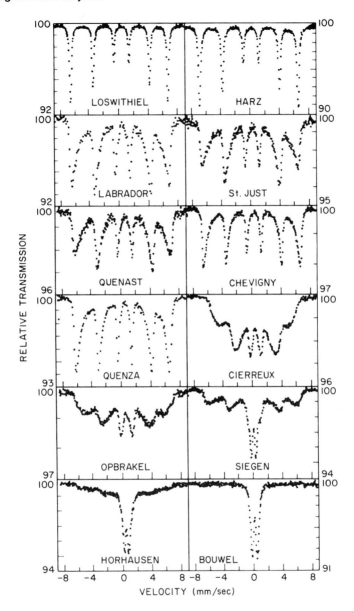

Fig. 7. Mössbauer spectra obtained at room temperature of some naturally occurring goethite (α-FeOOH) samples. The different spectral shapes are due to different particle-size distributions (Govaert *et al.*, 1976).

Pobell and Wittman (1966) studied iron-containing phases in samples of. Portland cement. They found that the iron is present in a microcrystalline calcium aluminum ferrite phase, and from the superparamagnetic behavior of these particles in different samples they were able to study the increase in particle size with increasing iron content.

Microcrystals of iron oxides and FeOOH (natural and synthetic) are used as pigments in paints. The applications of Mössbauer spectroscopy for studies of these materials have been reviewed by Keisch (1974, 1976).

Shinjo et al. (1975) observed that in the Mössbauer spectrum of a magnetic recording tape about 16% of the spectral area was due to superparamagnetic particles. These particles, which are useless for a recording purpose, are very difficult to observe by other methods.

Ferritin, which is important for iron storage in living species, consists of a core of ferric hydroxide in a protein shell. Several Mössbauer studies of ferritin, which demonstrate a superparamagnetic behavior, have been published. This subject has been reviewed by Oosterhuis and Spartalian (1976).

Microcrystals of several other inorganic magnetic compounds have been studied. Shinjo (1966) and van der Kraan and van Loef (1966) obtained Mössbauer spectra of synthetic α-FeOOH. An anisotropy constant of approximately 1.5×10^3 J m^{-3} was found (Shinjo, 1966). Dézsi and Coey (1973) studied ε-Fe$_2$O$_3$ particles. In other studies superparamagnetism has been found in microcrystalline particles of, for example, γ-Fe$_2$O$_3$ (Coey and Khalafalla, 1972; Boudart et al., 1975), Fe$_3$O$_4$ (McNab et al., 1968; Roggwiller and Kündig, 1973; Krupyanskii and Suzdalev, 1974a, b; Topsøe et al., 1974; Mørup et al., 1976), various ferrites (Gonser et al., 1968; McNab and Boyle, 1968), FeCl$_3$ (Litterst et al., 1974), granular Fe–SiO$_2$ (Dormann et al., 1976, 1977), and metallic iron (Boudart et al., 1977).

Superparamagnetic relaxation has also been observed in several noncrystalline samples. For example, Bukrey et al. (1974) found microcrystalline precipitates of iron oxides in a Na$_2$O·Li$_2$O·Fe$_2$O$_3$ glass system, and Popma and van Diepen (1974, 1975) observed a superparamagnetic behavior in noncrystalline Y$_3$Fe$_5$O$_{12}$.

Iron-doped polymers have been studied by Imshennik et al. (1976) and by Meyer (1976). In these investigations it was found that iron is present in small amorphous iron-rich clusters.

Other examples of superparamagnetism have been found in proton-irradiated compounds (Kopcewicz and Kotlicki, 1975, 1977) and in iron oxides formed by thermal decomposition (Gallagher and Kurkjian, 1966). Studies of solid-state reactions by Mössbauer spectroscopy have in several cases revealed the presence of superparamagnetic particles. This subject has been reviewed by Gallagher (1976). Examples of superparamagnetism in corrosion products are discussed in the review by Simmons and Leidheiser (1976).

Investigation of superparamagnetism in compounds that do not contain a Mössbauer isotope may be achieved through doping with a Mössbauer isotope such as ^{57}Fe (or ^{57}Co). Due to the exchange interaction of the ^{57}Fe atoms with the other magnetic ions, the ^{57}Fe atoms reflect the magnetic fluctuation in the microcrystals.

Several studies of ^{57}Co-doped microcrystals have been reported. For example, particles of NiO (Kündig et al., 1967b; Ando et al., 1967), Ni (Lindquist et al., 1968; Clausen et al., 1979; Mørup et al., 1979), and Co_3O_4 (Kündig et al., 1969) have been investigated in this way. Detailed studies of β-Co precipitates in Cu have also been carried out in this way.

Such small particles may be produced by precipitation from a supersaturated alloy. Nasu et al. (1967) made particle-size distribution studies of Co precipitates in a Cu alloy containing 1.97% Co. After annealing the samples at 1170 K, quenching into water, and aging for 10 hr at 870 K, they obtained Mössbauer spectra between 95 and 335 K. The spectra showed a typical superparamagnetic behavior, therefore making the determination of particle size distribution possible. The average size estimated in this way was 10 nm, which is in good agreement with the results from electron microscopy. Krop et al. (1971, 1974, 1975) made a detailed study of the superparamagnetic relaxation time of similar precipitates. They found that the temperature dependence of the relaxation time estimated from the Mössbauer spectra was not in agreement with the simple Néel formula [Eq. (12)], but could be fitted with Aharoni's theoretical model for particles with cubic anisotropy (Aharoni, 1973). It was also found that the magnetic field dependence of the superparamagnetic relaxation time did not follow the high energy barrier approximation by Brown [Eq. (13)], but rather followed the numerical values of Aharoni for particles with uniaxial anisotropy.

An unusual volume dependence of the superparamagnetic relaxation time has been found in FeNi alloy particles with 37% Ni (Afanas'ev et al., 1970) and metallic iron particles (Amulyavichus and Suzdalev, 1973). In both cases, a minimum in the relaxation time was found at a critical particle volume. The effect may be explained by the onset of the curling mode for spin rotation at this critical volume. Figure 8 shows some experimental results for metallic iron particles together with the theoretical values for the relaxation time for spin rotation in unison and spin rotation by curling, as calculated by Eisenstein and Aharoni (1976b). It is shown that the measured size dependence of the relaxation time is in qualitative accordance with theory, but the calculated critical radius differs from the one found experimentally by a factor of about two.

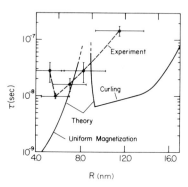

Fig. 8. Theoretical relaxation time of spherical Fe particles as a function of their radius R. For $R \lesssim 8$ nm only rotation in unison takes place. For 8 nm $\lesssim R \lesssim 9$ nm the relaxation time decreases with increasing particle size due to the onset of spin rotation by curling. Experimental results of Amulyavichus and Suzdalev (1973) at $T = 300$ K are shown for comparison (Eisenstein and Aharoni, 1976b).

Many other studies in which superparamagnetic particles have been observed could be mentioned. However, the examples discussed above demonstrate the wide range of different materials in which the effect has been found.

IV. Experimental Studies of Collective Magnetic Excitations in Microcrystals

In addition to the superparamagnetic relaxation effects discussed in Section III, many studies of magnetic microcrystals have revealed that the magnetic hyperfine splitting below the blocking temperature is smaller than that found in larger crystals (e.g. van der Kraan, 1972; Kündig et al., 1966; Hobson and Gager, 1970, Topsøe and Boudart, 1973; Simopoulos and Kostikas, 1973; Kubsch and Schneider, 1973; Mørup and Topsøe, 1976). To account for this observation many explanations involving, for example, surface effects and changes in magnetic ordering temperature have been proposed. Such effects may be present. However, as pointed out by Mørup and Topsøe (1976), collective magnetic excitations are probably the most important cause for the reduction in the hyperfine fields. They studied the effect in microcrystals of magnetite (Fe$_3$O$_4$). Since the hyperfine field also depends on temperature due to spin–wave excitations, it is necessary to compare the hyperfine splitting in small particles with that of large crystals at the same temperature. Spectra were therefore obtained from samples with average particle diameters of 6, 10, 12, and > 200 nm. Figure 9 shows the value of $h(V, T) = H(V, T)/H(V = \infty, T)$ as a func-

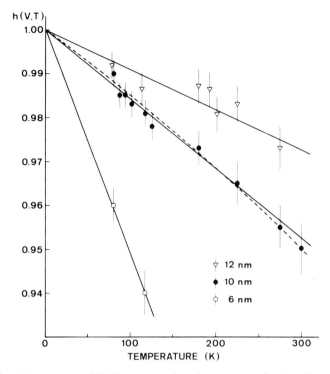

Fig. 9. The parameter $h(V,T)$ as a function of temperature for 6-, 10-, and 12-nm particles of Fe_3O_4. The solid lines indicate the best fit to the experimental results using Eqs. (23) and (26). The dotted line indicates the best fit to the results for the 10-nm particles using Eq. (22) (Mørup and Topsøe, 1976).

tion of temperature for the three samples of small particles. According to Eqs. (23) and (26) this quantity should approximately be proportional to $1 - kT/2KV$. It is seen from the figure that the results are, in fact, in accordance with a linear temperature dependence of $h(V, T)$. The values of K determined from the slopes of the best-fitted straight lines were in good agreement with those found from superparamagnetic relaxation.

In a more refined analysis the particle size distribution was also taken into account (Mørup et al., 1976). Since the reduction in the hyperfine field depends on the particle volume, the Mössbauer lines from particles of different sizes will not be shifted to the same extent. This gives rise to a broadening of the Mössbauer lines reflecting the particle-size distribution. Figure 10 shows some experimental and theoretical spectra of 10-nm Fe_3O_4 particles at various temperatures. The spectrum of Fe_3O_4 above the Verwey transition temperature, $T_V (= 119$ K) normally consists of two six-line patterns arising from the tetrahedral and octahedral sites with spectral area ratio close to $1:2$. In the spectra of 10-nm particles at low temperatures (e.g., at 180 K) the two components are, in fact, clearly distinguishable as in large crystals. However, at high temperatures the spectral shape is severely distorted. The theoretical spectra, which exhibit exactly the same temperature dependence, were calculated on the basis of (i) the Mössbauer parameters of bulk Fe_3O_4, (ii) a decrease in hyperfine splitting according to Eqs. (23) and (26), and (iii) a particle-size distribution consistent with that found from electron micrographs of the sample. Thus, the influence of collective magnetic excitations can explain the anomalous line shapes at high temperatures.

A similar analysis of Mössbauer spectra of ferritin has been made by Danson et al. (1977) and Williams et al. (1978). These authors calculated the influence of collective magnetic excitations on the value of the measured quadrupole shift, and their model was used to determine the particle-size distribution of FeOOH in ferritin. In a more recent study Nalovic and Janot (1979) used the same method to estimate the particle-size distribution in the synthetic α-FeOOH samples.

In general, the line broadenings and line shifts in Mössbauer spectra caused by collective magnetic excitations may not be easily distinguished from those arising from other effects, such as superparamagnetic relaxation and surface effects. However, it has been shown that the line shifts caused by superparamagnetic relaxation are generally far too small to explain the observed small magnetic hyperfine splitting in Mössbauer spectra of microcrystals (Mørup and Topsøe, 1976), and the line shifts and line broadenings caused by superparamagnetic relaxation may, in many cases, be neglected. Surface and interface magnetism is discussed in Section VIII and Chapter 2 (Topsøe et al., 1980). In short, however, at very low temperatures (e.g., 4.2 K) the hyperfine fields of atoms near the surface may be either larger or smaller than that of bulk atoms, depending on the material covering the surface. At higher temperatures the hyperfine field of surface atoms has been found to decrease more rapidly with increasing temperature than that of bulk atoms. In any case, the surface effects only affect a few atomic layers (unless the temperature is very close to the Curie temperature) and therefore surface effects are generally not important for particles with dimensions larger than about 10 nm.

Haneda and Morrish (1977) investigated the origin of the size-dependent hyperfine fields in small particles of γ-Fe_2O_3. In order to estimate the hyperfine field of surface ions, they obtained a spectrum of γ-Fe_2O_3 particles with the surface ions enriched in [57]Fe. The size of the surface hyperfine field was found to be very close to the bulk value. Therefore, they were able to conclude that the decrease in hyperfine fields with

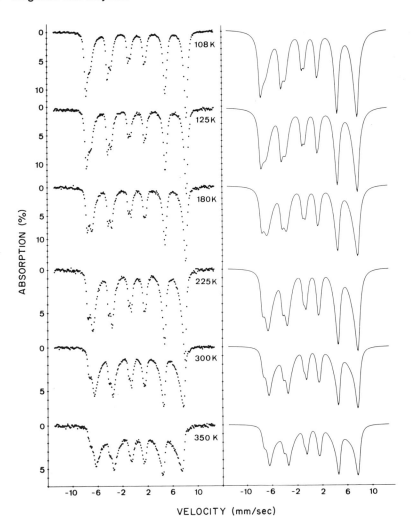

Fig. 10. Experimental and theoretical Mössbauer spectra of 10-nm particles of Fe_3O_4 at various temperatures. The line distortions in the theoretical spectra appear as a result of a particle-size dependent reduction in the hyperfine field due to collective magnetic excitations, in connection with the particle-size distribution (Mørup et al., 1976).

decreasing particle size is not due to surface effects, but can be explained by collective magnetic excitations.

When the particles are influenced by superparamagnetic relaxation one has, in fact, to consider the combined effect of collective magnetic excitations and superparamagnetic relaxation. In a rigorous model one should calculate Mössbauer spectra in the presence of fluctuations of the hyperfine field in all directions in space with a probability distribution given by, for example, Eq. (11). However, since the transition rate is high for transitions between states that are not separated by an energy barrier (states close to the same energy minimum), the calculations may be simplified considerably by considering only the two "thermally averaged states" (defined as thermal averages of the states near the two minima at $\theta = 0$ and $\theta = \pi$) with hyperfine fields given by Eq. (26). Such a model has been used to describe the spectra of 6-nm particles of Fe_3O_4 at various temperatures (Mørup et al., 1976).

V. Magnetic Field Dependence of Mössbauer Spectra of Small Particles

Studies of superparamagnetic particles in the presence of applied magnetic fields may, as discussed in Section II, C, be very useful since particle-size information can be obtained directly. This is in contrast to studies of superparamagnetic relaxation times and collective magnetic excitations, in which the observed effects depended on the product of the particle volume and the anisotropy energy constant. Combined studies of the temperature and magnetic field dependence will, of course, be especially useful since both volume and magnetic anisotropy information can be obtained. Important information about the spin structure may also be obtained by applying magnetic fields (Section VII).

According to Eq. (27) the hyperfine splitting of a superparamagnetic particle is restored for $\mu H/kT \gtrsim 1$. The magnetic moment of a ferromagnetic particle with a diameter of about 5 nm is typically of the order of several thousand Bohr magnetons. Therefore, magnetic saturation may be obtained in moderate magnetic fields even at room temperature. Eibschütz and Shtrikman (1968) observed this effect in 9.7-nm particles of $NiFe_2O_4$. The field dependence of the Mössbauer spectra of ^{57}Fe in Ni particles with an average diameter of 6.0 nm was studied at several temperatures by Lindquist et al. (1968). They found that the spectrum at 298 K changed from a single- to a six-line pattern, when the field was increased from a zero applied field to 5.9 kG. Similar results have been obtained for small particles of Fe_3O_4 (Roggwiller and Kündig, 1973), γ-Fe_2O_3 (Coey and Khalafalla, 1972); and $MgFe_2O_4$ (Gonser, 1968/69).

Van der Kraan (1971, 1972) found a pronounced increase in the area of the magnetically split component in the spectra of 7.0-nm α-Fe_2O_3 particles when a magnetic field was applied. Although α-Fe_2O_3 is antiferromagnetic (with a small canting of the spins above the Morin transition), small particles may still have a magnetic moment due to uncompensated surface spins. This may be the explanation of the observed field dependence.

As mentioned in Section II, B, both the superparamagnetic relaxation time and the populations of the energy minima are changed when a magnetic field is applied. In studies of Fe_3O_4 (Roggwiller and Kündig, 1973) and Co (Krop *et al.*, 1975), both effects have been taken into account and good agreement with theory has been found.

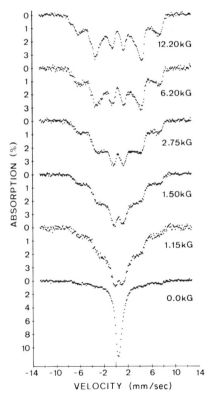

Fig. 11. Mössbauer spectra of 6-nm particles of Fe_3O_4 at 260 K in various applied magnetic fields (Mørup and Topsøe, 1977).

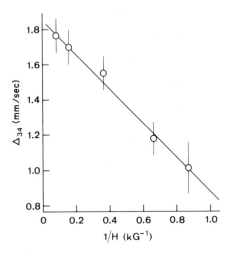

Fig. 12. Magnetic splitting of lines 3 and 4 in the spectra shown in Fig. 11 plotted as a function of the reciprocal applied magnetic field. From the slope of the line a particle size of 6.4 nm was estimated (Mørup and Topsøe, 1977).

Above the blocking temperature it is possible to determine the average value of the magnetic moments of ferro- or ferrimagnetic particles from the field dependence of the Mössbauer spectra. According to Eqs. (20) and (27) this is conveniently done by plotting $H_{hf} - H$ as a function of $1/H$. A straight line with the slope $H_{hf}(V = \infty, T)kT/\mu$ is obtained for $\mu H/kT \gg 1$ and, as a result, $\mu = M_s V$ can be determined. If the magnetization M_s of the particles is known (equal, for example, to the value for the bulk compound), then the average volume of the particles can be determined. Such determinations of particle size have been made on samples of small particles of Fe_3O_4 (Mørup and Topsøe, 1977; Lipka et al., 1977a, b). Figure 11 shows some spectra of small Fe_3O_4 particles at 260 K. The iron atoms at the octahedral and the tetrahedral sites in Fe_3O_4 have different hyperfine fields with opposite directions, and the corresponding Mössbauer lines are not completely resolved when the external magnetic field is small. However, an average splitting of the inner lines Δ_{34} can be obtained from the spectra. Figure 12 shows a plot of Δ_{34} as a function of $1/H$. The experimental points are fitted with a straight line, whose slope can be used to determine an average particle size of 6.4 nm. This value is in excellent agreement with x-ray diffraction and electron microscopy results. The same approach was used by Mørup et al. (1979) to determine the particle size of Ni particles doped with ^{57}Co.

In such studies the presence of a particle-size distribution gives rise to broad lines because particles with different values of μ exhibit different magnetic splitting. It may, in fact, be possible to determine the distribution in magnetic moments (and therefore also the particle-size distribution) from the line shape of the Mössbauer spectra.

VI. The Anisotropy Energy Constant in Small Particles

Sections III and IV showed that it is possible to determine the average value of KV and the distribution in KV from Mössbauer studies of superparamagnetic relaxation and collective magnetic excitations. If the volume V is known, e.g., from the applied magnetic field dependence of the Mössbauer spectra above T_B (as discussed in Section V) or from independent measurements (e.g., electron microscopy or x-ray diffraction), the value of the anisotropy energy constant K can be determined.

The values of the anisotropy energy constant in microcrystals obtained by using Mössbauer spectroscopy have, in many cases, been found to be considerably greater than those found in large crystals. Moreover, the values of K for a given compound may be different from system to system. For example, the value for unsupported α-Fe_2O_3 particles (van der Kraan, 1972) was found to be much greater than that found for small particles of α-Fe_2O_3 supported on a high surface area silica (Kündig et al., 1966, 1967a, b). Different values for the anisotropy energy constant in unsupported and supported γ-Fe_2O_3 microcrystals have also been found by Coey and Khalafalla (1972) and Boudart et al. (1975a).

For several reasons, the anisotropy energies of microcrystals are expected to be different from those of large crystals (see Section II, A). For example, the relative importance of surface anisotropy increases with decreasing particle size. This implies that the surface structure and the chemisorbed molecules on the surface of a microcrystal may have a significant influence on its anisotropy energy. When the particles are prepared on a high surface carrier (e.g., Al_2O_3 or SiO_2), the nature of the particle-support interface may also give contributions to the surface anisotropy. Moreover, the support interaction may affect the stress state and the shape of the particles and may thus give rise to stress and shape anisotropy. If the support is magnetic, exchange anisotropy may also be present. Finally, if the particles are not well separated, interaction anisotropy may also contribute to the total anisotropy energy. For these reasons the measured anisotropy energy constants of microcrystals are expected to depend critically both on particle size and preparation method.

Mössbauer studies of samples of small Fe_3O_4 particles with different particle sizes indicated that the anisotropy energy constant increases with decreasing particle size (Mørup and Topsøe, 1976). The absolute values of

K were also considerably greater than those found in macroscopic crystals, indicating the possible presence of a significant surface anisotropy. In accordance with this explanation, it was found that the exposure of 6.0-nm Fe_3O_4 particles to different molecules, such as acetone and oleic and stearic acid, resulted in different superparamagnetic relaxation times (Mørup et al., 1976, Lipka et al., 1977a, b).

Evidence of the presence of surface anisotropy for small particles of metallic iron has also been found using both Mössbauer spectroscopy and magnetic susceptibility (Boudart et al., 1975b, 1977). With the former technique, the presence of chemisorbed hydrogen was found to decrease the area of the hyperfine split spectrum and to increase the area of the paramagnetic component. This effect of chemisorbed hydrogen can be explained by either a decrease in the anisotropy energy barrier or a cancellation of magnetic moments of surface atoms. Using magnetic susceptibility, however, the presence of chemisorbed hydrogen was found to have no effect on the measured saturation magnetization at temperatures above the blocking temperature. Thus the combined results of Mössbauer spectroscopy and magnetic susceptibility show that the primary effect of chemisorbed hydrogen is to decrease the magnetic anisotropy energy barrier.

In these studies by Boudart et al. the hydrogen was chemisorbed and desorbed at a relatively high temperature (683 K) and it was therefore suggested that the observed changes in the magnetic anisotropy energy constant might be explained by changes in the surface structure of the particles which may occur at this temperature. In a more recent study hydrogen was chemisorbed at 300 K on silica supported iron particles with an average diameter of 6.3 nm (Mørup et al., 1980). In this case hydrogen chemisorption resulted in an effect opposite to that observed by Boudart et al., namely an increase in the magnetic anisotropy energy constant from $(1.0 \pm 0.1) \cdot 10^5 \, J\,m^{-3}$ in the absence of chemisorbed molecules to $(1.2 \pm 0.1) \cdot 10^5 \, J\,m^{-3}$ when hydrogen was chemisorbed. It was also found that chemisorption of CO at 200 K resulted in an increase in the magnetic anisotropy energy constant to $(1.3 \pm 0.1) \cdot 10^5 \, J\,m^{-3}$. It is not likely that surface reconstruction during chemisorption takes place at T ≤ 300 K and this may explain why hydrogen chemisorption at low and at high temperatures leads to opposite changes in the magnetic anisotropy energy constant.

The influence of hydrogen chemisorption on the superparamagnetic relaxation time of 4-nm Ni particles has also been investigated (Clausen et al., 1979; Mørup et al., 1979). The particles were doped with ^{57}Co and used as a source in a ^{57}Fe Mössbauer experiment. Figure 13 shows spectra obtained at 78, 25, 16, and 4.2 K in vacuum and with hydrogen chemisorbed at the surface. At 78 K the spectra essentially consist of a single line

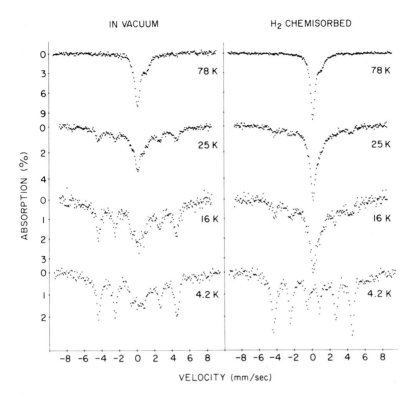

Fig. 13. Mössbauer spectra of 4-nm Ni particles at 78, 25, 16, and 4.2 K in vacuum and with hydrogen chemisorbed at the surface (Mørup *et al.*, 1979).

(the small "shoulder" at about 1 mm/s can be interpreted as Fe^{2+} in a nonmetallic phase). At 4.2 K the spectra are magnetically split with hyperfine fields of 277 kG in vacuum and 275 kG in H_2 (the Fe^{2+} component gives rise to a broad background in hydrogen and to a somewhat narrower line in vacuum). At 16 and 25 K the spectra contain both paramagnetic and ferromagnetic components. It is clearly seen that the area ratio of the ferromagnetic and paramagnetic components is significantly smaller in hydrogen than in vacuum. Numerical analyses of the spectra indicated that the anisotropy constant decreases from about 6.0×10^4 J m^{-3} in vacuum to about 3.0×10^4 J m^{-3} in hydrogen. These values are greater than the bulk value for magnetocrystalline anisotropy by a factor of about ten. Thus the results indicated that surface anisotropy gives a significant contribution to the total anisotropy energy constant and that $K_{surface}$ is very sensitive to hydrogen chemisorption.

During passivation (slow oxidation) of small particles of metallic iron formation of a thin surface layer, consisting of microcrystals of γ-Fe$_2$O$_3$ and Fe$_3$O$_4$, has been observed (Haneda and Morrish, 1978; 1979). Mössbauer studies of such partly oxidized iron particles showed that the hyperfine fields of the oxide microcrystals were 30–50 kG smaller than those of the bulk oxides. This was explained by the influence of collective magnetic excitations (Haneda and Morrish, 1979). It is interesting that in such systems a considerable exchange interaction is present at the metal-oxide interface. Haneda and Morrish therefore concluded that the magnetic anisotropy of the small oxide particles is mainly determined by the contribution from exchange anisotropy. The authors also pointed out that the expressions normally used for calculation of the reduction in the magnetic hyperfine splitting caused by collective magnetic excitations [Eqs. (22) and (23)] are not valid in this case, because exchange anisotropy is not uniaxial, but unidirectional as seen from Eq. (7). The reduction in the magnetic hyperfine splitting may, however, in this case be calculated by use of the general expression given by Eq. (24).

In a recent study of disordered alloys of Fe$_3$Pt it was observed that these alloys behave superparamagnetically, and the relaxation time was found to increase when the samples were subjected to an uniaxial elastic tensile stress (Sakharov and Kuz'min, 1976). Figure 14 shows Mössbauer spectra of one of the samples, which was quenched from 1028 K to room temperature. Spectrum A was obtained without any stress, and spectrum B was obtained with a stress of $(4.1 \pm 0.3) \times 10^8$ N m^{-2}. Here C and D indicate decompositions of the spectra A and B into "ferromagnetic" and "paramagnetic" components. It is clearly seen that the application of stress increases the relative area of the ferromagnetic component. These measurements thus seem to give evidence of induced stress anisotropy in magnetic clusters in the disordered alloys.

Mössbauer spectroscopy can also give information about the orientation of the magnetization and the anisotropy energy constant in particles that are so large that the Mössbauer spectra are not influenced by superparamagnetic relaxation or collective magnetic excitations. In this case the orientation of the magnetization can be found from the area ratio of the Mössbauer lines. Thus information about the texture can be obtained (Keune and Sette Camara, 1975; Nagy *et al.*, 1975), and the anisotropy energy constant determined from the magnetic field dependence of the area ratio of the Mössbauer lines (Beckmann *et al.*, 1968).

Haneda and Morrish (1976) investigated the orientation of the magnetization of γ-Fe$_2$O$_3$ particles in recording tapes in this way. They found that the degree of orientation of the magnetization depended on the concentra-

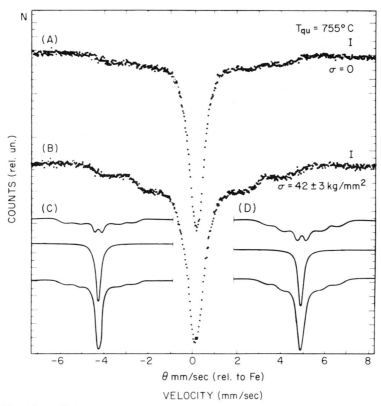

Fig. 14. Mössbauer spectra of a Fe_3Pt alloy quenched from 1028 K. Spectrum A was obtained without applied stress, whereas spectrum B was obtained with a uniaxial stress of $(4.1 \pm 0.3) \times 10^8$ N m^{-2}. C and D illustrate the decomposition of the spectra A and B into ferromagnetic and superparamagnetic components. The influence of the stress on the superparamagnetic fraction is clearly seen indicating that stress anisotropy has been induced (Sakharov and Kuz'min, 1976).

tion of γ-Fe_2O_3 particles, but was less dependent on the strength of the magnetic field applied during coating of the dispersed particles onto the plastic substrate.

VII. Magnetic Order in Small Particles

The magnetic order in a microcrystal may well depend on the particle size. For example, the Curie or Néel temperature may show particle-size dependence. If the magnetic ordering temperature decreases with decreasing particle size, the observed hyperfine field of a microcrystal will be smaller than that of a macroscopic crystal. The effect will be in the same direction as if collective magnetic excitations were present (Section IV). There are, however, several results which indicate that small particles have magnetic ordering temperatures close to the values found in large crystals. Mössbauer spectra of iron particles with a particle size below 10 nm have been recorded *in situ* up to about 700 K without observing hyperfine fields significantly different from those of larger particles (Boudart *et al.*, 1977). The effect of collective magnetic excitations was small since very large anisotropy constants were present. Magnetization measurements of small particles of Co, Ni, and γ-Fe_2O_3 have also shown that the Curie temperature is little affected by the particle size (Jacobs and Bean, 1963).

Other types of magnetic transitions may also be affected by particle size changes. In α-Fe_2O_3 the spin orientation changes at about 260 K, which is the Morin transition temperature T_M. Below T_M, the spins are parallel to the (001) direction, and above T_M the spins are perpendicular to this direction (with a small canting angle). This transition gives rise to changes in the sign and magnitude of the quadrupole coupling and in the size of the hyperfine field and is therefore easily observed by using Mössbauer spectroscopy. T_M has been observed to decrease when the particle size decreases (van der Kraan, 1972, 1973; Nakamura *et al.*, 1964; Kündig *et al.*, 1966). Schroeer and Nininger (1967) found a Morin temperature of 166 ± 10 K in particles with an average diameter of 575 Å. The same authors later published a detailed study of the particle size dependence of the Morin transition (Nininger and Schroeer, 1978). Krupyanskii and Suzdalev (1975) studied the Morin transition in supported and unsupported α-Fe_2O_3 particles. They found that the presence of the support had a marked effect on T_M.

Large crystals of Fe_3O_4 show a magnetic transition at 119 K, the so-called Verwey transition temperature T_V. This transition has also been studied in small particles by using Mössbauer spectroscopy. Krupyanskii and Suzdalev (1974a, b) found that T_V increases with decreasing particle size. However, this result contradicts the measurements by Topsøe *et al.* (1974), who also showed the necessity of carrying out the experiments *in situ*, since otherwise nonstoichiometry easily results in this type of samples. A later investigation of the Verwey transition in 10-nm particles yielded a transition temperature of about 100 K (Topsøe and Mørup, 1975), indicating a decrease in T_V with decreasing particle size.

Several Mössbauer studies have been performed in order to investigate the magnetic order in small particles of γ-Fe_2O_3 (Coey and Khalafalla, 1972; Coey, 1971; Morrish and Clark, 1974; Morrish *et al.*, 1976). When a large magnetic field is applied parallel to the γ-ray direction, it is expected that the second and the fifth lines in the spectrum disappear, due to the ferrimagnetic order normally found in γ-Fe_2O_3. However, in the studies of small particles is was found that even in magnetic fields of 50 and 90 kG, these lines had a significant intensity. These results may be explained by a noncollinearity of the spins which increases with decreasing particle size or by a noncollinearity which is large near the surface. Morrish *et al.* (1976) measured the effect in particles of γ-Fe_2O_3 which were enriched in [57]Fe at

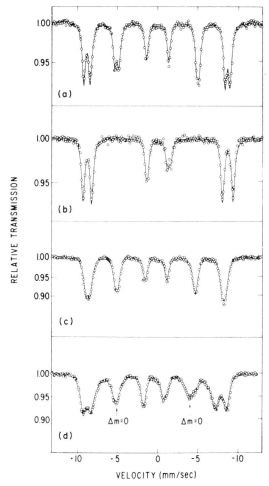

Fig. 15. Mössbauer spectra of (a), (b) bulk $NiFe_2O_4$ and (c), (d) oleic acid-coated $NiFe_2O_4$ particles with and without a 68.5 kG field applied parallel to the gamma ray direction: (a) $T = 30$ K, $H = 0$ kG; (b) $T = 23$ K, $H = 68.5$ kG; (c) $T = 23$ K, $H = 0$ kG; (d) $T = 23$ K, $H = 68.5$ kG. In the spectrum of the bulk sample lines 2 and 5 disappear when the field is applied. This is not the case for small coated particles. The difference is explained by a strong pinning of the surface spins due to the coating (Berkowitz *et al.*, 1975).

the surface. This study showed that the latter hypothesis is correct, i.e., the surface has a noncollinear magnetic structure.

A similar result was obtained by Berkowitz *et al.* (1975). These authors studied the magnetization of 8-nm $NiFe_2O_4$ particles coated with organic molecules such as oleic acid. The coated particles reached only 75% of the normal saturation magnetization in an applied magnetic field of 200 kG. Particles in alcohol did not show such a deviation from saturation magnetization. Mössbauer spectra were obtained from bulk $NiFe_2O_4$ and small particles coated with oleic acid or in alcohol. The spectra were taken in a zero applied field and a 68.5-kG field applied parallel to the gamma-ray direction. The spectra of the first two samples are shown in Fig. 15. For the bulk sample it is seen that lines 2 and 5 disappear in the applied field, while this is not the case for the oleic acid coated small particles. For the particles in alcohol, lines 2 and 5 disappeared in the presence of the applied field. The results were interpreted in terms of a pinning of the surface spins when the surface was covered by molecules of oleic acid.

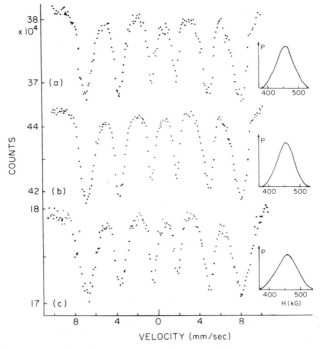

Fig. 16. Mössbauer spectra of an amorphous ferric gel at 4.2 K in fields of (a) 0 G, (b) 10 kG, and (c) 50 kG applied parallel to the gamma ray direction. The insensitivity of the relative areas of lines 2 and 5 on the value of the field is explained by a statistical distribution of the spin directions (sperimagnetism). The insets show the hyperfine field distributions (Coey and Readman, 1973b).

In studies of thin ferromagnetic films such as metallic iron films, the shape anisotropy is normally dominant, so that the magnetization direction is in the plane of the film (Shinjo, 1976; Shinjo *et al*., 1977). However, in studies of very thin films of metallic iron deposited between layers of Cu, Ag, or MgO, it has been found that the magnetization is preferentially oriented perpendicular to the film plane (Keune *et al*., 1975, 1979; Shinjo *et al*., 1979). This result also suggests that the spin orientation is influenced by the surface effects. A theoretical expression for the orientation of the magnetization at the surface of a thin film has been derived by Quach *et al*. (1978).

Coey and Readman (1973b) studied the spin structure in an amorphous ferric gel. They obtained Mössbauer spectra in a zero applied field and fields of 10 and 50 kG parallel to the gamma-ray direction (Fig. 16). It was found that the relative intensity of lines 2 and 5 was not affected by the magnetic field. From this result and magnetization measurements they concluded that the spins have no preferred direction, instead they have random directions relative to each other. This type of magnetic ordering was called "sperimagnetism."

VIII. The Hyperfine Field of Surface Atoms

In Section VII we discussed results showing that the orientation of the magnetic moments may depend on the distance from the surface. In this section we shall discuss Mössbauer studies of the size of the hyperfine field of atoms near the surface. Mössbauer studies of surface and interface magnetism are discussed in more detail in Chapter 2 (Topsøe *et al*., 1980).

Liebermann *et al*. (1969, 1970) proposed on the basis of magnetization measurements on thin films of ferromagnetic metals that the surface layers may be nonmagnetic (magnetically "dead layers"). Magnetic susceptibility measurements on small particles of Ni (Selwood, 1975) and Fe_3O_4 (Kaiser and Miskolczy, 1970) have also shown a lower magnetization than expected from studies of macroscopic crystals. Mössbauer spectroscopy may contribute to the interpretation of such results since surface layers may give rise to separate components in the Mössbauer spectra.

In order to study surface atoms by Mössbauer spectroscopy, a large fraction of the Mössbauer nuclei examined must be at the surface. This can be achieved, for example, by studying very small particles or very thin films, enriching samples with the Mössbauer isotope at the surface, or using conversion electron Mössbauer spectroscopy (Bäverstam *et al*., 1974; Petrera *et al*., 1976; Dumesic and Topsøe, 1977). In all cases, of course, either *in situ* studies must be carried out, or the surface must be covered by a coating that prevents attack of the surface by gas phase molecules (e.g., oxidation).

The hyperfine field of ^{57}Fe surface atoms in thin films of metallic iron, cobalt, and nickel have been studied in detail (Shinjo et al., 1973, 1974, 1977, 1979; Shinjo, 1976). These experiments showed that a distribution of hyperfine fields is present near the surface and the actual distribution depends on the material covering the surface. However, the average value is only slightly different from the bulk value. This result is obviously inconsistent with the proposed existence of "dead layers."

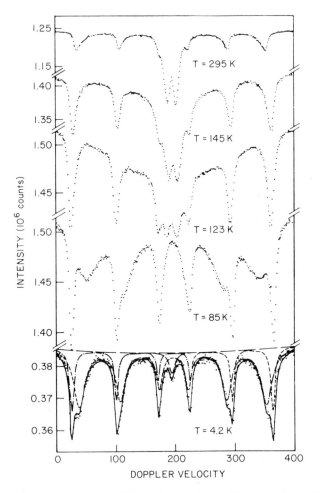

Fig. 17. Mössbauer spectra at different temperatures of 7-nm particles of α-Fe_2O_3, enriched in ^{57}Fe in the surface. For the spectrum obtained at 4.2 K is shown a decomposition into the surface component and the "bulk" component (van der Kraan, 1973).

Microcrystals of α-Fe_2O_3, enriched in [57]Fe at the surface, have been studied by van der Kraan (1971, 1972, 1973). The particles were prepared by precipitation in an aqueous solution. Some of the experimental results for 7-nm particles are shown in Fig. 17. It can be seen that below the blocking temperature the spectra can be described in terms of two different hyperfine fields. The large one is similar to that of macroscopic crystals and probably arises from the atoms in the interior of the particles, whereas the smaller hyperfine field can be attributed to the surface atoms. The magnitude of the latter hyperfine field was found to decrease faster with increasing temperature than the former.

Haneda and Morrish (1977) studied γ-Fe_2O_3 particles enriched in [57]Fe at the surface. They found that in this system the hyperfine fields for the surface layer are only about 1% smaller than those of the interior of the particles. This difference is considerably smaller than that found for α-Fe_2O_3.

The diverging results for α-Fe_2O_3 and γ-Fe_2O_3 may be due to the different crystal structures. However, different sample preparation methods may give rise to, e.g., different amounts of water molecules at the surface, and this may also explain the results.

Mössbauer studies of 6-nm Fe_3O_4 particles in oleic acid and acetone have also shown that the surface atoms in these samples have hyperfine fields similar to those of the atoms in the interior of the particles (Lipka *et al.*, (1977a, b)). Thus, the low value of the magnetization of such Fe_3O_4 particles observed by Kaiser and Miskolczy (1970) cannot be due to "dead layers," but may be explained by a noncollinear spin arrangement near the surface similar to that found in γ-Fe_2O_3 and in $NiFe_2O_4$.

A general result of these Mössbauer studies of surface atoms is that "dead layers" have not been found. The studies have also shown that only very few surface layers exhibit hyperfine fields different from those of the interior of the particles.

IX. Magnetic Fluctuations in Diamagnetically Substituted Iron Oxides

In several Mössbauer studies of diamagnetically substituted magnetically ordered iron oxides, spectra with broad and/or partially collapsed lines have been found. This indicates the presence of relaxation effects.

The spectra of these substances often show the following features: At very low temperatures normal magnetically split spectra are observed, but at higher temperatures the lines become broadened. Above a temperature T_p ($0 < T_p < T_c$, where T_c is the Curie temperature), the magnetic splitting (partially) collapses and a paramagnetic component is found. Finally, for $T > T_c$, only the paramagnetic component is present. In the temperature range $T_p < T < T_c$, the spectra often have similarities with those of superparamagnetic particles (Coey, 1972).

In fact, it has been shown that for certain concentrations of statistically substituted diamagnetic ions, the magnetic ions should be found in magnetic clusters of various size (Coey, 1972). It has been suggested that such clusters may behave superparamagnetically (Ishikawa, 1962, 1964). Coey (1972) studied some diamagnetically substituted garnets which show relaxation effects. In the region $T_p < T < T_c$ he found that the application of a magnetic field gives rise to a substantial broadening of the spectra. Some typical examples are shown in Fig. 18. These results were explained by the presence of small magnetic clusters containing 10^3-10^5 ions.

Similar results have been found in studies of, for example, the $Ga_2O_3-Fe_2O_3$ system (Nagarajan and Srivastava, 1977) and in solid solutions of $Fe_{2-2x}Mg_{1+x}Sn_xO_4$ (Basile and Poix, 1976).

NiZn–ferrites, $Zn_xNi_{1-x}Fe_2O_4$, which also show relaxation effects have been studied extensively. Daniels and Rosencwaig (1970) found

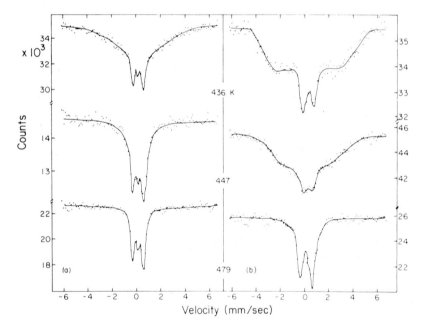

Fig. 18. Mössbauer spectra of $Y_3(Fe_{0.75}Sc_{0.25})_2[Fe]_3O_{12}$ at different temperatures (a) in the zero applied field and (b) in an applied field of 10 kG. It is seen that below the Curie temperature (450 K) the applied field greatly enhances the hyperfine splitting and reduces the relative area of the paramagnetic component. This is explained in terms of small magnetic clusters, which behave similar to superparamagnetic particles (Coey, 1972).

typical relaxation effects in samples with various concentrations of Zn. In a sample with $x = 0.70$, they found that the application of a field of 13.5 kG significantly reduced the relaxation effects at 77 K. Bhargava *et al.* (1976) also studied NiZn–ferrites with $x = 0.75$ at various temperatures in the zero applied field and in a 12-kG field. They found that the magnetic splitting disappeared far below the Curie temperature (which was found from neutron diffraction experiments). However, between 80 and 200 K Mössbauer relaxation spectra were observed and the splitting of the two inner lines (3 and 4) increased considerably when the 12-kG field was applied. The spectra were computer fitted with a relaxation model, and it was found that the values of $\langle S_z \rangle$ found from the Mössbauer spectra were considerably smaller than those found from neutron diffraction experiments, but the values of $\langle S_z \rangle$, derived from the Mössbauer spectra, increased when the field was applied. These results also suggest the presence of small magnetic clusters. However, it is noticeable that reasonably good fits of the spectra could be obtained assuming the same relaxation time for all the atoms. This is in contrast to the broad distribution in superparamagnetic relaxation times normally found in assemblies of magnetic microcrystals (see Section II, C).

In diamagnetically substituted materials of this type the magnetic interaction between the clusters is expected to be significant. Therefore, interaction anisotropy (Section II, A, 5) [and possibly also exchange anisotropy (Section II, A. 6)] is expected to be important. Thus, we have to consider magnetic fluctuations in a system in which the interaction energy [given, for example, by Eq. (6)] is significant. In this case the fluctuations of the magnetization of the individual clusters are strongly coupled and, consequently, the relaxation time will be less dependent on the cluster size than in a system with noninteracting magnetic particles. The narrow distribution in relaxation times may therefore be explained by the presence of magnetic interactions among clusters.

X. Concluding Remarks

Mössbauer spectroscopy seems to be extremely useful for the study of magnetic microcrystals. As this chapter illustrates, the technique can give information on magnetic fluctuations, magnetic order, surface magnetism, and several other properties.

The information obtained via the fluctuation effects is particularly useful since studies of superparamagnetic relaxation and collective magnetic excitations allow a determination of the distribution in KV for an assembly of particles. Moreover, in a sample with fast superparamagnetic relaxation it is possible to determine the magnetic moments of the particles from the field dependence of the spectra. Combination of these results may then lead to determinations of both V and K individually. It is of special interest that V and K can be measured under controlled atmosphere conditions. The value of K provides information regarding for example (i) shape, (ii) stress state, (iii) surface structure, (iv) interaction with the support, and (v) interaction with other particles. It is likely that many future investigations will concentrate on measurements of K and changes in K.

In many cases the surface anisotropy seems to be important. It is therefore highly desirable that more theoretical work is done in order to calculate the surface anisotropy energy constants of various materials and the dependence on the chemical state of the surface atoms.

Theoretical work on the influence of magnetic interactions among the particles on the magnetic fluctuation effects would also be valuable.

Future studies will probably include applications of other Mössbauer isotopes than ^{57}Fe. For example, several isotopes of the rare earths are well suited for Mössbauer spectroscopy, and these elements, as well as many of their compounds and alloys, have interesting magnetic properties. It is likely that Mössbauer spectroscopy will reveal new interesting phenomena in microcrystals of these materials.

The great sensitivity of Mössbauer spectroscopy to the magnetic properties of microcrystals has already appeared valuable in many different areas such as ceramics, glasses, alloys, catalysis, geology, biology, and ancient pottery. Further advances in the application and theoretical understanding of magnetic microcrystals will undoubtedly take place, and Mössbauer spectroscopy will play a significant role in these developments.

Acknowledgments

The authors are indebted to A. Aharoni for valuable discussions and to Haldor Topsøe and Nan Topsøe for critical reading of the manuscript and many valuable comments and suggestions.

References

Afanas'ev, A. M., and Onishchenko, E. V. (1976). *Zh. Eksp. Teor. Fiz.* **70**, 621 [*English transl.: Sov. Phys. JETP* **43**, 322 (1976)].

Afanas'ev, A. M., Suzdalev, I. P., Gen, M. Ia., Gol'danskii, V. I., Korneev, V. P., and Manykin, E. A. (1970). *Zh. Eksp. Teor. Fiz.* **58**, 115 [*English transl.: Sov. Phys. JETP* **31**, 65 (1970)].

Afanas'ev, A. M., Manykin, E. A., and Onishchenko, E. V. (1972). *Fiz. Tverd. Tela* **14**, 2502 [*English transl.: Sov. Phys.-Solid State* **14**, 2175 (1973)].

Aharoni, A. (1964). *Phys. Rev.* **135**, A447.

Aharoni, A. (1969). *Phys. Rev.* **177**, 793.

Aharoni, A. (1973). *Phys. Rev. B* **7**, 1103.

Aharoni, A., and Eisenstein, I. (1975). *Phys. Rev. B* **11**, 514.

Amulyavichus, A. P., and Suzdalev, I. P. (1973). *Zh. Eksp. Teor. Fiz.* **64**, 1702 [*English transl.: Sov. Phys. JETP* **37**, 859 (1973)].

Ando, K. J., Kündig, W., Constabaris, G., and Lindquist, R. H. (1967). *J. Phys. Chem. Solids* **28**, 2291.

Basile, F., and Poix, P. (1976). *Phys. Status Solidi (a)* **35**, 153.

Bäverstam, U., Ekdahl, T., Bohm, C., Liljequist, D., and Ringström, B. (1974). *In* "Mössbauer Effect Methodology" (C. Seidel and I. J. Gruverman, eds.), Vol. 9. Plenum Press, New York.

Beckmann, V., Bruckner, W., Fuchs, W., Ritter, G., and Wegener, H. (1968). *Phys. Status Solidi* **29**, 781.

Belozerskii, G. N., and Pavlyukhin, Y. T. (1977). *Fiz. Tverd. Tela (Leningrad)* **19**, 1279 [*English transl.: Sov. Phys. Solid State* **19**, 745 (1977)].

Belozerskii, G. N., Pavlyukhin, Y. T., and Gittsovich, V. N. (1976). *Zh. Eksp. Teor. Fiz.* **70**, 717 [*English transl.: Sov. Phys. JETP* **43**, 371 (1976)].

Berkowitz, A. E., Lahut, J. A., Jacobs, I. S., and Levinson, L. M. (1975). *Phys. Rev. Lett.* **34**, 594.

Blume, M., and Tjon, J. A. (1968). *Phys. Rev.* **165**, 446.

Bhargava, S. C., Mørup, S., Knudsen, J. E. (1976). *J. Phys.* **37**, C6-93.

Boudart, M., Delbouille, A., Dumesic, J. A., Khammouma, S., and Topsøe, H. (1975a). *J. Catal.* **37**, 486.

Boudart, M., Topsøe, H., and Dumesic, J. A. (1975b). *In* "The Physical Basis for Heterogeneous Catalysis" (E. Drauglish and R. I. Jaffee, eds.), p. 337. Plenum Press, New York.

Boudart, M., Dumesic, J. A., and Topsøe, H. (1977). *Proc. Nat. Acad. Sci. U. S.* **74**, 806.

Bozorth, R. M. (1951). "Ferromagnetism." van Nostrand-Reinhold, Princeton, New Jersey.

Brown, W. F., Jr. (1959). *J. Appl. Phys. Suppl.* **30**, 130S.

Brown, W. F., Jr. (1963a). *J. Appl. Phys.* **34**, 1319.

Brown, W. F., Jr. (1963b). *Phys. Rev.* **130**, 1677.

Bukrey, R. R., and Kenealy, P. F., Beard, G. B., and Hooper, H. O. (1974). *Phys. Rev. B* **9**, 1052.

Chikazumi, S., and Oomura, T. (1955). *J. Phys. Soc. Jpn.* **10**, 842.

Clausen, B. S., Mørup, S., and Topsøe, H. (1979). *Surf. Sci.* **82**, L589.

Coey, J. M. D. (1971). *Phys. Rev. Lett.* **27**, 1140.

Coey, J. M. D. (1972). *Phys. Rev. B* **6**, 3240.

Coey, J. M. D., and Khalafalla, D. (1972). *Phys. Status Solidi (a)* **11**, 229.

Coey, J. M. D., and Readman, P. W. (1973a). *Earth Planet. Sci. Lett.* **21**, 45.

Coey, J. M. D., and Readman, P. W. (1973b). *Nature (London)* **246**, 476.

Collins, D. W., Dehn, J. T., and Mulay, L. N. (1967). *In* "Mössbauer Effect Methodology" (I. J. Gruverman, ed.), Vol. 3, p. 103. Plenum Press, New York.

Constabaris, G., Lindquist, R. H., and Kündig, W. (1965). *Appl. Phys. Lett.* **7**, 59.

Corciovei, A., Costache, G., and Vamanu, D. (1972). *Solid State Phys.* **27**, 237.

Cullity, B. D. (1972). "Introduction to Magnetic Materials." Addison-Wesley, Reading, Massachusetts.

Daniels, J. M., and Rosencwaig, A. (1970). *Can. J. Phys.* **48**, 381.

Danon, J., Enriques, C. R., Mattievich, E., and Da c. de M. Coutinho Beltrão, Maria (1976). *J. Phys.* **37**, C6-866.

Danson, D. P., Williams, J. M., and Janot, C. (1977). *Proc. Int. Conf. Mössbauer Spectrosc., Bucharest, Romania* (D. Barb and D. Tarina, eds.), Vol. 1, p. 307.

Dattagupta, S., and Blume, M. (1974). *Phys. Rev. B* **10**, 4540.
Dézsi, I., and Coey, J. M. D. (1973). *Phys. Status Solidi (a)* **15**, 681.
Dormann, J. L., Gibart, P., and Renaudin, P. (1976). *J. Phys.* **37**, C6-281.
Dormann, J. L., Gibart, P., and Suran, G. (1977). *Physica* **86-88B**, 1431.
Dumesic, J. A., and Topsøe, H. (1977). *Adv. Catal.* **26**, 121.
Eibschüts, M., and Shtrikman, S. (1968). *J. Appl. Phys.* **39**, 997.
Eisenstein, I., and Aharoni, A. (1976a). *J. Appl. Phys.* **47**, 321.
Eisenstein, I., and Aharoni, A. (1976b). *Phys. Rev. B* **14**, 2078.
Eisenstein, I., and Aharoni, A. (1977a). *Phys. Rev. B* **16**, 1278.
Eisenstein, I., and Aharoni, A. (1977b). *Phys. Rev. B* **16**, 1285.
Eisenstein, I., and Aharoni, A. (1977c). *Physica* **86-88B**, 1429.
Eynatten, G. von, and Bömmel, H. E. (1977). *Appl. Phys.* **14**, 415.
Gallagher, P. K. (1976). *In* "Applications of Mössbauer Spectroscopy" (R. L. Cohen, ed.),
    Vol. 1, p. 199. Academic Press, New York.
Gallagher, P. K., and Kurkjian, C. R. (1966). *Inorg. Chem.* **5**, 214.
Gangas, N. H., Simopoulos, A., Kostikas, A., Yassoglou, N. J., and Filippakis, S. (1973).
    Clays Clay Minerals **21**, 151.
Gangas, N. H. J., Sigalas, I., and Moukarika, A. (1976). *J. Phys.* **37**, C6-867.
Giessen, A. A. van der (1967). *J. Phys. Chem. Solids* **28**, 343.
Gonser, U. (1968/69). *Mater. Sci. Eng.* **3**, 1.
Gonser, U., Wiedersic, H., and Grant, R. W. (1968). *J. Appl. Phys.* **39**, 1004.
Govaert, A., Dauwe, C., Plinke, P., Grave, E. de, and Sitter, J. de (1976). *J. Phys.* **37**, C6-825.
Haneda, K., and Morrish, A. H. (1976). *IEEE Trans. Magn.* **12**, 767.
Haneda, K., and Morrish, A. H. (1977). *Phys. Lett.* **64A**, 259.
Haneda, K., and Morrish, A. H. (1978). *Surf. Sci.* **77**, 584.
Haneda, K., and Morrish, A. H. (1979). *Nature (London)* **282**, 186.
Hanzel, D., and Sevsek, F. (1976). *J. Phys.* **37**, C6-277.
Helgason, O., Steinthorsson, S., Mørup, S., Lipka, J., and Knudsen, J. E. (1976). *J. Phys.* **37**,
    C6-829.
Herpin, A. (1968). "Theorie du Magnétisme." Presses Univ. de France, Paris.
Hill, V. G., Weir, C., Collins, R. L., Hoch, D., Radcliffe, D., and Wynter, C. (1978). *Hyperfine
    Interact.* **4**, 444.
Hobson, M. C., Jr., and Gager, H. M. (1970). *J. Catal.* **16**, 254.
Housley, R. M., Grant, R. W., Muir, A. H., Blander, M., Abdel-Gawad (1971). *Proc. Lunar
    Sci. Conf. Geochim. Cosmochim. Acta, 2nd, Suppl.* **2** 3, 2125.
Huffman, G. P., and Dunmyre, G. R. (1975). *Proc. Lunar Sci. Conf., 6th*, p. 757.
Imshennik, V. K., Suzdalev, I. P., Bogner, L., Ogrodnik, A., and Litterst, F. J. (1976). *J. Phys.*
    37, C6-751.
Ishikawa, Y. (1962). *J. Phys. Soc. Jpn.* **17**, 1877.
Ishikawa, Y. (1964). *J. Appl. Phys.* **35**, 1054.
Jacobs, I. S., and Bean, C. P. (1963). *In* "Magnetism" (G. T. Rado and H. Suhl, eds.), Vol. III.
    Academic Press. New York.
Janssen, M. M. P. (1970). *J. Appl. Phys.* **41**, 384.
Johnson, C. E., and Glasby, G. P. (1969). *Nature (London)* **222**, 376.
Kaiser, R., and Miskolczy, G. (1970). *J. Appl. Phys.* **41**, 1064.
Keisch, B. (1974). *J. Phys.* **35**, C6-151.
Keisch, B. (1976). *In* "Applications of Mössbauer Spectroscopy" (R. L. Cohen, ed.), Vol. 1, p.
    263. Academic Press, New York.
Keune, W., and Sette Camara, A. (1975). *Phys. Status Solidi (a)* **27**, 181.
Keune, W., Williamson, D. L., and Lauer, J. (1975). *Proc. Int. Conf. Mössbauer Spectrosc.,
    Cracow, Poland* (A. Z. Hrynkiewicz and J. A. Sawicki, eds.), Vol. 1, p. 39.

Keune, W., Lauer, J., Gonser, U., and Williamson, D. L. (1979). *J. Phys.* **40**, C2-69.
Kneller, E. (1969). *In* "Magnetism and Metallurgy" (A. E. Berkowitz and E. Kneller, eds.). Academic Press, New York.
Knudsen, J. E. and Mørup, S. (1980). *J. Phys.* **41**, C1-155.
Kodama, K., McKeague, J. A., Tremblay, R. J., Gosselin, J. R., and Townsend, M. G. (1977). *Can. J. Earth Sci.* **14**, 1.
Kopcewicz, B., and Kopcewicz, M. (1976). *J. Phys.* **37**, C6-841.
Kopcewicz, M., and Kotlicki, A. (1975). *Radiat. Eff.* **24**, 267.
Kopcewicz, M., and Kotlicki, A. (1977). *Proc. Int. Conf. Mössbauer Spectrosc., Bucharest, Romania* (D. Barb and D. Tarina, eds.), p. 87.
Kostikas, A., Simopoulos, A., and Gangas, N. H. (1974). *J. Phys.* **35**, C1-107.
Kostikas, A., Simopoulos, A., and Gangas, N. H. (1976). *In* "Applications of Mössbauer Spectroscopy" (R. L. Cohen, ed.), Vol. 1, p. 241. Academic Press, New York.
Kraan, A. M. van der (1971). *J. Phys. Suppl.* **32**, C1-1034.
Kraan, A. M. van der (1972). Thesis (unpublished).
Kraan, A. M. van der (1973). *Phys. Status Solidi (a)* **18**, 215.
Kraan, A. M. van der, and Loef, J. J. van (1966). *Phys. Lett.* **20**, 614.
Krop, K., and Williams, J. M. (1971). *J. Phys. F* **1**, 938.
Krop, K., Korecki, J., Zukrowski, J., Karaś, W., and Williams, J. M. (1973). *Proc. Int. Conf. Mössbauer Spectrosc., Bratislava, Czechoslovakia* (M. Hucl and T. Zemčik, eds.), Vol. 2, p. 462. Nuclear Information Centre, Prague.
Krop, K., Korecki, J., Zukrowski, J., and Karaś, W. (1974). *Int. J. Magn.* **6**, 19.
Krupyanskii, Y. F., and Suzdalev, I. P. (1974a). *J. Phys.* **35**, C6-407.
Krupyanskii, Y. F., and Suzdalev, I. P. (1974b). *Zh. Eksp. Teor. Fiz.* **67**, 736 [*English transl.: Sov. Phys. JETP* **40**, 364 (1975)].
Krupyanskii, Y. F., and Suzdalev, I. P. (1975). *Fiz. Tverd. Tela* **17**, 588 [*English transl.: Sov. Phys. Solid State* **17**, 375 (1975)].
Kubsch, H., and Schneider, H. A. (1973). *Proc. Int. Conf. Mössbauer Spectrosc., 5th, Bratislava, Czechoslovakia* (M. Hucl and T. Zemčik, eds.), p. 718. Nuclear Information Centre, Prague.
Kündig, W., Bömmel, H., Constabaris, G., and Lindquist, R. H. (1966). *Phys. Rev.* **142**, 327.
Kündig, W., Lindquist, R. H., and Constabaris, G. (1967a). "Colloque Ampere," Vol. XIV, p. 1029. North-Holland Publ., Amsterdam.
Kündig, W., Ando, K. J., Lindquist, R. H., and Constabaris, G. (1967b). *Czech. J. Phys.* **B17**, 467.
Kündig, W., Kobelt, M., Appel, H., Constabaris, G., and Lindquist, R. H. (1969). *J. Phys. Chem. Solids* **819**.
Liebermann, L. N., Fredkin, D. R., and Shore, H. B. (1969). *Phys. Rev. Lett.* **22**, 539.
Liebermann, L. N., Clinton, J., Edwards, D. M., and Mathon, J. (1970). *Phys. Rev. Lett.* **25**, 232.
Lindquist, R. H., Constabaris, G., Kündig, W., and Portis, A. M. (1968). *J. Appl. Phys.* **39**, 1001.
Lipka, J., Mørup, S., and Topsøe, H. (1977a). *Proc. Int. Conf. Soft Magn. Mater., 3rd, Bratislava, Czechoslovakia* p. 763.
Lipka, J., Mørup, S., and Topsøe, H. (1977b). Unpublished results.
Litterst, F. J., Bröll, W., and Kalvius, G. M. (1974). *J. Phys.* **35**, C6-415.
Manykin, E. A., and Onishchenko, E. V. (1976). *Fiz. Tverd. Tela (Leningrad)* **18**, 3203 [*English transl.: Sov. Phys. Solid State* **18**, 1870 (1976)].
Mathalone, Z., Ron, M., and Biran, A. (1970). *Solid State Commun.* **333** (1970).
McNab, T. K., and Boyle, A. J. F. (1968). *In* "Hyperfine Structure and Nuclear Radiations," p. 957. North-Holland Publ., Amsterdam.

McNab, T. K., Fox, R. A., and Boyle, A. J. F. (1968). *J. Appl. Phys.* **39**, 703.
Meiklejohn, W. H. (1962). *J. Appl. Phys. Suppl.* **33**, 1328.
Meiklejohn, W. H., and Bean, C. P. (1957). *Phys. Rev.* **105**, 904.
Meyer, C. T. (1976). *J. Phys.* **37**, C6-777.
Morrish, A. H. (1965). "The Physical Principles of Magnetism." Wiley, New York.
Morrish, A. H., and Clark, P. E. (1974). *Proc. Int. Conf. Magn. ICM-73* Vol. II, p. 180. Nauka, Moscow.
Morrish, A. H., Haneda, K., and Schurer, P. J. (1976). *J. Phys.* **37**, C6-301.
Mørup, S., and Topsøe, H. (1976). *Appl. Phys.* **11**, 63.
Mørup, S., and Topsøe, H. (1977). *Proc. Int. Conf. Mössbauer Spectrosc., Bucharest, Romania* (D. Barb and D. Tarina, eds.), p. 229.
Mørup, S., Topsøe, H., and Lipka, J. (1976). *J. Phys.* **37**, C6-287.
Mørup, S., Clausen, B. S., and Topsøe, H. (1979). *J. Pys.* **40**, C2-78.
Mørup, S., Clausen, B. S., and Topsøe, H. (1980). *J. Phys.* **41**, C1-331.
Nagarajan, R., and Srivastava, J. K. (1977). *Phys. Status Solidi (b)* **81**, 107.
Nagy, D. L., Kulcsár, K., Ritter, G., Spiering, H., Vogel, H., Zimmerman, R., Dézsi, I., and Pardavi-Horvath, M. (1975). *J. Phys. Chem. Solids* **36**, 759.
Nakamura, T., Shinjo, T., Endoh, Y., Yamamoto, N., Shiga, M., and Nakamura, Y. (1964). *Phys. Lett.* **12**, 178.
Nalovic, L., and Janot, C. (1979). *Rev. Phys. Appl.* **14**, 475.
Nasu, S., Shinjo, T., Nakamura, Y., and Murakami, Y. (1967). *J. Phys. Soc. Jpn.* **23**, 664.
Néel, L. (1949). *Ann. Geophys.* **5**, 99.
Néel, L. (1953). *C. R. Acad. Sci. Paris* **237**, 1468.
Néel, L. (1954). *J. Phys. Rad.* **15**, 225.
Nininger, R. C., and Schroeer, D. (1978). *J. Phys. Chem. Solids* **39**, 137.
Oosterhuis, W. T., and Spartalian, K. (1976). *In* "Applications of Mössbauer Spectroscopy" (R. L. Cohen, ed.), Vol. 1, p. 141. Academic Press, New York.
Petrera, M., Gonser, U., Hasmann, U., Keune, W., and Lauer, J. (1976). *J. Phys.* **37**, C6-295.
Pincus, P. (1960). *Phys. Rev.* **118**, 658.
Pobell, F., and Wittman, F. (1966). *Zh. Angew, Phys.* **20**, 488.
Popma, J. A., and Diepen, A. M. van (1974). *Mater. Res. Bull.* **9**, 1119.
Popma, J. A., and Diepen, A. M. van (1975). *AIP Conf. Proc.* **24**, 123.
Povitskii, V. A., Makarov, E. F., Murashko, N. V., and Salugin, A. N. (1976). *Phys. Status Solidi (a)* **33**, 783.
Quach, H. T., Friedmann, A., Wu, C. Y., and Yelon, A. (1978). *Phys. Rev. B* **17**, 312.
Rathenau, G. W., and Snoek, J. L. (1941). *Physica* **8**, 555.
Roggwiller, P., and Kündig, W. (1973). *Solid State Commun.* **12**, 901.
Sakharov, E. M., and Kuz'min, R. N. (1976). *Pis'ma Zh. Eksp. Teor. Fiz.* **24**, 279 [*English transl.: JETP Lett.* **24**, 248 (1976)].
Saraswat, I. P., Vajpei, A. C., and Garg, V. K. (1977). *Indian J. Chem.* **15A**, 493.
Schroeer, D. (1970). *In* "Mössbauer Effect Methodology" (I. J. Gruverman, ed.), Vol. 5, p. 141. Plenum Press, New York.
Schroeer, D., and Nininger, R. C. (1967). *Phys. Rev. Lett.* **19**, 632.
Selwood, P. W. (1975). "Chemisorption and Magnetization." Academic Press, New York.
Shinjo, T. (1966). *J. Phys. Soc. Jpn.* **21**, 917.
Shinjo, T. (1976). *IEEE Trans. Magn.* **12**, 86.
Shinjo, T., Matsuzawa, T., Takada, T., Nasu, S., and Murakami, Y. (1973). *J. Phys. Soc. Jpn.* **35**, 1032.
Shinjo, T., Matsuzawa, T., Mizutani, T., and Takada, T. (1974). *Jpn. J. Appl. Phys. Suppl. 2 Part 2*, 729.

Shinjo, T., Pfannes, H. D., and Gonser, U. (1975). *Proc. Int. Conf. Mössbauer Spectrosc.*, *Cracow, Poland* (A. Hrynkiewicz and J. A. Sawicki, eds.), p. 465.

Shinjo, T., Hine, S., and Takada, T. (1977). *Proc. Int. Vac. Congr., 7th, and Int. Conf. Solid Surf., 3rd, Vienna* p. 2655.

Shinjo, T., Hine, S., and Takada, T. (1979). *J. Phys.* **40**, C2-86.

Simmons, G. W., and Leidheiser, H. (1976). *In* "Applications of Mössbauer Spectroscopy" (R. L. Cohen, ed.), Vol. 1, p. 85. Academic Press, New York.

Simopoulos, A., and Kostikas, A. (1973). *Proc. Int. Conf. Mössbauer Spectrosc., 5th, Bratislava, Czechoslovakia* (M. Hucl and T. Zemčik, eds.), p. 759. Nuclear Information Centre, Prague.

Singh, A. K., Jain, B. K., and Chandra, K. (1977). *Phys. Status Solidi* **44**, 443.

Topsøe, H., and Boudart, M. (1973). *J. Catal.* **31**, 346.

Topsøe, H., and Mørup, S. (1975). *Proc. Int. Conf. Mössbauer Spectrosc., Cracow, Poland* (A. Z. Hrynkiewicz and J. A. Sawicki, eds.), Vol. 1, p. 321.

Topsøe, H., Dumesic, J. A., and Boudart, M. (1974). *J. Phys.* **35**, C6-411.

Topsøe, H., Dumesic, J. A., and Mørup. S. (1980). *In* "Applications of Mössbauer Spectroscopy" (R. L. Cohen, ed.), Vol. 2. Academic Press, New York.

West, F. G. (1961). *J. Appl. Phys. Suppl.* **32**, 2495.

Wickman, H. H. (1966). *In* "Mössbauer Effect Methodology" (I. J. Gruverman, ed.), Vol. 2, p. 39. Plenum Press, New York.

Williams, J. M., Danson, D. P., and Janot, C. (1978). *Phys. Med. Biol.* **23**, 835.

2

Catalysis and Surface Science

Henrik Topsøe

Haldor Topsøe Research Laboratories
Lyngby, Denmark

James A. Dumesic

Department of Chemical Engineering
University of Wisconsin
Madison, Wisconsin

Steen Mørup

Laboratory of Applied Physics II
Technical University of Denmark
Lyngby, Denmark

APPLICATIONS OF MÖSSBAUER
SPECTROSCOPY, VOL. II

I. Introduction

Mössbauer spectroscopy has many unique features that make this technique of special interest for catalysis and surface science studies. It is of particular interest that Mössbauer spectroscopy is one of the few techniques that allows UHV surface science studies, as well as *in situ* studies of catalysts under conditions typical of the catalytic reaction. Therefore, the number of applications keeps growing, and many new areas of application in catalysis and surface science are being explored. This chapter will discuss some of the most recent developments within these two fields.

Catalysis and surface science will be combined in this chapter since the two fields greatly overlap. Surface studies on well-defined systems are often used as models for understanding the complex surfaces of real catalysts. On the other hand, studies on small catalyst particles, in which a significant number of the atoms are surface atoms, can be equally useful for examining surface phenomena.

In Section II some of the experimental considerations special for catalysis and surface science studies will be given. *In situ* cells for measurements at high or low temperatures, with and without magnetic fields, will be described. The use of applied magnetic fields is especially important for obtaining particle size information. Possible ways of studying materials not containing Mössbauer isotopes are briefly elucidated. The backscattering techniques utilizing the finite penetration depth of conversion electrons or x rays are also discussed.

Surface studies will be discussed in Section III. The discussion will be devoted to characterization of structural, electronic, and magnetic properties of surfaces, as well as to the effect of chemisorption on these properties. The examples will be taken from studies on small particles and thin films. Studies where the adsorbate contains the Mössbauer active isotope will be exemplified by recent studies using Grafoil as the adsorbent. Section III also deals with the many recent studies of the magnetic properties of surfaces and interfaces. The backscatter studies, which have been especially useful when information regarding many surface layers is desired, will complete this section.

Section IV deals with the application of Mössbauer spectroscopy to catalysis. There will be subsections devoted to characterization of both unsupported and supported catalysts. The range of applications of Mössbauer spectroscopy will be illustrated by discussion of some catalyst systems selected out of those currently attracting much interest. The discussion will include multimetallic clusters and alloy catalysts, ammonia synthesis catalysts, hydrocarbon synthesis catalysts, ruthenium catalysts, partial oxidation catalysts, hydrodesulfurization catalysts, and catalysts for electrocatalysis. Section IV will also include discussions of particle size and dispersion measurements and properties of supported catalysts. The whole chapter will emphasize recent developments since the literature up to about 1975 has been reviewed elsewhere (Dumesic and Topsøe, 1977).

II. Methodology

A. *In situ Cells for Catalyst Studies*

A major advantage of Mössbauer spectroscopy is the fact that the technique allows *in situ* experiments under a great variety of conditions. The measurements can be carried out under experimental conditions with temperatures ranging from below that of liquid helium to above 1000 K and pressures ranging from UHV to above 100 atm.

Catalytic reactions typically occur in the temperature range 500–800 K. In addition, catalyst activation often requires similar temperatures. Thus the capability of high temperature measurements with the catalysts exposed to different gases is an important requirement for an *in situ* cell for catalyst studies. The possibility of performing Mössbauer experiments at high and low temperatures is also of great interest since the temperature dependence of the Mössbauer parameters often contains information that is essential for the detailed interpretation of the spectra.

For some isotopes with high γ-ray energies (e.g., Ru, Au, Ni), measurements must be carried out at low temperatures to obtain a measurable Mössbauer effect. Catalyst studies using such isotopes may be carried out by conducting the reactions at high temperatures and subsequently carrying out measurements at, for example, 4.2 K. The ideal Mössbauer cell therefore allows both high and low temperature studies. Nevertheless, when studying samples containing ^{57}Fe, a simple cell like the one shown in Fig. 1, allowing measurements at room temperature and above, is often sufficient (B. S. Clausen, 1976). The cell can be made of either quartz or

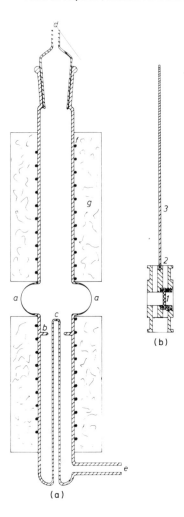

Fig. 1. *In situ* glass Mössbauer cell and stainless steel absorber holder for studies above room temperature: (a)   a, window for transmission of γ rays; b, basis for absorber holder; c, tube for temperature measurement and regulation; d, inlet of gas; e, outlet of gas; f, heating coil; g, insulating material. (b)   1, absorber; 2, channels for gas flow; 3, rod for removal of the absorber holder (B. S. Clausen, 1976).

Pyrex glass with its windows made thin enough to allow transmission of the γ rays.

It should be pointed out that in addition to collecting *in situ* Mössbauer spectra, it may be advantageous to measure simultaneously the catalytic reaction rate. Such studies could be carried out using a cell like the one shown in Fig. 1. Other solutions have been described elsewhere (see, e.g.,

C. A. Clausen and Good, 1975; Dumesic and Topsøe, 1977; Raupp and Delgass, 1979c).

In studies of catalysts consisting of magnetically ordered microcrystals, it is of great advantage to be able to apply magnetic fields [see Section II, B and Chapter 1 (Mørup et al., 1980b)]. The cell shown in Fig. 2 (B. S. Clausen et al., 1979a) allows such studies in the temperature range 78–725 K with the sample in ultrahigh vacuum or exposed to a desired gas or gas mixture.

A cell allowing in situ pretreatment at temperatures up to about 700 K and measurements down to liquid helium temperature in the same chamber was constructed by Bartholomew (1972). Such a cell is quite complicated and has many possible heat leaks. A simpler, but less flexible, approach described by C. A. Clausen and Good (1975) allowed pretreatment of the catalysts at high temperatures in a small, thin-windowed glass container. This container was sealed after the pretreatment and could subsequently be placed in a conventional liquid helium dewar.

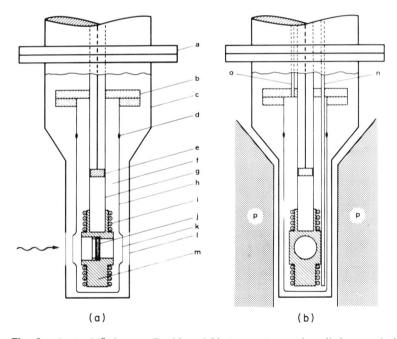

(a)   (b)

Fig. 2. In situ Mössbauer cell with variable temperature and applied magnetic field capabilities: a, flange with O-ring; b, Conflat flange with Cu gasket; c, vacuum and heat shield; d, glass to metal seal; e, movable piston; f, in situ chamber; g, liquid nitrogen tube; h, Pyrex glass; i, heating coil; j, Mössbauer sample wafer; k, Pyrex window; l, Mylar or Ni foil windows; m, Cu block; n, gas inlet tube; o, gas outlet tube; p, pole shoes. (B. S. Clausen et al., 1979a. J. Phys. E: Sci. Instrum. **12**, 439. Copyright of The Institute of Physics.)

Recently B. S. Clausen (1979) has described how high temperature pretreatments in a simple way can be combined with liquid helium temperature measurements. By use of a glass cell and a toploaded cryostat, the sealing of the cell can be avoided. This design was used for studies of chemisorption effects on nickel particles (B. S. Clausen et al., 1979b).

Other designs of in situ cells are described in the literature (Cohen and Wertheim, 1974; Delgass et al., 1976; and Dumesic and Topsøe, 1977). Finally, for in situ catalyst studies it is often convenient to connect the Mössbauer cells to a general purpose vacuum gas handling system which allows: (i) gases (or gas mixtures) to be passed over the sample, (ii) evacuation of gases adsorbed on the catalyst surface, and (iii) dosage of calculated amounts of gases to the catalyst.

B. Applied Magnetic Fields

In many cases the application of magnetic fields to the sample during collection of the Mössbauer spectrum yields valuable new information on catalysts and surfaces. Small fields (up to about 20 kG) are conveniently obtained by using an iron core electromagnet. Fields up to about 100 kG can be obtained by using a superconducting coil. However, since the superconducting coil must be kept at a temperature near liquid helium temperature, very large fields are not easily combined with in situ cells operating at high temperatures.

The application of magnetic fields can, for example, give information about the magnetic moment of the magnetic species (atoms or particles). In paramagnetic samples the magnetic splitting of the spectra as a function of the applied field is essentially proportional to a Brillouin function. Thus the spin state of the atoms can be estimated from the field dependence of the spectra (Chappert, 1974). For example, high- and low-spin states of iron atoms can be distinguished in this way. The type of magnetic order in magnetically ordered compounds may also be elucidated by studies in applied fields (see, e.g., Section III, F).

In studies of superparamagnetic particles the magnetic splitting of the spectra is essentially proportional to the Langevin function L ($\mu H/kT$), where μ is the magnetic moment of a particle. Since μ is often very large (of the order of 10^4 μ_B), magnetic saturation can be achieved in moderate applied fields even above room temperature. Thus the magnetic moment of the particles can be determined from the field dependence of the spectra; and, if the magnetization of the particles is known, the particle size can be estimated. This is discussed in more detail in Chapter 1 (Mørup et al., 1980b), and an application to Ni catalysts is given in Section IV, B. The cell shown in Fig. 2 allows in situ studies in the presence of an applied magnetic field.

C. Important Mössbauer Isotopes

Although the Mössbauer effect has been detected in about 100 nuclear transitions, few have practical importance for studies of surfaces and catalysts. For many γ transitions the high recoil energy makes cooling of both the source and the absorber necessary, which, of course, limits the possibility of doing *in situ* studies at reaction temperatures. Other transitions have a very short lifetime resulting in very broad lines, which makes it difficult to obtain chemical information from the spectra. Also, other drawbacks limit the use of certain isotopes, e.g., a short lifetime of the parent nuclide, need of access to an accelerator to produce the source, or high internal conversion coefficients. A detailed discussion of the feasibility of using the different Mössbauer isotopes for catalyst and surface studies has been given by Dumesic and Topsøe (1977).

The most important Mössbauer isotope in catalyst and surface studies has turned out to be ^{57}Fe. This is due to the ease with which it can be employed and the relatively large number of iron-containing systems. ^{119}Sn, ^{121}Sb, and ^{151}Eu are also important Mössbauer isotopes for studies of catalysts and surfaces. In spite of difficulties in performing experiments with isotopes like ^{99}Ru and ^{197}Au, several interesting Mössbauer catalytic studies have appeared (see, e.g., Sections IV, D and IV, G).

D. Doping with Mössbauer Isotopes and Source Experiments

Catalysts and surfaces that do not contain Mössbauer atoms can be studied by doping the sample with a suitable Mössbauer isotope. This can be done either by doping the sample with a stable isotope (e.g., ^{57}Fe) or with a radioactive mother isotope (e.g., ^{57}Co). In the former case the sample can be used in normal transmission geometry. In the latter case the sample is used as a source and the Mössbauer spectrum is generated using a standard single-line absorber. Such experiments are often referred to as *source experiments* or *emission Mössbauer spectroscopy measurements*. When the sample is studied in an *in situ* cell, the required Doppler velocity is most conveniently imparted to the standard absorber.

When a sample is doped with a stable Mössbauer isotope and then used as the absorber in transmission geometry, a sufficient number of Mössbauer atoms must be added to the sample in order to produce a spectrum in a reasonably short time. This quantity may be so great that the solubility limit is exceeded, or the structural, chemical, and catalytic properties of the sample as a whole may be affected. On the other hand, when the sample is doped with source atoms, a much smaller number of atoms are necessary. However, in source experiments, the decay process may give rise to formation of long-lived changes in the chemical surroundings of the Mössbauer nuclei. For instance, when a sample is doped with ^{57}Co, the electron capture process preceding the population of the excited Mössbauer level will result in formation of unstable electronic configurations of the ^{57}Fe atoms. (For a detailed discussion of such *aftereffects*, see, e.g., Wertheim, 1971; Friedt and Danon, 1972; Seregin *et al.*, 1978.)

Most of the unstable electronic configurations will decay in a time that is short compared to the lifetime of the excited Mössbauer state and will, therefore, not be detected. In conductors, the high electron mobility ensures that unstable charge states of the Mössbauer atoms have very short lifetimes, and source experiments then give useful information about the stable state of the Mössbauer atoms. In insulators, however, some of the radioactive atoms may be found in unstable valence states with lifetimes longer than the lifetime of the Mössbauer nucleus. In these cases the observed valences may, therefore, not be those of the stable Mössbauer atom, and this makes the interpretation of the spectra difficult. In spite of this, information on the structure, the dynamics, the chemical state, and magnetic properties can often be obtained.

E. Backscatter Experiments

After a Mössbauer nucleus has resonantly absorbed a γ quantum, it decays to its ground state with the reemission of a γ quantum or with the emission of an atomic s electron. The latter process is called internal conversion, and the internal conversion coefficient α is defined as the ratio of number of emitted internal conversion electrons to the number of reemitted γ rays. After emission of a conversion electron, the hole in the s-electron shell is filled by one of the outer electrons; this takes place with the emission of an x-ray quantum or an Auger electron. Thus, by using a scattering geometry instead of the usual transmission geometry, a Mössbauer spectrum can be obtained by counting either the γ rays, x rays, or electrons. For ^{57}Fe, the value of α is approximately 10, and it is, therefore, advantageous to detect the emitted *internal conversion electrons* or *x rays*.

One advantage of conversion electron Mössbauer spectroscopy (CEMS) is that the escape depth of the electrons (≈ 100 nm) is several orders of magnitude smaller than for γ or x rays. Thus, even for massive samples, the Mössbauer spectrum becomes somewhat surface sensitive (see Chapter 3).

The detection of electrons instead of γ quanta imposes some restrictions on the design of the spectrometer. In particular, it must be noted that the electrons cannot penetrate the thin windows of beryllium or aluminum normally used in proportional counters. Therefore, the sample must be placed inside the counter during the measurements, thus limiting the conditions under which experiments can be carried out. *In situ* studies of surfaces exposed to different gases generally cannot be carried out by the use of this method since the gases may hinder the functioning of the counter.

The essential features of a simple Mössbauer counter for CEMS (a so-called resonance counter) are shown in Fig. 3. This particular counter was constructed by Fenger (1973) for studies of surfaces of massive flat samples. The sample is sealed to the counter with a thick layer of vacuum grease and is used as a combined absorber and cathode in the counter, which contains three 0.1-mm stainless steel anode wires. The 14.4-keV γ radiation emitted from a ^{57}Co source, which is moved by a conventional transducer, is resonantly absorbed in the sample, and the emitted conversion electrons are detected at the anodes in the counter, which is filled with a 1–10% methane in helium gas mixture.

Similar resonance counters have been designed for studies of samples small enough to be placed inside the counter (Fenger, 1969; Sette Camara and Keune, 1975; Tricker *et al.*, 1976a; Tricker, 1977). Weyer (1976) constructed a resonance counter with a flat anode plate parallel to the cathode, instead of using anode wires. The counter had small dimensions and mass and could easily be moved by a conventional Mössbauer drive system and used for Mössbauer experiments with a stationary source outside the counter. A counter that can be used for low temperature studies has been described by Sawicki *et al.* (1976).

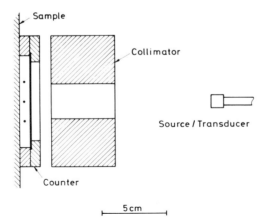

Fig. 3. Mössbauer resonance counter for studies of surfaces of massive samples (Fenger, 1973. *Nucl. Instrum. Methods* **106**, 203).

It has been shown that the sensitivity of resonance counters is so high that even a monolayer of ^{57}Fe on a support can be detected (Petrera *et al.*, 1976). The conversion electrons emitted from different depths in the sample will have different energies. The electrons emitted from the outer surface layers will possess the full internal conversion energy ($\approx$ 7 keV), whereas those emitted from subsurface layers will leave the sample with a lower energy (Krakowski and Miller, 1972). Thus, by collecting Mössbauer spectra using different detected electron energies, a nondestructive depth profile of the sample may be carried out (Bonchev *et al.*, 1969). This technique has been referred to as *energy selective conversion electron Mössbauer spectroscopy*. The technique requires good energy resolution of the electron detector, which is achieved by using electron multiplier tubes for detection of electrons in connection with a β-ray spectrometer (Bonchev *et al.*, 1969; Bäverstam *et al.*, 1974a, b; Schunck *et al.*, 1975; Oswald and Ohring, 1976; Jones *et al.*, 1978). The essential features of such a spectrometer are shown in Fig. 4 (Jones *et al.*, 1978). In this system the current in the coils of the spectrometer determines the energy of electrons that are detected by the electron multiplier tube. Using such a system, a depth resolution of about 5 nm can be obtained (Bäverstam *et al.*, 1974a, b).

Recently Tricker *et al.* (1979) demonstrated a different method for obtaining depth selective backscatter Mössbauer spectra. This method is based on the fact that there is a component in the electron spectrum

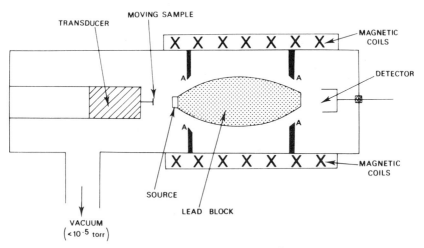

Fig. 4. β-ray spectrometer for conversion electron Mössbauer spectroscopy (CEMS) surface studies. The radius of the magnetic coils is 10 cm and the sample to detector distance 26 cm; A = lead apertures (Jones *el al.*, 1978. *Appl. Surf. Sci.* **1**, 338).

arising from photoelectrons produced by backscattered γ and x rays from atoms within the sample (Tricker et al., 1977). This component can be enhanced by evaporating an inert overlayer of a few tens of nanometers onto the surface of the sample. It is interesting that the signal from the deeper layers increases with increasing thickness of the overlayer made, for example, of gold. Thus, studies with different thickness of the overlayer give information on the depth distribution of Mössbauer atoms in different chemical and magnetic environments.

Backscatter Mössbauer spectroscopy by detection of x rays can be used for studies of surface layers of 10–100 μm. It is mainly used for nondestructive studies of surfaces of massive (thick) samples. The experimental technique and its applications for studies of steel surfaces have been discussed by Collins (1968). A simple experimental arrangement for ^{57}Fe x-ray backscatter Mössbauer spectroscopy, utilizing a proportional counter with a high efficiency for 6.3-keV x rays and a low efficiency for 14.4-keV γ rays, has been described by Østergaard (1970).

F. Analysis of Mössbauer Spectra

Mössbauer parameters are usually obtained by computer fitting the data to a number of Lorentzian lines and using one of the many available fitting routines. For well-crystallized samples this is done without major difficulties, but in catalyst and surface studies the Mössbauer atoms will often be located in many different sites, each with a different set of Mössbauer parameters. As a result, the experimental spectra often consist of broad lines, reflecting variations in the surroundings of the Mössbauer atoms, and computer fitting with a small number of Lorentzian lines may not give meaningful Mössbauer parameters.

Computer programs that can be used for determinating distribution in Mössbauer parameters have recently been developed (Window, 1971; Hesse and Rübartsch, 1974; Mangin et al., 1976) and will undoubtedly be applied in future catalyst and surface studies.

III. Surface Studies

A. Feasibility of Surface Studies

Mössbauer spectroscopy can conveniently be used for surface studies when a significant fraction of the Mössbauer atoms in the sample are present at the surface. Two different cases can be imagined for studies of this type: (i) The Mössbauer atoms are present in small particles or thin films and (ii) the resonant atoms are present in chemical species adsorbed (adsorbate) on surfaces of a solid. Both cases will be discussed in this section, together with possible ways of obtaining surface information from samples having only a small fraction of the Mössbauer atoms located at the surface.

In the case of iron Mössbauer spectroscopy, the 14.4-keV γ rays have a nonresonant penetration depth into metallic iron of approximately 10 μm (where the penetration depth is the distance from the surface to the point in the sample where the γ-ray intensity is reduced by photoelectric absorption to half of its initial value). This depth corresponds to approximately 10^5 atomic layers. Therefore, if the Mössbauer experiment is carried out in the conventional transmission geometry (i.e., detection of γ rays that have passed through the sample), the contribution to the observed spectrum from the surface of bulk samples will be imperceptibly small. It is only in samples in which a significant fraction of the resonant atoms (e.g., ^{57}Fe) is actually present at the surface that the technique is sensitive to surface phenomena. Surface information may be obtained from transmission geometry measurements on *small particles* and *thin films*. For example, if the resonant atoms are located uniformly throughout a small spherical particle with a diameter of less than about 10 nm, more than 10% of the resonant atoms will be at the surface. In the present discussion, it is appropriate to speak of the "resonant atom dispersion," which is defined as the fraction of resonant atoms in the sample present at the surface. Correspondingly, resonant atom dispersions higher than ≈ 0.01 are necessary for "conventional" Mössbauer spectroscopy surface studies using transmission geometry.

Surface information may be obtained for samples with a small total atom dispersion by enriching the surface in the Mössbauer isotope. As an example of the *surface-enrichment* approach, consider the case of iron. The actual total atom dispersion of iron in the sample may be small, but by enriching the surface with ^{57}Fe while maintaining the natural abundance of ^{57}Fe (2.2%) in the bulk, the resonant atom dispersion (of ^{57}Fe) in the sample can be made much larger (by a factor of ≈ 40) than the actual total iron dispersion (Van der Kraan, 1973; Haneda and Morrish, 1977; Lauer *et al.*, 1977).

Detailed information about surface structures of such particles is not easily obtained since their surface structure may be complex. In order to obtain information about surface structures it may, therefore, be advantageous to perform experiments on systems with well-defined surface structures (e.g., single crystals). Such studies would be of particular interest when combined with other surface investigations of the same sample, using such techniques as Auger electron spectroscopy and x-ray and ultraviolet photoelectron spectroscopies. The combined use of Mössbauer spectroscopy and electron spectroscopy has, in fact, been reported in the literature (Tricker *et al.*, 1974; Clausen, C. A., and Good, 1977c). From the viewpoint of Mössbauer spectroscopy, however, studies on well-defined surfaces are not straightforward, owing to the difficulty in generating the surface sensitive Mössbauer spectrum within a reasonable span of time (e.g., in less than several days). For example, if a monolayer of ^{57}Fe atoms is deposited on an iron-free surface, the relative γ-ray absorption would be very low when using the transmission geometry, and a counting time of the order of weeks may be required to obtain a spectrum with reasonably good signal-to-noise ratio.

In order to reduce the counting time and yet retain the conventional transmission geometry for Mössbauer spectroscopy, it is necessary to use several surface monolayers in the γ beam (Keune and Gonser, 1971; Keune *et al.*, 1974). Alternatively, a few atomic layers of the resonant isotope can be evaporated onto a substrate, followed by evaporating an "inert" covering layer over the resonant isotope. Repeated evaporation of the resonant isotope and inert covering layers leads to a "*sandwich arrangement*" of these materials, and in this way a single sample with many "surface" layers can be prepared (Shinjo, 1976; Shinjo *et al.*, 1977; Lauer *et al.*, 1977).

The preparation of such "sandwich" samples has been discussed by Marchal and Janot (1972). It should be noted that "sandwich"-type samples are ideal for studying solid–solid interfacial phenomena, whereas other configurations must be used for studies of solid–gas or solid–vacuum problems.

A significant reduction in the counting time can also be achieved by generating the Mössbauer spectrum through detection of resonantly scattered radiation using the *scattering geometry*, instead of the transmission geometry. If the case of a sample consisting of a monolayer of ^{57}Fe atoms is considered again, the advantage of going from transmission to conversion electron backscatter Mössbauer spectroscopy lies in a reduction of the counting time by about two orders of magnitude. This method has been discussed in Section II, E.

Another possibility of collecting a Mössbauer spectrum from a monolayer of ^{57}Fe is based on the incorporation of the parent Mössbauer nuclide (^{57}Co) into the surface under study (Burton and Godwin, 1967; Shinjo *et al.*, 1973; Shinjo, 1979). In this case, the Mössbauer spectrum is generated in the conventional transmission geometry using the sample as the radioactive source and using a single-line iron-containing material (e.g., $K_4Fe(CN)_6 \cdot 3H_2O$) as an absorber. If one considers a sample consisting of a monolayer of ^{57}Co atoms, the advantage of this type of *source experiment* will be a reduction in the counting time by about three orders of magnitude, relative to the counting time for transmission experiments with a ^{57}Fe monolayer sample.

B. Surface Information from Mössbauer Spectra

In this section the different types of surface information that can be obtained from a Mössbauer spectrum are discussed. In the analysis of the spectrum of a sample with a high resonant atom dispersion, the identification of surface components is not always a straightforward procedure. In many cases a comparison with spectra of similar samples with smaller resonant atom dispersions (e.g., "bulk samples") can assist in the interpretation. It may also be useful to change selectively the chemical state of the surface atoms (e.g., by chemisorption). In this case the spectral component from the surface atoms is altered, while the component from all other Mössbauer atoms remains unaffected. When the surface component in the spectrum has been identified, an analysis of the Mössbauer parameters of this component can give detailed information about the physical and chemical properties of the surface atoms, which is briefly discussed in what follows below.

The spectral area of the surface component is closely related to the f factor, and studies of this parameter elucidate the vibrational state of the surface atoms and the corresponding strength of bonding of these atoms.

The isomer shift is most useful for determining the surface chemical state. It reflects the distribution of electrons surrounding the nucleus, and thus gives information about both the valence of the surface atoms and the extent of electron transfer among atoms at the surface. This may allow examination of both the intrinsic difference between the electronic structure of bulk and surface atoms and also the influence of chemisorption on the electronic surroundings of surface atoms.

The quadrupole splitting is particularly useful for surface structure measurements. The quadrupole splitting is determined by the electric field gradient at the nucleus of the resonant atom. Thus it depends on the symmetry of the charge surrounding the nucleus through effects due to crystal field splitting, charge transfer between the central atom and its neighbors, and the distribution of charges over the lattice sites. The quadrupole splitting may therefore serve to identify the site in which the Mössbauer atom is located.

The magnetic hyperfine interaction may also give information relevant to surface studies. In magnetic samples the hyperfine field of surface atoms gives information about their electronic structure reflecting, for example, the chemical bonds of atoms at surfaces and interfaces. This is discussed further in Section III, D.

From measurements of superparamagnetic relaxation, the particle size (dispersion) and the magnetic anisotropy energy constant may be obtained (Mørup *et al.*, 1980b). The latter parameter has been shown to depend both on the detailed geometric arrangement of the surface atoms and on the chemical state of these atoms.

Mössbauer spectra of paramagnetic samples may be affected by relaxation effects, and in such cases the spin–spin and spin–lattice relaxation times may be estimated. The spin–spin relaxation time depends on the local concentration of magnetic atoms (Mørup *et al.*, 1976; Bhargava *et al.*, 1979) and may, therefore, be useful in studies of adsorbate structures. For example, it may be possible to distinguish the situation where the adsorbed molecules are arranged randomly from the situation where they are arranged in islands. The spin–lattice relaxation time is determined by the interaction between the phonons and the spins of the paramagnetic ions and may thus provide information about the phonon spectrum of the surface atoms.

Finally, the diffusion constant of surface atoms may be estimated from the diffusional broadening of the Mössbauer lines. An example of such a study is given in Section III, C.

C. Diffusional and Vibrational Motion of Surface Atoms

1. Adsorbate experiments

It was mentioned earlier that Mössbauer spectroscopy surface studies can be carried out with the resonant atoms present in adsorbed chemical species. Such adsorbate experiments can give information on the vibrational motion of Mössbauer atoms in the adsorbed molecules. Moreover, in cases where surface diffusion takes place, this motion is reflected in the Mössbauer spectra as a line broadening, which increases with increasing diffusion constant. Very rapid liquidlike diffusion results in the disappearance of the Mössbauer signal.

Adsorbate Mössbauer studies can be divided into two groups depending on whether the adsorbents are polycrystalline or have preferred orientation of certain surfaces. A series of interesting studies using Grafoil as an adsorbent belong to this latter group. Grafoil is the trade name for a type of high surface area graphite sheets where the crystallites are all aligned such that their basal planes are oriented (to within ± 15°) parallel to the plane of the sheets. Since Grafoil exhibits both high surface area and preferred orientation, it can be used for studies that normally would have to be done on single crystal surfaces.

The following example will illustrate how information on the aniso-tropic vibrational motion of iron and tin compounds adsorbed on Grafoil can be obtained from Mössbauer spectroscopy studies. The structural information obtained from these studies will be discussed further in Section III, D.

Bukshpan *et al.* (1975) studied the temperature dependence of the recoil-free fraction f for $Sn(CH_3)_4$, $SnCl_4$, and SnI_4 adsorbed on Grafoil at temperatures T between 77 and 300 K. For these species, plots of $\ln f$ versus T (shown in Fig. 5) were linear. The authors treated the vibrating tin atoms as harmonic oscillators. They were then able to calculate the frequency of vibration from the slope of the $\ln f$ versus T plot. Provided that the internal modes of vibration for each molecular unit [e.g., $Sn(CH_3)_4$] have frequencies much greater than the frequency of vibration of the adsorbed species (as a unit) on the surface, the temperature dependence of the recoil-free fraction is primarily due to the latter mode of vibration.

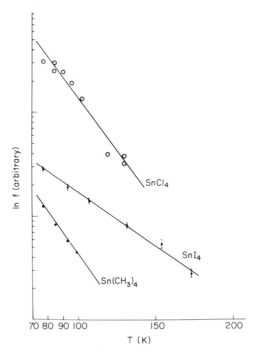

Fig. 5. Temperature dependence of the recoil-free fraction $f(T)$ for (a) $Sn(CH_3)_4$, (b) $SnCl_4$, (c) SnI_4 (Bukshpan *et al.*, 1975. *Surf. Sci.* **52**, 466).

With this assumption, the following frequencies were found for the vibration of the three tin species adsorbed on Grafoil: 4.6×10^{12}, 5.4×10^{12}, 6.7×10^{12} s^{-1} for $Sn(CH_3)_4$, $SnCl_4$, and SnI_4, respectively.

The above vibration frequencies were then used to estimate the strength of the chemical bond between the Grafoil surface and the adsorbed species. A two-parameter equation was used to express the shape of the potential energy well for adsorption, namely: $E(z) = Az^{-9} - Bz^{-3}$, where E is the potential energy, z is the distance between the adsorbed species and the surface, and A and B are constants. The two constants in this equation were evaluated from the experimentally determined frequency of vibration and the estimated equilibrium separation between the adsorbed species and the surface, the latter obtained from published values of bond lengths and ionic radii of the various components. Finally, from the estimated values of the two constants in the potential energy equation, the potential energy minimum (binding energy) for the adsorbed species on the surface was calculated. For the $Sn(CH_3)_4$, $SnCl_4$, and SnI_4 species on Grafoil, the binding energies were 16, 23, and 106 kcal/mole, respectively.

The adsorption of $FeCl_2$ on Grafoil was later studied by Shechter et al. (1976a). These authors prepared samples by heating the $FeCl_2$ and Grafoil to 1023 K (which is above the sublimation temperature of $FeCl_2$), thereby allowing the $FeCl_2$ vapor to deposit on the carbon surface. The amount of $FeCl_2$ deposited on the Grafoil was restricted to surface coverages less than a monolayer. Mössbauer spectra were then collected at various temperatures (between 80 and 300 K) for samples containing different amounts of $FeCl_2$. While the Mössbauer spectrum of bulk, crystalline $FeCl_2$ is a quadrupole-split doublet, the spectra of fractional monolayer "films" of $FeCl_2$ on Grafoil show two distinct quadrupole-split doublets. One spectral doublet, called B, had Mössbauer parameters that were quite similar to those of bulk $FeCl_2$, whereas the other doublet, called L, had a significantly larger quadrupole splitting. For low $FeCl_2$ coverages on Grafoil (e.g., 0.2 monolayer coverage) the B doublet was the dominant spectral component, and at high $FeCl_2$ coverages (e.g., 0.9 monolayer coverage) the L doublet was dominant in the Mössbauer spectra. This result suggests that at high coverages the $FeCl_2$ (B doublet) is present as a "film" with nearly the bulk structure (a layered Cl-Fe-Cl arrangement), whereas at low coverages this bulk structure is not achieved and the Fe^{2+} (L doublet) is present in sites of symmetry different from those in the bulk.

The authors determined the temperature dependence of the recoil-free fractions for the B and L doublets. From these values they calculated the binding energy of these species on Grafoil, using the same procedure as that mentioned earlier in this section; for the B doublet the binding energy was ≈ 80 kcal/mole and for the L doublet ≈ 60 kcal/mole. Both of these values are significantly greater than the heat of fusion (10 kcal/mole) and the heat of evaporation (30 kcal/mole) of $FeCl_2$, indicating that the iron is strongly bonded to the Grafoil surface.

When an adsorbed species is only weakly bound to the surface (as in physical adsorption), it may in some cases be appropriate to treat the adsorbed layer as a two-dimensional fluid, and an important aspect of this treatment is the possible existence of a two-dimensional melting point. This problem was investigated by Shechter *et al.* (1976b) for 1–3, butadiene iron tricarbonyl (BIT) adsorbed on Grafoil. The strategy of this study was to use ^{57}Fe Mössbauer spectroscopy to measure the recoil-free fractions both perpendicular and parallel to the surface. This was done by collecting Mössbauer spectra for different orientations of the Grafoil sheets in the γ-ray beam. The recoil-free fraction perpendicular to the surface $f_\perp$ is a measure of the binding energy of the adsorbed species to the surface. As shown in Fig. 6, the value of $f_\perp$ was observed to decrease smoothly with

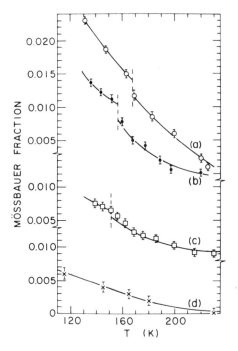

Fig. 6. Mössbauer resonant absorption fractions of butadiene iron tricarbonyl adsorbed on Grafoil, with the γ-ray direction normal to the average surface plane direction. Fractional monolayer coverages x, are curve (a), $x = 0.3$; curve (b), $x = 0.6$; curve (c), $x = 1.0$, and curve (d), bulk solid. The breaks in the curves occur at the two-dimensional melting points of the films and influence $f_\perp$ because of misaligned crystallites (Shechter *et al.* 1976b).

increasing temperature in the range from 100 to 230 K (except for minor discontinuities at about 150 K, which will be discussed later).

The binding energy was again estimated from the temperature dependence of the recoil-free fraction $f_\perp$, and a value of ≈ 9 kcal/mole was thereby found. This low value is consistent with the BIT being physically adsorbed on the Grafoil surface.

The recoil-free fraction parallel to the surface $f_\parallel$ depends on the interaction between adsorbed species. For BIT on Grafoil, the value of $f_\parallel$ (shown in Fig. 7) decreases smoothly with increasing temperature up to ≈ 150 K, after which further temperature increase leads to an abrupt decrease in $f_\parallel$. This dramatic decrease in $f_\parallel$ takes place over a 2 K temperature range; above this transition range $f_\parallel$ is approximately zero. The results indicate a substantial increase in the mobility of the Mössbauer atoms in directions parallel to the Grafoil plane. Thus below ≈ 150 K the BIT is present as a two-dimensional solid, and the interactions between neighboring species lead to finite values of $f_\parallel$, whereas above ≈ 150 K the BIT is present as a two-dimensional liquid. As mentioned earlier, the crystallographic planes in Grafoil are not perfectly oriented. This is presumably the reason for the small discontinuities observed in $f_\perp$ at about 150 K.

In a related study using the ^{119}Sn Mössbauer resonance, Shechter (1977) reported a similar behavior for adsorbed tetramethyl tin $Sn(CH_3)_4$.

Information about the vibrational motion of adsorbates has also been obtained from studies using high surface area polycrystalline adsorbents. Suzdalev et al. (1966, 1968) studied the motion of SnO and $SnO_2 \cdot nH_2O$ on

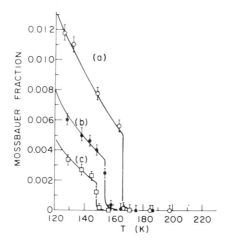

Fig. 7. Mössbauer resonant absorption fractions of butadiene iron tricarbonyl adsorbed on Grafoil with the γ-ray direction parallel to the Grafoil surface. Sample coverages are as in Fig. 6 (Shechter et al., 1976b).

high surface area silica powder using ^{119}Sn Mössbauer spectroscopy. These experiments showed that the spectral area of the $SnO_2 \cdot nH_2O$ component decreased more rapidly than that of the SnO component. The vibrational motion of the tin atoms was assumed to have two contributions, one arising from the internal vibrations of the tin atoms within the compound and another arising from the vibrations of the tin compound on the silica support. The latter contribution was found to increase dramatically above 230 K. Another interesting effect was observed, namely, a broadening of the Mössbauer lines of the $SnO_2 \cdot nH_2O$ compound, which increased with increasing temperature. This broadening was attributed to diffusion of the Mössbauer atoms and it was, in fact, possible to estimate a value of 12 kJ for the activation energy of the diffusion, assuming that the diffusion takes place via a jump mechanism. The authors found that the mobility could be decreased by decreasing the pore size of the silica adsorbent.

2. Vibrational Motion of Surface Atoms

In the studies discussed above, the Mössbauer atoms were present in species adsorbed *on* surfaces, and it is, therefore, not surprising that the motion of these atoms differed considerably from atoms in bulk compounds. It must also be realized, however, that the motion of atoms *in* the surface of bulk phases may differ from that of atoms inside the bulk.

The vibrational properties of surface atoms can (as discussed in Section III, A) be investigated by studying small particles or thin films having a significant fraction of the resonant atoms at the surface. The f factor of small particles has been the subject of numerous investigations. Such studies may not, however, directly give information about the vibration of surface atoms since the f factor for small particles is influenced by a number of different factors. For example, the usual Debye approximation for the phonon spectrum has to be modified when considering small particles because the longest phonon wavelength that can be present in a small crystal is of the order of the particle size. This is usually referred to as a low-frequency cutoff. Since the lowest frequencies give a relatively large contribution to the amplitude of the thermal vibrations of the Mössbauer atom, this effect should give rise to an increase in the f factor of small particles as compared to the bulk value. On the other hand, surface vibrational modes may become important in small particle studies, and this will lead to a decrease in the f factor. Moreover, for very small particles, the motion of the particles relative to the support or the matrix in which they are embedded may not be negligible. This type of vibration also gives rise to a decrease in the f factor relative to the bulk value.

The early studies (Marshall and Wilenzik, 1966; Roth and Hörl, 1967; Suzdalev et al., 1967) of the particle-size dependence of the f factor have given somewhat conflicting indications on the relative importance of the different effects. Von Eynattan and Bömmel (1977) studied the f factor for metallic iron microcrystals, embedded in paraffin wax, in the range 7.0–45.0 nm. They observed a decrease in the f factor with decreasing particle size. The f factor was also found to depend on the coupling of the microcrystals to their surroundings, in agreement with earlier findings (Van Wieringen, 1968). The results therefore suggest that the small f factors in the microcrystals may originate from vibrations of the particles. However, the authors also considered the possible contribution from surface phonons, and they were not able to exclude that these may contribute to the observed f factors. Viegers and Trooster (1977) found no evidence for surface modes in an [197]Au Mössbauer spectroscopy study of gold microcrystals. These authors measured the temperature dependence of the f factor for several samples of gold particles, with average diameters of 3–17 nm, embedded in gelatin. At 4.2 K the f factor was close to that of bulk gold, but at high temperatures the f factor decreased with decreasing particle size. They analyzed the results in terms of a Debye model that included the presence of both a low-frequency cutoff and surface modes, but this model did not give a satisfactory fit of the results. The authors were therefore led to the conclusion that the vibration of the particles in the gelatin as a whole plays a significant role in determining the f factor.

The above studies, the studies of the mobility of $SnO_2 \cdot nH_2O$ mentioned in Section III, C.1, and many studies of supported catalysts have shown that the f factor for small particles is influenced by the vibration of the particles relative to the surroundings. It may, therefore, be difficult, in general, to unravel the contribution of surface vibrations to the f factor of small particles.

Van der Kraan (1972, 1973) devised an experimental approach whereby the above two effects may be distinguished. He prepared 7 nm particles of α-Fe_2O_3 with and without surface enrichment with [57]Fe. The temperature dependences of the spectral areas for the two samples were then compared. It was found that the f factor of the surface-enriched particles decreased more rapidly with increasing temperature than that of the unenriched particles. This suggests that the amplitude of vibrations of surface atoms is enhanced by the presence of surface phonons. It should, however, be mentioned that the composition of the surface layers may be different from that of the interior of the particles, and this may also influence the results.

A different approach to studies of vibrations of surface atoms is based on the sensitivity of paramagnetic Mössbauer spectra to the spin–lattice relaxation time. Since the spin–lattice relaxation is mainly sensitive to the vibrational modes that give rise to relative displacements of the Mössbauer atom and its nearest neighbors, such studies may give information about parts of the vibrational spectrum that are not influencing the f factor significantly (Knudsen and Mørup, 1977). Studies of spin–lattice relaxation of Fe^{3+} ions in ion-exchange resins have been carried out by Suzdalev et al. (1968). The iron concentration in the sample was very low (1.6 wt%) in order to suppress spin–spin relaxation, which otherwise would be so fast that spin–lattice relaxation could not be studied in detail. In the study it was found that adsorption of water greatly increased the spin–lattice relaxation time of Fe^{3+} ions.

D. Structure and Electronic Properties

The structure and the electronic properties of small particles, thin films, and adsorbed layers are often different from those of bulk samples. In many cases this is due to interactions with the support, but intrinsic surface and interface effects may also be present. Structural and electronic information is mainly obtained from the isomer shift and the quadrupole splitting of Mössbauer spectra.

The quadrupole splitting is proportional to the electric field gradient (EFG) at the Mössbauer nucleus, and this parameter is therefore very sensitive to structural changes as discussed in Section III, B. It is, however, worth noting that the texture of phases containing the Mössbauer nuclei can also be determined from quadrupole-split Mössbauer spectra. This information is derived from the area ratio of the Mössbauer lines (see Chapter 2 of Volume I). Consider, for example, ^{57}Fe Mössbauer spectra. Samples with a random orientation of the EFG of the iron-containing species yield symmetric doublet spectra (for simplicity we neglect here the possible anisotropy in the recoil-free fraction, which may lead to Gol'danskii–Karyagin effect). However, in absorbers with a preferred orientation of EFG of the iron-containing species, the area ratio of the two lines in the spectrum depends strongly on the orientation of the sample relative to the γ-ray beam. For example, the spectrum of a sample with axially symmetric EFG at the Mössbauer nuclei shows area ratios of 3 : 1 or 3 : 5 when the symmetry direction is parallel or perpendicular to the γ-ray beam, respectively. Thus preferred orientations may easily be determined. This has, for example, been demonstrated in studies of thin films and layers of adsorbed molecules on Grafoil.

1. Studies of Adsorbates on Grafoil

In some of the Mössbauer studies of chemisorbed molecules on Grafoil discussed in Section III, C, interesting structural information was obtained. For example, in the studies of adsorbed $FeCl_2$ by Shechter et al. (1976a), two quadrupole doublets, B and L, were observed. The former closely resembled that of bulk $FeCl_2$. The L doublet had a larger quadrupole splitting and dominated the spectra at low coverages. In addition, it was found that the two lines of each doublet were of unequal intensity and that the spectral area asymmetry for each doublet was dependent on the $FeCl_2$ coverage. These observations, as well as the large binding energy (see Section III, C), show that the iron did not simply crystallize as bulk $FeCl_2$ during the sample preparation, but instead formed a two-dimensional structure with the Grafoil surface. Additional evidence for the formation of a surface compound was that the asymmetry of the spectral doublets changed with different orientations of the Grafoil in the γ-ray beam. These Mössbauer results led the authors to propose a number of two-dimensional surface structures (at different coverages) as possible origins of the L doublet. These structures included $FeCl_2$ monomers, dimers, and larger clusters adsorbed in registry with the hexagonal structure of the Grafoil surface. While it is not possible to uniquely determine surface structure from the quadrupole splitting alone, the results of Shechter et al. illustrate the utility of Mössbauer spectroscopy for detecting and studying surface structure changes.

The structure of $(CH_3)_3SnCl$ molecules adsorbed on Grafoil has also recently been studied by Bukshpan (1977). Mössbauer spectroscopy and x-ray diffraction studies of crystalline $(CH_3)_3SnCl$ suggest that the tin ion is coordinated to five nearest neighbors: three methyl groups and two chloride ions. (The methyl groups define a plane, and the two chloride ions lie above and below this plane, respectively.) The $(CH_3)_3SnCl$ molecules thereby form a chainlike structure when they crystallize to form a three-dimensional solid, with the primary axis of the EFG along the (Cl-Sn-Cl) chains. The Mössbauer spectra of fractional monolayer coverages (≈ 0.30 -0.90 monolayers) of $(CH_3)_3SnCl$ on Grafoil showed isomer shifts and quadrupole splitting similar to those of solid $(CH_3)_3SnCl$.

Thus the chainlike arrangement of $(CH_3)_3SnCl$ molecules also seems to be present in the structure of the two-dimensional layer adsorbed on the Grafoil. The primary difference between the Mössbauer spectra of adsorbed and solid $(CH_3)_3SnCl$, however, was that the two peaks of the quadrupole-split doublet were of unequal intensity for the adsorbed species, whereas they were of equal intensity for the solid phase. This indicates that the primary axis of the EFG at each tin ion in the adsorbed layer has a preferred orientation with respect to the γ-ray beam. During the measurements, the Grafoil surface was placed in the γ-ray beam with the basal planes of the graphite microcrystals approximately perpendicular to the γ-ray beam. The spectral area ratio of the two peaks forming the quadrupole doublet for the adsorbed tin was used to calculate the orientation of the EFG with respect to the Grafoil surface (basal planes). In this way it was shown that the chains of $(CH_3)_3SnCl$ were oriented parallel to the surface.

2. Thin Film Studies

When thin metallic iron films are prepared on various substrates, the films normally consist of α-Fe (bcc), which is the stable phase of iron below about 1183 K. However, Keune et al. (1977a, b) have demonstrated that epitaxial growth of metallic iron on (001) Cu planes may lead to the formation of iron films having the γ structure (fcc). Sandwich samples were prepared using the following metal desposition sequence: (i) 200 nm of Cu was condensed onto a NaCl crystal held at 573 K, (ii) a 1.8-nm ^{57}Fe film was deposited on the Cu film, (iii) this ^{57}Fe film was covered with a 100-nm Cu layer, (iv) the ^{57}Fe–Cu deposition sequence was repeated three more times, and (v) the final ^{57}Fe film was covered with another 200-nm Cu film followed by removal of the original NaCl substrate with water. Electron diffraction studies of the multilayer sample showed only the fcc diffraction spots of Cu with the (001) Cu planes parallel to the film surfaces. The absence of the bcc diffraction spots of α-Fe was suggestive of an epitaxial growth of fcc γ-Fe on the copper. In contrast to the ferromagnetic α-Fe, the fcc form of iron is antiferromagnetic with a Néel temperature of about 80 K. Mössbauer spectra obtained at different temperatures are shown in Fig. 8. At 295 K the main line is a fairly narrow singlet with an isomer shift characteristic of γ-Fe. At 77 and 4.2 K a significant broadening of the singlet is observed. The latter behavior is also consistent with the presence of γ-Fe since the small hyperfine field for antiferromagnetic γ-Fe (≈ 24 kG) does not result in a resolved six-peak pattern. The antiferromagnetic ordering of the 1.8-nm ^{57}Fe films was further verified by collecting a Mössbauer spectrum at about 13 K in the presence of a 49-kG applied magnetic field (Fig. 8d). For a ferromagnet, the net magnetic moment will align with the applied field, and sharp Mössbauer peaks will be observed with a magnetic splitting approximately equal to the value of the applied field plus or minus the magnitude of the magnetic hyperfine field. (The plus and minus signs are used depending on whether the hyperfine field is positive or negative, respectively.) However, an antiferromagnetic sample has no net magnetic moment, and the application of an external magnetic field should lead to a broadening of the Mössbauer spectrum due to a distribution of magnetic splittings. Indeed, as seen in Fig. 8d, the applied 49-kG field leads to such a broad, magnetically split spectrum, indicative of antiferromagnetism.

The presence of γ-Fe has also been detected in the Mössbauer spectra of very thin iron films on polycrystalline Cu substrates (Keune et al., 1979)

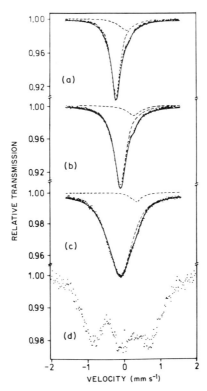

Fig. 8. Mössbauer spectra of ~ 1.8-nm Fe on (001) Cu at (a) 295 K, (b) 77 K, (c) 4.2 K, (d) ~ 13 K and with an applied longitudinal field of 49 kG. (Adapted from Keune *et al.*, 1977a. *J. Appl. Phys.* **48**, 2976.)

and in supersaturated Cu–Fe alloys (Williamson *et al.*, 1976). In the latter study, information about the surface atoms of the γ-Fe precipitates was obtained.

Interesting results regarding surface structures have also been obtained by Aggarwal and Mendiratta (1977) who studied thin films of FeTe. They obtained [57]Fe Mössbauer spectra of such films deposited on SiO substrates and found that the quadrupole splitting increased and the isomer shift decreased with decreasing film thickness. These results were interpreted in terms of changes in the electronic properties with film thickness. The spectra also showed an asymmetry in the spectral area ratio of the quadrupole doublet, indicating a preferred orientation of the FeTe phase relative to the substrate. Measurements of FeTe films grown on an Al substrate resulted in completely different spectra showing only a spectral singlet with isomer shift different from that observed on the SiO substrate.

The authors suggested that the disappearance of the quadrupole splitting could be due to a structure of these films, which is similar to that of the cubic Al substrate.

A different type of substrate effect has been observed in studies of thin films of α-Fe sandwiched between layers of Ag (Duncan et al., 1978; Keune et al., 1979) and Cu (Keune et al., 1979). In these studies it was found that the isomer shift was slightly larger than that of bulk iron, thus indicating a smaller electron density at the ^{57}Fe nuclei. This effect is probably due to a change in the conduction electron density in the iron film caused by the presence of the metallic substrate.

The properties of thin iron oxide films and passivation layers have been studied by several authors, and it has been found that different oxide and hydroxide phases can be formed, depending on the conditions under which the oxidation takes place. Keune and Gonser (1971) evaporated ^{57}Fe onto an Al substrate at a rate of ≈ 50 nm hr^{-1}. The vacuum during this evaporation was only 10^{-6} Torr, and the temperature of the aluminum substrate was 300 K. The film so produced had a thickness of 33 nm, and it was subsequently cut into several pieces that were stacked in the γ-ray beam. At 4.2 K the Mössbauer spectrum was magnetically split with an internal hyperfine field of 460 kG, while at 77 K the spectrum contained only a quadrupole-split doublet. The isomer shifts of these two spectra were indicative of Fe^{3+}. In fact, γ-FeOOH has a Néel temperature of 73 K, and at 4.2 K its magnetic hyperfine field is 460 kG. Thus deposition of an iron film in a relatively poor vacuum (i.e., 10^{-6} Torr) can lead to complete oxidation of the iron to Fe^{3+}, and for the conditions used in the study of Keune and Gonser, the oxidized product was γ-FeOOH.

Keune and Gonser (1971) also studied the influence of the substrate temperature by evaporating iron onto a Be substrate held at 4.2 K. The residual gas pressure was the same as that for the previous experiment, and the thickness of the film was 110 nm. For this sample, however, the Mössbauer spectrum at 4.2 K was a quadrupole doublet characteristic of Fe^{3+}. Heating the sample to room temperature followed by recooling to 4.2 K then led to the appearance of magnetic splitting ($\approx 475 \pm 25$ kG) in the Mössbauer spectrum. These results suggest that the sample prepared by evaporation onto a cold (4.2 K) substrate was amorphous, and heating to room temperature resulted in crystallization, probably to γ-FeOOH.

Passivation layers on iron surfaces in a K_2CrO_4 solution were studied by Pritchard and Dobson (1973). In order to obtain a large surface sensitivity, they electrodeposited ^{57}Co on the surface of a disk of mild steel. In this case the passivation layer was found to be nonmagnetic even at 77 K and, in spite of the reducing conditions maintained during and after the passivation procedure, only Fe^{3+} was observed. It was suggested that the film is either superparamagnetic or consists of a mixed (Cr, Fe) oxide with a low Néel temperatue (see also Chapter 3 of Volume I). Metallic iron surfaces rapidly oxidize when exposed to air. The nature of such iron oxide layers on small particles of metallic iron has been studied extensively, and the results are of great importance for the understanding of passivation and corrosion. Suzdalev and Amulyavichus (1972) studied oxide films with thicknesses from 1.5 to 5 monolayers at the surface of metallic iron particles formed by oxidation in air. They found that these very thin oxide layers give rise to magnetically split components with broad lines. Topsøe et al. (1973) and Van Diepen et al. (1977) have studied passivated iron particles. After room temperature passivation, the passivation layer was found to be about 20 atomic layers deep (Topsøe et al., 1973). In both these studies the room temperature spectra of the passivated samples showed, besides the α-Fe component, a broad background resulting from Fe^{3+} and possibly also Fe^{2+} phases. Using x-ray diffraction, Van Diepen et al. (1977) found the presence of a spinel phase, and it was suggested that the results can be explained by an inhomogeneous oxide layer that is iron rich ("FeO") near the metal–oxide interface and oxygen rich ("γ-Fe_2O_3") at the outside. Such an inhomogeneous layer would have a large number of magnetically inequivalent sites, which can explain the broad lines in the Mössbauer spectrum.

More recently Haneda and Morrish (1978) studied partly oxidized iron particles with an overall average particle size of 29.3 nm. In this case it was possible to identify the oxide phases as a mixture of microcrystalline γ-Fe_2O_3 and Fe_3O_4.

Studies of oxide surface layers utilizing conversion electron Mössbauer spectroscopy are discussed in Section III, G.

3. Small Particle Studies

Atoms at the surface of small particles are in a less symmetric environment than atoms in the bulk. Therefore, the Mössbauer atoms at surfaces are expected to show quadrupole splittings different from those of bulk atoms. This was examined theoretically by Hrynkiewicz et al. (1971) who made computer simulations of ^{57}Fe Mössbauer spectra of small Fe^{3+}-containing particles. The calculations were carried out for various sizes and shapes of the particles. In theoretical spectra of small particles ($\lesssim 10$ nm) with simple cubic crystal structure, Fe^{3+} surface components with large quadrupole splittings (1.5–2.0 mm s^{-1}) were present. The calculations also showed that the Mössbauer spectra of small particles may critically depend on the particle shape.

The authors also calculated the electric field gradient (EFG) for atoms at various distances from the surface in spherical particles with a spinel structure. This investigation showed that for atoms within a surface shell of about 0.5 nm in thickness, large disturbances of the EFG were present. Below this depth the EFG is only slightly different from the bulk value.

The calculations were carried out assuming that the crystal structure and lattice spacings of small particles are independent of the particle size. While the first assumption is often very good, the latter may not always be valid. For example, x-ray studies of small particles of α-Fe_2O_3 have shown that the lattice parameter increases with decreasing particle size (Schroeer and Nininger, 1967). This is expected to give rise to a dependence of the quadrupole splitting on the particle size. Mössbauer studies of small particles of α-Fe_2O_3 above the superparamagnetic blocking temperature showed that the quadrupole splitting increases almost linearly with reciprocal particle diameter (Kündig et al., 1967). This result was interpreted by the authors in terms of a "core and shell" model with a quadrupole splitting of 1.38 mm s^{-1} for atoms near the surface (the shell) and a quadrupole splitting of 0.79 mm s^{-1} for atoms within the core of the particles. When the particle size decreases, the relative contribution from the former increases, resulting in an increase in the observed average value of the splitting. Schroeer (1968, 1970) noticed that the changes in quadrupole splitting are comparable to those occurring under thermal expansion and suggested that the increased splitting may be explained by the increase in the lattice parameter with decreasing particle size. The origin of the particle-size dependence of the quadrupole splitting was further elucidated by Van der Kraan (1972, 1973). In short, it was found that enrichment of the surface in ^{57}Fe (as discussed in Section III, C, 2) considerably increased the quadrupole splitting. This result strongly supports the "core and shell" model by Kündig et al. (1967).

E. Chemisorption Studies

The contributions of surface atoms to the Mössbauer spectra of thin films and small particles may be identified because they may have isomer shifts, quadrupole splittings, and magnetic hyperfine fields that are different from the bulk values. When the surroundings of surface atoms are changed by chemisorption, the spectral components due to these atoms are expected to be altered. Observation of these effects in adsorbents requires, of course, that a large fraction of the Mössbauer atoms are present very close to the surface. In this section, a few illustrative studies will be presented, with more given throughout Section IV. Studies on adsorbates containing Mössbauer atoms were described in the two previous sections.

Hobson and Gager (1970) and Gager et al. (1973) studied chemisorption of different molecules on very small silica-supported particles of α-Fe_2O_3.

Figure 9 shows some typical results of the effect of chemisorption of different gases. Spectrum B-1 was obtained after heating the sample in flowing hydrogen followed by outgassing at 875 K. The three lines in the spectrum were interpreted as a superposition of two ferrous doublets, designated as Fe^{2+} and Fe^{2*}. Adding water vapor to the sample (spectrum B-2) changed the spectrum into a new ferrous quadrupole doublet with a rather large splitting. Outgassing at 773 K restored the spectrum to its original state. Quite different changes were observed when H_2S was adsorbed (spectrum B-3). It could also be concluded that the H_2S was chemisorbed rather than physically adsorbed because pumping on the sample at room temperature did not alter the spectrum.

These spectra clearly show that the changes in the environments of the Mössbauer atoms upon chemisorption depend on the type of molecules

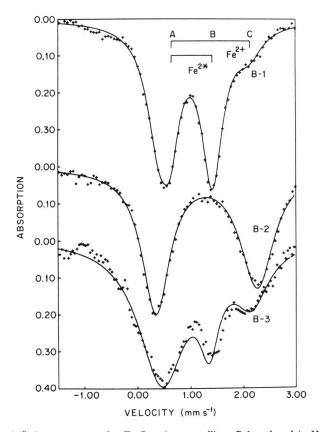

Fig. 9.   Mössbauer spectra of α-Fe_2O_3 microcrystallites. B-1, reduced in H_2 and out-gassed at 873 K; B-2, H_2O adsorbed (23 Torr); B-3, H_2S adsorbed (85 Torr) (Gager *et al.*, 1973. *Chem. Phys. Lett.*, **23**, 386).

that are chemisorbed. In many cases it is this type of information that is desired in studies of real catalysts, and a detailed analysis may elucidate important parameters such as catalyst dispersion, reactivity of gases with the catalyst, changes in the chemical state of the catalyst during the reactions, etc.

Cation-exchanged zeolites may have all the cations located at the internal surfaces of the zeolite structure, and as such may be interesting model systems for studies of chemisorption. Many such studies have appeared (see, e.g., Delgass, 1976 and Dumesic and Topsøe, 1977) and have helped elucidate both the nature of the cationic exchange sites and the structure of the chemisorption complexes. Rees (1976) has, for example, studied a ferrous-exchanged A-zeolite. The location of the Fe^{2+} ions in the zeolite was traced through adsorption of methanol, acetonitrile, and acetylene molecules, which can enter only into the large cavities in the zeolite. All three gases were observed to interact with all of the ferrous ions. This result shows that these ions must be located on or near the walls of the large cavities. Before adsorption of gases, the ferrous ions showed a spectrum with a relatively low value of the room temperature isomer shift (1.08 mm s^{-1} with respect to sodium nitroprusside) and a small and almost temperature-independent quadrupole splitting (0.46 mm s^{-1} at 300 K). The low isomer shift is indicative of the ion being at a site of low coordination number. The ferrous ions were assigned to sites in the window between the large cavities and the sodalite cage. In these sites the ferrous ions will have three nearest oxygen neighbors. Simple crystal field calculations for the ferrous ions in this type of site were in agreement with the observed small and temperature-independent quadrupole splittings. In Fig. 10 the effect of adsorbing different amounts of water is shown. The zeolites contain 2.66 iron atoms per unit cell. The area of the original quadrupole-split ferrous doublet was observed to decrease linearly with increasing amount of water admitted, until it disappeared when about 5.3 water molecules per unit cell were admitted. This suggests that two molecules of water are coordinated to each ferrous ion. The existence of two Fe^{2+} doublets (both different from the original ferrous doublet) in the samples containing even more water (6.5 molecules per unit cell) was suggested to be due to dissociation of one of the water ligands. For less than 10 water molecules per unit cell, the motion of the proton resulting from the dissociation must be slower than $\sim 10^{-7}$ s for the two different sites to be seen individually.

Very interesting results about the mobility of ethylene in the zeolite structure were also found. On cooling a sample containing about 1.5 ethylene molecules per unit cell to below 200 K, the spectrum changed from a single quadrupole-split doublet into a spectrum with two doublets. This was taken as evidence that at high temperature (> 200 K) the

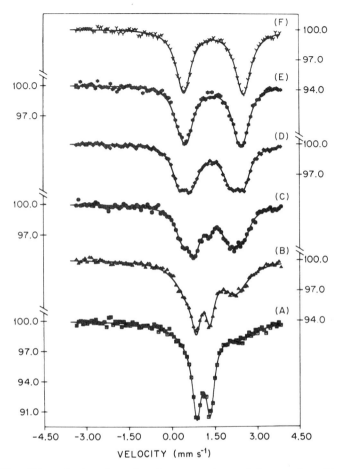

Fig. 10.   Effect of adsorption of water on spectrum of ferrous–A zeolite. Water molecules per unit cell: A, 0.80; B, 2.84; C, 4.75; D, 6.50; E, 10.5; F, 28. [Rees, L.V.C. in "Magnetic Resonance in Colloid and Interface Science" (H. A. Resing and C. G. Wade, eds.), ACS Symposium Series No. 34; American Chemical Society, Washington, D. C., 1976, p. 308.]

ethylene moves freely in the zeolite structure, whereas at lower temperatures the ethylene "residence time" on the ferrous ions becomes of the order of 10^{-7} s.

In addition to the changes in isomer shift and quadrupole interaction that are illustrated by the above examples, it has been found that the magnetic properties of surface atoms may also change on chemisorption. This is discussed in Sections III, F, IV, E, and Chapter 1 (Mørup et al., 1980b).

F. Surface and Interface Magnetism

1. Introduction

The magnetic properties of atoms at surfaces and interfaces have recently attracted much attention. The objective of many studies has been to test the finding by Lieberman *et al.* (1969, 1970) that the magnetic moment of atoms at the surface of thin films of Fe, Co, and Ni apparently vanishes. Such nonmagnetic layers have been referred to as "magnetically dead layers."

Mössbauer spectroscopy has been used extensively for studies of surface and interface magnetism (see, e.g., Shinjo, 1979). The surface sensitivity has been obtained by doping the surface with ^{57}Co, by enriching it in ^{57}Fe, or by studying very thin films or very small particles.

The measured hyperfine fields are, to a first-order approximation, proportional to the magnetic moment of the Mössbauer atoms. Therefore, the information obtained from the magnetic splitting of the spectra is related to the size of the magnetic moments of surface atoms. Furthermore, the orientation of the magnetic moments of the surface atoms relative to the γ-ray direction is reflected in the area ratio of the Mössbauer lines in the magnetically split spectrum.

Mössbauer studies of iron alloys have shown that the hyperfine field depends on the local environments of the iron atoms (see, for example, Schwartz, 1976). Therefore, one might expect that the hyperfine field of surface and interface atoms will depend on the positions and the type of neighboring atoms. Specifically, different fields will be expected for surface atoms at a solid–vacuum interface, surface atoms with chemisorbed molecules, and atoms located at a solid–solid interface.

The temperature dependence of the hyperfine field of surface atoms is also of great interest. Owing to the small number of neighboring magnetic atoms, the magnetization of surface atoms may decrease more rapidly with increasing temperature than does the bulk magnetization. Monte Carlo calculations carried out by Binder and Hohenberg (1974) showed that well below the Curie temperature this decrease in magnetization is significant only for one or two surface layers. However, when the Curie temperature is approached, several (> 10) layers show a decreased magnetization as compared to the bulk value. Some results calculated on the basis of the Heisenberg model are shown in Fig. 11.

In the above calculation, it was assumed that the exchange interaction between a surface atom and its neighbors is identical to the interaction among the bulk atoms. Binder and Hohenberg (1974, 1976) and Hohenberg and Binder (1975) have also discussed situations where these exchange interactions are different. In the extreme case where the surface

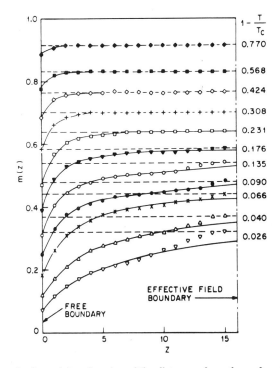

Fig. 11. Magnetization $m(z)$ as function of the distance z from the surface. The results were obtained from Monte Carlo calculations on a Heisenberg system. Here z is the distance from the surface given as number of layers. (Adapted from Binder and Hohenberg, 1974.)

and the bulk exchange interactions have different signs, the magnetic order of the surface layer and the bulk may be different (for example, "dead layers" or surface antiferromagnetism may coexist with bulk ferromagnetism). Interesting situations may also arise when the surface exchange constant is numerically greater than the bulk exchange constant. This may lead to magnetically "live layers," i.e., magnetically ordered surface layers, on top of a paramagnetic bulk phase.

2. Source Experiments

One method of achieving information about the surface magnetic properties of a bulk sample is to prepare a surface containing ^{57}Co and then to use the sample as a source in a Mössbauer experiment. Examples of this approach are discussed in this section.

Shinjo et al. (1973) electrodeposited $\approx 10^2$ atomic layers of cobalt onto a polished surface of metallic cobalt, followed by deposition of about one-tenth of a monolayer of ^{57}Co. In addition, a second sample was prepared by depositing 10^3 atomic layers of cobalt onto a copper surface, followed by deposition of one-tenth of a monolayer of ^{57}Co. In both preparations, the samples were rapidly cooled to 77 K after ^{57}Co deposition. During the measurements the samples had the ^{57}Co surface layer covered with ice. At 4.2 K, the Mössbauer spectrum (obtained with a standard stainless steel absorber) of the first sample was a magnetically split pattern with an isomer shift characteristic of zero valent ^{57}Fe in metallic cobalt. The six peaks, however, were broadened due to the presence of a distribution of magnetic hyperfine fields (roughly from 260 to 350 kG) at the different ^{57}Co atoms. (The value of the hyperfine field for ^{57}Fe in bulk metallic cobalt is 330 kG at 4.2 K.) A similar Mössbauer spectrum was obtained with the sample at 77 K, but during the collection of the spectrum at room temperature, oxidation of the metallic cobalt took place, evidenced by the presence of an Fe^{2+} spectral pattern. This behavior is consistent with the location of the ^{57}Co atoms near the surface of the sample. The Mössbauer spectra of the copper-supported cobalt were similar to those just described for the cobalt deposited on the polished cobalt surface.

It is interesting that for all the metallic surfaces no paramagnetic components were observed in the Mössbauer spectra. Since the magnetic hyperfine field tends to be proportional to the local magnetic moment in ferromagnetic alloys, it was concluded that although a distribution of magnetic moments is indeed present, the magnetic moments of ^{57}Fe atoms at cobalt surfaces are not zero.

Shinjo et al. (1974) also prepared samples of ^{57}Co electrodeposited onto iron surfaces using a procedure similar to that described above. At 4.2 K, the Mössbauer spectrum was again a six-peak metallic iron pattern with symmetric but broad peaks. The distribution in the magnetic hyperfine field, calculated from the broadening of the spectral peaks, was from 240 to 340 kG, with an average value of 290 kG. This should be compared with a value of 340 kG for bulk metallic iron. Thus, as in the case of a cobalt surface, the magnetic moments of ^{57}Fe atoms in metallic iron surfaces are not zero, but only slightly smaller than the value for bulk metallic iron.

3. Thin Film Studies

Information regarding surface magnetic properties may also be obtained from thin film (this section) and small particle studies (the following section). Thin films and small particles are usually in contact with a substrate or support, which may cause strain, charge transfer, magnetic coupling, and chemical reaction between the solid phases. Furthermore, the small dimensions of such systems may induce changes in the properties. For a discussion of the magnetic properties of thin films and some of the above problems, see, e.g., Binder and Hohenberg (1974), Gradmann (1974), Bergmann (1978), Göpel (1978) and Alvarado (1979).

In absorber Mössbauer experiments it is generally necessary to stack a number of identical films in order to achieve sufficiently large resonant absorptions. By analogy with those studies of α-Fe_2O_3 particles discussed in Sections III, C, 2 and III, D, 3, the surface sensitivity of the Mössbauer spectrum may be enhanced by enrichment of the film surfaces in ^{57}Fe. For example, Shinjo (1976) and Lauer et al. (1977) have prepared samples consisting of layers of 100-nm Cu, 10-nm natural iron, and 0.5-nm ^{57}Fe, which were successively vacuum deposited on a mylar substrate at 77 K. To facilitate collection of Mössbauer spectra, this Cu–Fe–^{57}Fe–Cu evaporation sequence was repeated nine times. By this procedure, approximately one-third of the ^{57}Fe atoms are in the first surface layer, and beneath this surface a metallic iron support still remains. The Mössbauer spectra of this sample at 4.2, 77, and 295 K are magnetically split with broad peaks due to a distribution of magnetic hyperfine fields. Through computer fittings of these spectra, this distribution was calculated. The range of magnetic fields was from ≈ 210 to 340 kG with two high probability fields of 290 and 340 kG. The latter is equal to the value for bulk metallic iron and is undoubtedly due to those ^{57}Fe atoms below the surface layers. Thus the former value is interpreted as being due to the ^{57}Fe surface atoms. This interpretation is supported by the results for ^{57}Co source atoms, electrodeposited on an Fe surface and coated with ice, showing that the surface magnetic moments are slightly smaller than bulk magnetic moments.

It might be argued that in preparing the above sandwich structures, switching from evaporation of Fe to ^{57}Fe may cause a discontinuous growth of the iron lattice. In order to study this problem, Lauer et al. (1977) prepared a sample with the following evaporation sequence repeated 10 times: 50-nm Cu, 10-nm natural Fe, 0.5-nm ^{57}Fe, 10-nm natural Fe, and 50-nm Cu. If ^{57}Fe–Fe interface effects were present, these should be clearly visible in the spectra. However, the Mössbauer spectra of the ^{57}Fe layers sandwiched between layers of natural iron showed parameters identical to those of bulk metallic iron. Hence ^{57}Fe–Fe interface effects can be neglected with the preparation technique used by the authors. The effects observed in the Cu–Fe–^{57}Fe–Cu sample must therefore originate from the ^{57}Fe–Cu interface.

Keune et al. (1979) prepared very thin films of ^{57}Fe (down to 0.2 nm in thickness) between layers of Cu. At 4.2 K the hyperfine field of a 3-nm film was equal to that of bulk α-Fe, whereas at room temperature the hyperfine field was 315 $\pm$ 3 kG, i.e., about 5% smaller than the bulk value. For films with thickness less than 1.8 nm, magnetically split spectra with very broad lines were obtained at 4.2 K. The average hyperfine field was about 290 kG, i.e., about 15% smaller than that of bulk α-Fe at 4.2 K. In the spectrum of a 1.8-nm Fe film obtained at 4.2 K, two magnetically split components could be resolved. One of these showed a hyperfine field of about 306 kG and was attributed to surface atoms, whereas the hyperfine field of the other component was very close to the value for bulk α-Fe.

In the above investigations the iron surfaces were covered with ice or copper. In a later study, Shinjo et al. (1977) have prepared sandwich samples using MgF_2 instead of Cu as the material covering the ^{57}Fe surfaces. The MgF_2 layers were 20-nm thick, and the ^{57}Fe layer thickness was varied from 6 to 1.6 nm for the different samples prepared. Mössbauer spectra obtained at 4.2 and 300 K are shown in Figs. 12 and 13, respectively. The Mössbauer spectra of the 6-nm ^{57}Fe layers were identical to those of bulk metallic iron. However, as the ^{57}Fe layer thickness de-

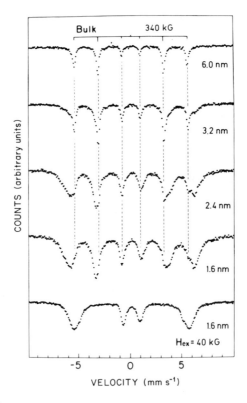

Fig. 12. Mössbauer absorption spectra at 4.2 K for the MgF_2-coated Fe films: 6.0, 3.2, 2.4, and 1.6 nm without external field and 1.6 nm with an external field of 40 kG. (Adapted from Shinjo et al., 1977.)

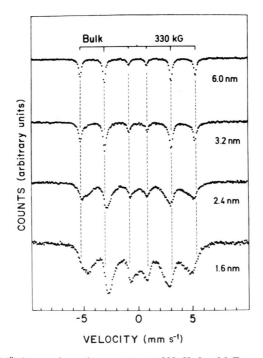

Fig. 13. Mössbauer absorption spectra at 300 K for MgF_2-coated thin Fe films. (Adapted from Shinjo *et al.*, 1977.)

creased, deviations from the bulk behavior were observed. Specifically, at 300 K the magnetic hyperfine field decreased with decreasing film thickness, while at 4.2 K the magnetic splitting increased as the film thickness decreased. A spectrum obtained at 4.2 K for the 1.6-nm film, in the presence of a 40 kG field, is also shown in Fig. 12. The complete disappearance of lines 2 and 5 in this spectrum indicates that the magnetization is perfectly oriented parallel to the direction of the external field. This can be taken as evidence for the sample being completely ferromagnetic. The increased splitting for the thin films (compared to bulk samples) at 4.2 K suggests that the surface magnetic moment of the ^{57}Fe atoms is increased by the presence of MgF_2. The decreased magnetic splitting for the thin films at 300 K was interpreted as resulting from the smaller number of magnetic neighbors of surface atoms compared to those of the inside of the bulk. The latter interpretation is in accordance with the theoretical calculations by Binder and Hohenberg (1974) (see Fig. 11).

In related studies, thin iron films coated with MgO and Sb have been studied (Shinjo *et al.*, 1979, Hine *et al.*, 1979). At 4.2 K a coating of Sb was found to decrease the hyperfine field of the surface atoms, whereas MgO gives rise to a hyperfine field of about 380 kG for the surface atoms, which is considerably higher than the bulk value (340 kG).

Duncan *et al.* (1978) studied the temperature dependence of the spectra of thin, metallic iron films coated with Ag. For these samples it was not possible to decompose the spectra into a surface and a bulk component, although the spectra of a 2-nm film showed a small line broadening corresponding to a spread in the hyperfine field of about 6 kG both at 4.2 K and at room temperature. The size of the average hyperfine fields as a function of temperature for films of various thicknesses is shown in Fig. 14. At 4.2 K the 2.0-nm film shows a magnetic splitting that is larger than the bulk value by about 2–3%. The thicker films show hyperfine fields closer to the bulk value. At about 80 K the hyperfine fields of all the films are very close to the bulk value, but at higher temperatures the hyperfine fields of the thin films decreased more rapidly than in bulk. This effect was most pronounced in the thinnest films.

Duncan *et al.* (1978) suggested that the increase in hyperfine field at 4.2 K for thin films relative to the bulk value may be due to some interaction between the conduction electrons of the Fe atoms at the film surface and the outer electrons of the layers of silver covering the film. The decrease in

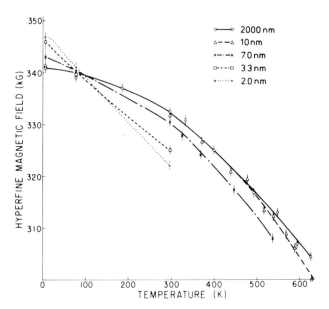

Fig. 14. A plot of the hyperfine field of thin Fe films coated with Ag as a function of temperature (Duncan *et al.*, 1978. *Hyperfine Interactions* **4**, 886).

hyperfine field with decreasing film thickness at high temperatures was attributed to enhanced spin–wave excitations in thin films.

Keune *et al.* (1979) studied a 0.5-nm film of metallic iron deposited between 30-nm-thick layers of Ag. In this case it was possible to resolve the 4.2 K spectrum into a bulk and a surface component, the latter showing a hyperfine field that was 5.3% larger than the bulk value. These results are in qualitative agreement with the results of Duncan *et al.* (1978).

In order to determine the magnetic hyperfine field as a function of depth in Ag-coated iron films, Owens *et al.* (1979) prepared a series of films using isotopically pure ^{56}Fe, which shows no Mössbauer effect. During the preparation a thin layer of ^{57}Fe was deposited at varying depths in the films to serve as a probe of the magnetic hyperfine field. Figure 15 shows the magnetic hyperfine field at 4.2 K for films of about 20 atomic layers in thickness, with the ^{57}Fe layers at different depths. The center of the film shows a hyperfine field close to the bulk value at 4.2 K, but when the surface is approached, the hyperfine field increases by about 8 kG. This elegant experiment clearly shows that the departures from bulk behavior are associated with the Ag/Fe interface. Figure 16 shows the hyperfine fields at various depths at room temperature. It can be seen that the hyperfine field near the surface is now smaller than the bulk value by about 5 kG. This result qualitatively agrees with the calculations by Binder and Hohenberg (1974) (Fig. 11).

In the discussion so far only studies of iron surfaces located at iron–solid interfaces have been mentioned. The studies show that the magnetic

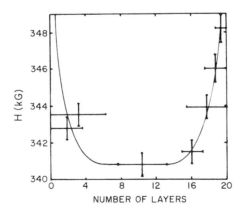

Fig. 15. Magnetic hyperfine fields at 4.2 K of probe ^{57}Fe layers located at different depths in a ∼ 20 atomic-layer thick ^{56}Fe film covered with Ag. The vertical error bars indicate uncertainties in determination of the average hyperfine field, while the horizontal bars indicate the thickness of the ^{57}Fe layers (Owens *et al.*, 1979).

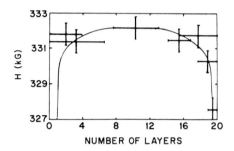

Fig. 16. Magnetic hyperfine fields at 298 K of probe [57]Fe layers located at different depths in a ∼ 20 atomic-layer thick [56]Fe film covered with Ag. The error bars have the same meaning as in Fig. 15 (Owens *et al.*, 1979).

properties of the surface iron atoms depend strongly on the material covering the surface. However, direct information about the intrinsic surface magnetic properties of iron is not obtained from such studies, but may be obtained from iron–vacuum interface studies.

Varma and Hoffman (1972) reported studies of thin iron films on a glass substrate with the iron surface exposed to UHV. These authors found an average hyperfine field of 298 ± 3 kG for a 1.2-nm film at 15 K. Although the iron–glass interface also may affect this result, the measurement suggests that the magnetic moments for iron atoms at the iron–UHV surface are somewhat smaller than in bulk.

The values of the hyperfine field of iron surface atoms with different coverings are summarized in Table I. Although the iron films were prepared in different ways and the spectra also have been analyzed in different ways, there is good agreement between the results obtained by different groups on the effect of a given covering. It is generally found that the hyperfine field of iron is reduced by about 10–15% at the Fe–Cu interface and increased by about 5% at the Fe–Ag interface. Since the work function of Fe is smaller than that of Cu and larger than that of Ag (Michaelson, 1977), it is likely that the different effects of Cu and Ag covering on the hyperfine field of iron interface atoms may be explained by differences in the charge transfer at Fe–Cu and Fe–Ag interfaces. However, no quantitative explanation has yet been given to explain the observed effects.

In all the studies, the hyperfine fields of surface iron atoms differ from the bulk value by less than about 15%. This clearly shows that "magnetically dead layers" are not present.

The discussion in this section has so far been centered around thin films of metallic iron. The magnetic properties of cobalt and nickel films have also received much attention (Liebermann *et al.*, 1970; Fulde *et al.*, 1973;

TABLE I

Surface Hyperfine Field of Metallic Iron at 4.2 K

Coverage	Hyperfine field (kG)	Reference
Fe	340[a]	Shinjo et al. (1977)
Cu	290[b]	Lauer et al. (1977)
Cu	306 ± 8[b]	Keune et al. (1979)
Cu	303[b]	Shinjo (1976)
Ag	358[b]	Keune et al. (1979)
Ag (111) plane	347[c]	Duncan et al. (1978)
Ag (111) plane	349[b]	Owens et al. (1979)
Sb (0001) plane	320[b]	Hine et al. (1979)
MgF_2	375[b]	Shinjo et al. (1977)
MgO	380[b]	Shinjo et al. (1979)
SiO	320[d]	Lee et al. (1964)
H_2O	290[e]	Shinjo (1976)
UHV/glass	298 ± 3[f]	Varma and Hoffman (1972)

[a] Bulk hyperfine field.
[b] Surface component.
[c] Average hyperfine field for 2.0-nm film.
[d] Average hyperfine field for 0.46-nm film.
[e] [57]Co experiment, average hyperfine field at the surface.
[f] Average hyperfine field for 1.2-nm film at 15 K.

Bergmann, 1978, 1979; Landolt and Campagna, 1978). As discussed in Section III, F, 2, Co surfaces may be studied by doping with [57]Co, and conducting source experiments. This approach was used by Gradmann et al. (1977) in the study of thin fcc Co (111) films grown on Cu (111). At 4 K the Mössbauer spectrum of a thin film containing only four atomic layers could be analyzed in terms of two magnetically split components. One of these components had a hyperfine field of 320 kG and was attributed to the two inner layers, while the other one had a hyperfine field of 228 kG and was attributed to the surface layers. A film containing only two atomic layers yielded a spectrum that could be analyzed in terms of only one hyperfine field equal to 262 kG. In a similar fashion, Ni films may be studied by [57]Fe Mössbauer spectroscopy by doping the films with [57]Co or [57]Fe. For example, information relevant to the surface magnetic behavior of Ni was obtained by Shinjo et al. (1977) in a study of thin Ni-rich NiFe alloy films (95% Ni, 5% Fe) sandwiched between MgF_2 layers. The magnetic splitting of the [57]Fe atoms in these films was equal to the bulk value for the alloy and was independent of the alloy film thickness from 1.6 to 4.8 nm. Hence the presence of the alloy/MgF_2 interface appears to have only a small influence on the surface magnetic properties of the NiFe alloy. The result also indicates that the hyperfine magnetic field at a Ni surface will be only slightly different from that of the bulk.

The measurements discussed above provide information about the magnitude of the hyperfine field, and thereby the magnetic moments, of the surface atoms. A complete understanding of surface magnetism, however, requires the simultaneous determination of the orientation of the magnetic moments. This information can, in fact, also be obtained by using Mössbauer spectroscopy. Specifically, the hyperfine field is parallel or antiparallel to the magnetic moment of the Mössbauer atom, and the direction of the hyperfine field relative to the γ-ray direction is reflected in the relative areas of the six lines of a magnetically split spectrum. For example, if the hyperfine field is perpendicular to the γ-ray direction, the relative areas of the lines in a ^{57}Fe spectrum are equal to $3:4:1:1:4:3$, and if the hyperfine field is parallel to the γ-ray direction, the relative areas are $3:0:1:1:0:3$, i.e., lines 2 and 5 disappear in this case. For random orientations of the hyperfine fields of the Mössbauer atoms in the sample, the relative intensities are $3:2:1:1:2:3$.

In most of the Mössbauer studies of thin films and surface atoms, relative area ratios close to $3:4:1:1:4:3$ have been observed. Since the γ-ray direction in these measurements was perpendicular to the surfaces, it can be concluded that the magnetization direction lies in the plane of the film. This is consistent with the shape anisotropy for thin ferromagnetic films. However, in very thin films the surface anisotropy may play an important role (Néel, 1954). The lower symmetry of surface atoms may, for example, favor a spin orientation perpendicular to the surface. This was observed by Shinjo et al. (1979) in 0.8-nm iron films covered with MgO. They found an area ratio of $3:1:1:1:1:3$, which is close to the ratio expected for the magnetization perpendicular to the film plane. Furthermore, Keune et al. (1979) found an area ratio of $3:0.8:1:1:0.8:3$ in 0.5 and 0.2-nm iron films covered with Cu, and in a 0.5-nm iron film covered with Ag, a ratio of $3:0.45:1:1:0.45:3$ was observed. In thicker films, other contributions to the magnetic anisotropy such as shape and magnetocrystalline anisotropies become dominating because their sizes are proportional to the volume of the film, whereas surface anisotropy is proportional to the surface area.

In connection with the above results, it should be noted that the contribution from the demagnetizing field to the magnetic splitting of a Mössbauer spectrum of a thin film depends on the magnetization direction (Knudsen and Mørup, 1980). When a film is magnetized perpendicular to the film plane, the demagnetizing field takes its maximum value (about 21 kG in metallic iron). When the magnetization lies in the film plane, the demagnetizing field is zero. Some of the differences in the hyperfine fields of thin films with different coverings may, therefore, partly be explained by differences in the magnetization direction relative to the film plane, which gives rise to different contributions from the demagnetizing field.

It should be noted that other experimental methods, such as magnetic susceptibility measurements, yield the total magnetic moment of the sample. In such experiments one may, for example, observe a magnetization of small particles or thin films that is smaller than the bulk value. However, it is difficult to decide unambiguously about the origin of this effect since a number of possible explanations exist. For instance, the magnetic moments of all the atoms may be reduced, "magnetically dead layers" may exist, or a noncollinear spin arrangement may be present (e.g., surface antiferromagnetism). Mössbauer spectroscopy is an alternative method which may allow distinction among these possibilities.

4. Small Particle Studies

As discussed above, surface studies using thin films are usually carried out using sandwich structures, i.e., the surfaces are covered with other solid phases. Therefore, studies of these samples are well suited for investigation of interfaces, but they do not directly elucidate the intrinsic surface magnetic properties. Small particles are usually present on a support, and solid–solid interface effects may, as discussed in the last section, also be important for such systems. However, a large fraction of the atoms in small particles are, in fact, exposed to the gaseous environment over the sample, and the possible effects of chemisorption on surface magnetism may conveniently be estimated from such samples. It must be noted, however, that in studies of small particle systems it may often be difficult to separate surface effects from size effects. The intrinsic magnetic properties may no longer be equal to those of the bulk as the size becomes very small. For example, superparamagnetism and collective magnetic excitations are dependent on particle size and may not always be easily distinguished from surface effects (Mørup et al., 1980b).

For comparison with the many studies on metallic iron films, it is interesting to look at some experiments on small, supported iron particles. Boudart et al. (1975a, b, 1977) studied metallic iron particles supported on magnesium oxide. For particle sizes below 10 nm, the average magnetic hyperfine field was observed over a wide temperature range to be equal (within 1%) to that of bulk iron. The main effect of the large fraction of atoms present at the surface or iron–magnesium oxide interface was a broadening of the lines, which increased as the particle size decreased (Dumesic et al., 1975a, b). Furthermore, the lines were observed to be asymmetrically broadened toward the center of the spectrum, with the asymmetry increasing with decreasing particle size. A possible explanation is the presence of slightly smaller hyperfine magnetic fields for atoms at the surface, but the results of Topsøe et al. (1979a) indicate that the line broadening may be due to collective magnetic excitations.

The experiments on thin metallic films showed that the hyperfine field at solid–solid interfaces does not differ greatly from the bulk value. The studies of small metal particles indicate that this result also holds for surfaces with and without chemisorbed molecules (Boudart et al., 1977; B. S. Clausen et al., 1979b; Mørup et al., 1979a; Mørup et al., 1980a).

Several studies on different small particle oxide systems (α-Fe_2O_3: Van der Kraan, 1972, 1973; γ-Fe_2O_3: Haneda and Morrish, 1977; Fe_3O_4: Mørup and Topsøe, 1976; Mørup et al., 1976; Lipka et al., 1977) also show that magnetically dead layers are absent and the hyperfine magnetic fields at the surface are close to the bulk value.

Information about the magnetic order of surface atoms in small particles has been obtained in several studies. For example, in Mössbauer investigations of small particles of γ-Fe_2O_3 in large external fields parallel to the γ-ray direction, a finite intensity of lines 2 and 5 was observed (Coey, 1971). These lines disappear completely in similar experiments with large crystals, owing to the ferrimagnetic order in such samples. The magnetic moment of the small particle sample was also found to be smaller than that expected from the bulk properties of γ-Fe_2O_3. These results were interpreted in terms of a noncollinear spin arrangement near the surface of the small particles. In a similar study Morrish et al. (1976) obtained Mössbauer spectra of γ-Fe_2O_3 particles enriched in [57]Fe at the surface, and found for these samples a significant intensity of lines 2 and 5. This result unambiguously shows that the noncollinearity is a surface phenomenon. It is likely that the effect can be explained by the lower symmetry of the surface atoms, which may favor specific spin orientations for the various types of surface sites, as discussed in Section III, F, 3 with reference to the thin film results by Shinjo et al. (1979) and Keune et al. (1979). Since the finite intensity of lines 2 and 5 in the above studies is due to the surface atoms, the positions of these lines can be used to determine the hyperfine field for surface atoms. By using this method, Haneda and Morrish (1977) found that the hyperfine field of surface atoms of γ-Fe_2O_3 particles was only 1 or 2% smaller than that of the atoms in the interior of the particles.

Berkowitz et al. (1975) studied small particles of $NiFe_2O_4$ coated with organic molecules such as oleic acid. Particles coated in this way reached only 75% of the saturation magnetization in an applied field of 200 kG, whereas samples in alcohol showed no such decreased saturation magnetization. Mössbauer spectra, obtained with a 68.5 kG magnetic field applied parallel to the γ-ray direction, showed zero intensity of lines 2 and 5 for a bulk sample and a sample exposed to alcohol, while the sample exposed to oleic acid showed a significant intensity of these lines. This result indicates a noncollinear spin arrangement of the iron atoms near the surface induced by the oleic acid molecules, similar to that observed in γ-Fe_2O_3 particles. This pinning of the surface spins may be regarded as a strong magnetic anisotropy that is induced by the chemisorbed molecules. That is, by affecting the orientation of the surface spins, this pinning may also affect the energy required to change the magnetization direction for the particle as a whole, thereby contributing to the total magnetic anisotropy energy constant. This was, indeed, found by Mørup *et al.* (1976) in a study of 6.0-nm Fe_3O_4 particles exposed to acetone, oleic acid, and stearic acid. These authors found that the superparamagnetic relaxation time depends on the organic species covering the surface, thus indicating a change in the magnetic anisotropy energy constant.

A similar effect was observed in studies of MgO-supported iron particles (Boudart *et al.*, 1977), small silica-supported Ni particles (B. S. Clausen *et al.*, 1979b; Mørup *et al.* 1979a), and silica-supported Fe particles (B. S. Clausen, 1979; Mørup *et al.* 1980a). In these investigations it was found that the superparamagnetic relaxation time changed considerably when particles were exposed to hydrogen. These magnetic effects are discussed in Chapter 1 (Mørup *et al.* 1980b).

G. Surface Studies by Use of Backscattering Techniques

In most of the studies discussed above, the surface effects have concerned only a few atomic layers. The detection of surface effects could be achieved because a large fraction of the resonant atoms were present at the surface, or because the surface influenced the properties of atoms well below the surface. The effect of hydrogen chemisorption on the superparamagnetic relaxation time of Ni particles (B. S. Clausen *et al.*, 1979b) is an example of the latter type of surface information.

In contrast, this section will be concerned with surface studies of bulk samples through use of backscattering techniques, i.e., experiments in which the limited penetration depth of x rays and conversion electrons is utilized to obtain surface sensitivity. The methodology for such experiments was discussed in Section II, E.

The penetration depth of ^{57}Fe x rays and conversion electrons is of the order of 10–100 μm and 10–100 nm, respectively. Therefore, the surface layers discussed with these techniques are considerably thicker than those discussed in the previous sections.

Conversion electron Mössbauer spectroscopy (CEMS) has been extensively used for studies of corrosion of iron surfaces (see, for example,

Fenger, 1973; Toriyama *et al.*, 1974; Tricker *et al.*, 1974; Sette Camara and Keune, 1975; Carbucicchio, 1977; Graham *et al.*, 1978, and Tricker *et al.*, 1979). A review of corrosion studies has been given by Simmons and Leidheiser (1976). As an interesting example of such studies, Fig. 17 shows spectra of energy analyzed conversion electrons from the surface of a ^{57}Fe-enriched iron foil, which was oxidized by heating in air for 10 min at 623 K (Tricker *et al.*, 1976a). Spectrum A is obtained with low energy

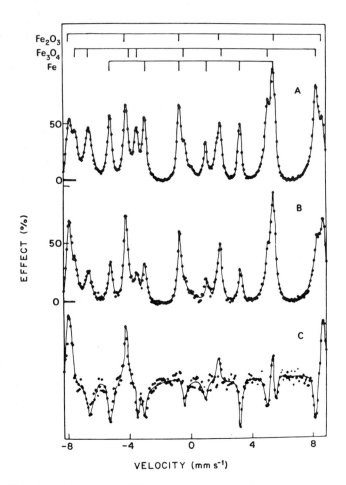

Fig. 17. Conversion electron Mössbauer spectroscopy (CEMS) spectra of an oxidized 90% ^{57}Fe-enriched iron foil obtained with A, low energy electrons; B, high energy electrons; C, difference spectrum. Note the marked enhancement of the Fe_2O_3 contribution in spectrum B. This is clearly seen by noting the increase in intensity of the Fe_2O_3 line at high positive velocities and in the difference spectrum (Tricker *et al.*, 1976a *Nucl. Instrum. Methods* **135**, 117).

electrons, spectrum B with high energy electrons, and C shows the corresponding difference spectrum. The line positions of α-Fe_2O_3, Fe_3O_4, and metallic Fe are indicated by bar diagrams above the spectra. The low energy electrons mainly arise from layers deep below the surface ($\sim$ 100 nm), whereas the high energy electrons are those emitted from the outer layers. The difference spectrum indicates that the outermost layer consists of α-Fe_2O_3, whereas the inner oxide layer mainly consists of Fe_3O_4.

Surface analysis of steels is another field of great technological importance (see Chapter 3 of Volume I and Chapter 3 of this volume). The amount of austenite in steel can be easily detected by Mössbauer spectroscopy since this phase is paramagnetic and, therefore, yields a component in the spectra that can be easily distinguished from magnetic α-iron and carbide phases (see Chapter 2 of Volume I and Chapter 7 of this volume). Swartzendruber and Bennett (1972) studied the amount of austenite in an Fe–C alloy of eutectoid composition that had been subjected to light surface grinding. Some spectra are shown in Fig. 18. The velocity scale was restricted to the central region and, therefore, only the two inner lines of the magnetic component are seen: (a) and (b) are spectra of an α-Fe foil and the Fe-C alloy, respectively, with the 14.4-keV γ rays counted; (c) and (d) show the spectra of the Fe–C alloy obtained by counting 6.3-keV x rays and the conversion electrons, respectively. It is evident that the central component, which is due to austenite, is significantly enhanced when conversion electrons are counted. This indicates that the cutting process raises the temperature of a thin surface layer into the austenitic range. This austenite is retained during rapid cooling by the cutting fluid and the substrate.

A metallic iron surface subjected to a strain (within the elastic limit) was studied by Mercader and Cranshaw (1975) by use of CEMS. In order to improve the signal-to-noise ratio, the surface was enriched in ^{57}Fe, and spectra were obtained both with the surface under tension and the surface under compression. The surface was strained by bending the bar. Small changes in isomer shifts, quadrupole couplings, and hyperfine fields relative to an unstrained surface were detected. The relative volume changes were estimated from the deformation of the sample, and the Mössbauer results were found to be in accordance with those obtained from iron subjected to hydrostatic pressure (Pound et al., 1961; Litster and Benedek, 1963; Pipkorn et al., 1964; Southwell et al., 1968) and for iron films subjected to a uniaxial tensile stress (Kjeldgård et al., 1975). With improvements of the technique such measurements may have technological applications.

In the field of geochemistry, the surface sensitivity of CEMS makes this technique valuable to mineral weathering studies. For example, Jones et al.

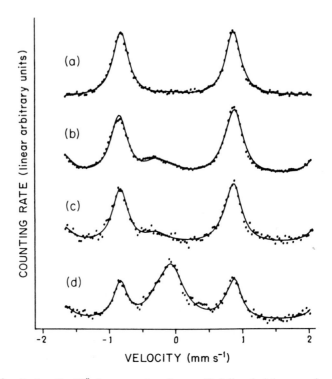

Fig. 18. Backscatter Mössbauer spectra of a pure Fe foil and of the ground surface of a spheroidized Fe–C alloy of the eutectoid composition (NBS Standard Reference Material 493). The velocity range has been restricted to the central region of the spectrum. (a) Fe foil, 14.4 keV γ rays counted; (b) Fe–C alloy, 14.4 keV γ rays counted. (c) Fe–C alloy, 6.3 keV x rays counted. (d) Fe–C alloy; conversion electrons counted. The large central peak due to retained austenite on the ground surface is clearly evident in (d). (Swartzendruber and Bennett, 1972. *Scripta Met.* **6**, 737. Copyright 1972. Reprinted with permission from Pergamon Press, Ltd.)

(1978) were able to identify the oxidation products on siderite surfaces as α-FeOOH. In a similar study, the oxidation mechanism of biotite has been investigated (Tricker *et al.*, 1976b).

An interesting application of CEMS was made by Forester (1973) who studied lunar samples in order to explore the hypothesis that the surface of these samples contains metallic iron particles produced by exposure to the solar wind. They observed, in fact, a component in the spectra that could be attributed to superparamagnetic iron particles.

A conversion electron study with the 3.5-keV conversion electrons reemitted after resonance absorption of 6.2-keV γ quanta of ^{181}Ta was made by Salomon *et al.* (1977). Spectra of tantalum single crystals and foil surfaces were compared to transmission Mössbauer spectra, and it was found that the foil surfaces (probed by the conversion electrons) showed larger line widths than bulk tantalum. The ^{181}Ta Mössbauer line is extremely sensitive to small changes in electron density at the ^{181}Ta nucleus, and it was suggested that the line broadening observed was due to absorbed residual gases at the surface giving rise to a distribution in isomer shifts. These absorbed species were evidently present in spite of sample degassing at 2673 K in UHV (10^{-9} Torr).

IV. Catalyst Studies

A. General Remarks

Mössbauer spectroscopy is an ideal spectroscopic technique for catalyst studies since real catalysts can be examined *in situ* at reaction conditions. The γ rays used in the experiments have a negligible interaction with gases and penetrate deeply into a solid catalyst, allowing both exterior and interior atoms to be observed. The use of Mössbauer spectroscopy under *in situ* reaction conditions has given new insight into many catalyst systems. It has, for example, been observed that properties such as surface composition, surface structure, and surface chemical state, may depend on the gaseous environment. Therefore, the possibility of performing *in situ* Mössbauer studies under well-defined conditions of adsorption or reaction is of great importance. The possibility of measuring simultaneously Mössbauer spectral parameters and catalytic behavior is particularly valuable. It should be added that many Mössbauer studies have shown that the results obtained from *in situ* experiments may differ from those obtained with the sample exposed to air.

Many studies of real catalysts are complicated by the presence of small particles, limiting the use of such techniques as x-ray diffraction. Particle size restrictions are, however, not important in Mössbauer spectroscopy. On the contrary, Mössbauer spectroscopy may be used to obtain information on particle size and even particle-size distribution (see Section IV, B). Furthermore, for well-dispersed catalysts a large fraction of the resonant atoms are surface atoms, and insight into the properties of these surface atoms may be obtained from Mössbauer spectroscopy. The way to obtain this information was discussed in Sections II and III, and many examples dealing with surface studies of real catalysts will be given in the sections that follow.

The conditions often used during the preparation of catalysts may result in structures that differ from the well-defined, ordered structures of ideal crystals. Mössbauer spectroscopy can be used to study such systems since the resonance does not require the atoms to be in an ordered lattice.

Sections IV, B and IV, C will deal with two problems that are important in many catalyst studies: measurements of particle size and properties of supported catalysts. Sections IV, D–J will give reviews of recent Mössbauer spectroscopic studies for selected catalyst systems. The areas selected are alloy catalysts, ammonia catalysts, Fischer–Tropsch catalysts, partial oxidation catalysts, hydrodesulfurization catalysts, and electroless- and electrocatalysis.

Inevitably, in making such a selection, many important Mössbauer studies will not be discussed here (for studies before $\sim$ 1975 see Dumesic and Topsøe, 1977). Although a few applications of Mössbauer spectroscopy to homogeneous catalyst systems have appeared, these will not be discussed in detail here. In fact, Mössbauer spectroscopy is not an ideal technique for studying liquid phase systems, but homogeneous catalysts may be studied after freezing the samples.

B. Particle Size and Dispersion Measurements

Knowledge of the particle size and particle-size distribution of the catalytic active phase is important in catalyst studies. For some reactions the activity is found to be proportional to the surface area of the active phase (i.e., constant turnover frequency) irrespective of particle size of this phase, whereas for other reactions the turnover frequency is dependent on the particle size of the catalyst. Reactions belonging to the first category have been termed structure insensitive, whereas reactions of the second type are referred to as being structure sensitive.

Many of the Mössbauer parameters such as the recoil-free fraction, isomer shift, quadrupole splitting, and magnetic hyperfine field may, for a given system, depend on the particle size. However, the problem is that the quantitative dependence of the Mössbauer parameters on the particle size is rather complex and in most cases not yet established. The present discussion, however, will deal mainly with two instances where the particle size and the fraction of atoms exposed at the surface (i.e., the degree of dispersion) can, in fact, be determined directly.

The particle size of magnetic microcrystals has often been estimated by use of Mössbauer spectroscopy from measurements of temperature dependence of the superparamagnetic component in the spectra or, more recently, by analysis of the line shifts and line shapes of the spectra below the blocking temperature (collective magnetic excitations) (see Mørup et al., 1980b). However, by using these methods, only the product of the anisotropy constant K and the volume of the particle V can be determined. If K is known, V can then be determined. However, as discussed in

Chapter 1 (Mørup *et al.* 1980a), the effective K may depend on such parameters as the particle shape, surface structure, chemisorbed gases, and the presence of stresses. It may, therefore, be difficult to estimate K, and as a result, particle-size determinations become uncertain. Before discussing how the particle size may be determined directly, it may be worthwhile mentioning that for studies of supported catalysts, K may be dependent on the nature of the support and the particle size. For example, the shape of supported microcrystals depends on the relative surface energies of (i) the microcrystal surface, (ii) the support surface, and (iii) the microcrystal–support interface.

Recently Mørup and Topsøe (1977) showed that the particle size of magnetically ordered microcrystals can, in fact, be directly determined from the dependence of the spectra on the strength of the applied magnetic field at temperatures above the superparamagnetic blocking temperature. Under these conditions, the magnetic splitting of the spectra is essentially proportional to the Langevin function $L(\mu H/kT)$, where μ is the magnetic moment of the particle. The particle size can then be conveniently determined from a plot of the magnetic hyperfine field as a function of T/H for spectra at high fields. See also Chapter 1 (Mørup *et al.* 1980a).

Figures 19 and 20 show an example of such a determination of μ for a Ni catalyst supported on silica (Mørup *et al.*, 1979a). The Ni particles were doped with [57]Co, and the experiments were carried out as source experiments. Figure 19 shows Mössbauer spectra at 78 K and in various applied fields. The substantial increase in the magnetic splitting with increasing field is typical of superparamagnetic particles. Figure 20 shows the value of the magnetic splitting of the spectra (corrected for the contribution due to the applied field) as a function of T/H. From the slope of the straight lines, a particle diameter of 4 nm was found. It was assumed that the magnetization is equal to the bulk value, that the magnetic anisotropy is negligible, and that the particles have spherical shape. The result obtained in this way agreed nicely with independent x-ray and chemisorption results (B. S. Clausen, 1979).

If the above type of measurement is combined with measurements of superparamagnetic relaxation or collective magnetic excitations, information about the prevailing magnetic anisotropies can also be directly obtained, and it is possible to determine not only distributions in KV, but also particle-size distributions. The possibility of measuring particle-size distributions has been useful in understanding the relative importance of different preparation parameters in the genesis of supported iron catalysts (Topsøe *et al.* 1979a).

Particle-size determinations of nonmagnetic materials may also be carried out if the Mössbauer parameters of the surface atoms are different

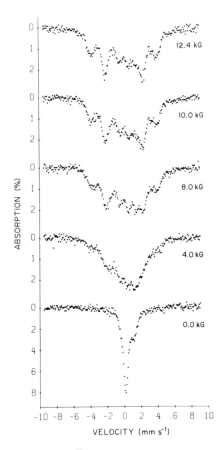

Fig. 19. Mössbauer spectra of ^{57}Co-doped silica-supported Ni particles in vacuum at 78 K in various applied magnetic fields. The spectra were obtained with a movable absorber of $K_4Fe(CN)_6 \cdot 3H_2O$ (Mørup *et al.*, 1979a).

from those of the atoms in the bulk. These differences may be intrinsic or may be induced by chemisorption ("titration"). Such measurements may directly give information about the dispersion (defined as the surface atom fraction of all the atoms) of the Mössbauer isotope-containing phase.

Some of the many examples of titration experiments used to establish the dispersion of iron-containing phases are those of iron ion-exchanged zeolites (e.g., Delgass *et al.*, 1969), silica-supported iron catalysts (Hobson and Gager, 1970), and alumina-supported catalysts (Topsøe and Mørup, 1975). In the last of these studies, a calcined Co–Mo/Al$_2$O$_3$ catalyst, doped with ^{57}Fe, was studied at 300 K. Figure 21 shows the spectra obtained in air (spectrum a) and in a H$_2$/H$_2$S gas mixture (spectrum b). It is seen that the exposure to H$_2$/H$_2$S at 300 K gives rise to substantial

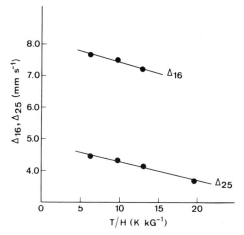

Fig. 20.   Magnetic splitting $\triangle_{16}$ and $\triangle_{25}$ of lines 1 and 6 and lines 2 and 5, respectively, plotted as a function of T/H for the spectra of the ^{57}Co-doped Ni particles shown in Fig. 19 (Mǿrup *et al.*, 1979a).

changes in the spectrum. Since diffusion of the gases into the oxidic phases is not likely to occur at this temperature, it can be concluded that a large fraction of the ^{57}Fe atoms are located very close to the surface. A quantitative analysis of spectrum b indicated that the chemical state of about 37% of the iron atoms was changed from Fe^{3+} to Fe^{2+}. Although the spectral components were not analyzed in detail, these measurements

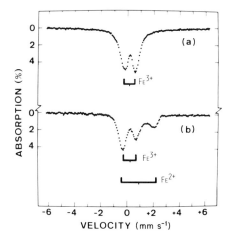

Fig. 21.   Room temperature Mössbauer spectra of ^{57}Fe in $Co-Mo/Al_2O_3$ catalysts after different treatments: (a) after calcination, (b) after room temperature sulfiding in H_2/H_2S ($\sim$ 2% H_2S). (Adapted from Topsǿe and Mǿrup, 1975.)

show the usefulness of Mössbauer spectroscopy for such surface titration studies.

Section IV, D will give several examples of the use of intrinsic differences or titration-induced differences to establish the degree of dispersion and surface composition of supported bimetallic catalysts.

C. Supported Catalysts

Many catalysts consist of a so-called active component located on a support. The support may have several functions: it may "create" and stabilize a large surface area (i.e., high dispersion) for the active components, it may stabilize structures which would not otherwise exist, and it may also interact (e.g., electronically) with the active components and thus form an integrated part of the catalyst system. Mössbauer spectroscopy has contributed greatly to elucidating the role of the support in supported catalyst systems.

This section will concentrate on two important problems in supported catalysts: preparation (reduction) of supported metallic iron catalysts (Section IV, C, 1) and the possible modification of the electronic structure of the catalyst caused by the support (Section IV, C, 2). Different support effects are discussed in other sections. Sections III, D and IV, D, G, H, and I present examples of how the chemical and structural forms of supported species may depend on the support, whereas Section IV, J gives some examples of how the reactivity of the supported phases are influenced by the support.

1. Supported Metallic Iron Catalysts

Iron is usually placed in contact with a support by processes such as impregnation, ion exchange, or coprecipitation, followed by drying and high temperature oxidation (calcining). The metallic state of iron is then achieved by hydrogen reduction of these oxidic precursors. Pure iron oxides readily reduce to metallic iron in hydrogen at temperatures as low as 573 K. If, however, iron interacts and/or reacts with the support, then a surface or bulk compound may form, which is more resistant to reduction than the pure iron oxides themselves.

In order to describe the preparation of supported metallic iron catalysts, it is convenient to define the reducibility of such catalysts as the fraction of the total amount of iron that can be reduced to the metallic state.

Silica has been used extensively as a support for both iron oxides and metallic iron. Table II summarizes literature data for the reducibility of iron catalysts supported on silica. The reducibility is not constant, but

TABLE II

Reducibility of SiO$_2$ Supported on Iron Catalysts

wt% Fe	Preparation[a]	Pretreatment	Reduction	Reducibility[b]	Fe^0 particle size (nm)	References
0.1	Fe(NO$_3$)$_3$ imp.	{ vac., 748 K, 8 hr / air, 773 K, 3 hr }	973 K, 2 hr	~0	—	Garten (1979b)
0.17	ion exchange	vac., 683 K, 3 hr	673 K, 15 hr	~0	—	Huang and Anderson (1975)
0.5	Fe(NO$_3$)$_3$ imp.	O$_2$, 573 K	973 K, 4 hr	0.2	—	Dézsi et al. (1979)
0.8	Fe(NO$_3$)$_3$ imp.	air, 523 K	723 K, 8 hr	0.15	—	Yoshioka et al. (1970)
1.4	Fe(NO$_3$)$_3$ imp.	O$_2$, 773 K, 24 hr	~773 K	~0	—	Hobson and Gager (1968)
2	Fe(NO$_3$)$_3$ imp.	O$_2$, 648 K, 1 hr	723 K, 8 hr	~0	—	Tachibana et al. (1969)
3	Fe(NO$_3$)$_3$ imp.	O$_2$, 773 K, 16 hr	723 K	~0	—	Hobson and Cambell (1967)
3	Fe(NO$_3$)$_3$ imp.	air, 773 K, 3 hr	703 K, 48 hr	0.4[c]	~6[d]	B. S. Clausen et al. (1979a)
3.9	Fe(NO$_3$)$_3$ imp.	air, 473–573 K	773 K	0.56	—	Sigg and Wicke (1977)
3.9	Fe(NO$_3$)$_3$ imp.	air, 473–573 K	873 K	0.76	—	Sigg and Wicke (1977)
3.9	Fe(CO)$_5$	air, 473–573 K	773 K	1.0	—	Sigg and Wicke (1977)
4.94	Fe(NO$_3$)$_3$ imp.	air, 723 K, 4 hr	628 K, 12 hr	0.67	—	Amelse et al. (1978)
4.94	Fe(NO$_3$)$_3$ imp.	air, 723 K, 4 hr	628 K, 24 hr	>0.9	13	Amelse et al. (1978)
5	Fe(NO$_3$)$_3$ imp.	—	673 K	~0.8	—	Arnold and Hobert (1968)
10	Fe(NO$_3$)$_3$ imp.	vac., 413 K, 16 hr	723 K, 24 hr	0.15	e	Raupp and Delgass (1979a)
10	Fe(NO$_3$)$_3$ imp.	O$_2$, 648 K, 12 hr	723 K, 8 hr	0.67	7.9[f]	Raupp and Delgass (1979a)
10	Fe(NO$_3$)$_3$ imp.	O$_2$, 723 K, 24 hr	723 K, 8 hr	1.0	10.3[f]	Raupp and Delgass (1979a)
10	Fe(NO$_3$)$_3$ imp.	{ vac., 413 K, ~16 hr / O$_2$, 723 K, ~24 hr }	723 K, ~24 hr	0.15	—	Raupp and Delgass (1979a)
> 5, < 20	imp.	air, 723 K	~673 K	~0.8	—	Hobert and Arnold (1971)

[a] imp.: impregnation.

[b] Fraction of metallic iron calculated assuming equal recoil-free fraction for Fe^0 and nonreduced iron; Fe^0 assumed ferromagnetic, unless otherwise stated.

[c] Taking superparamagnetic Fe^0 into account.

[d] From magnetic field dependence of spectra (B. S. Clausen, S. Mørup, and H. Topsøe, unpublished result).

[e] No observable x-ray diffraction peak.

[f] From x-ray line broadening on passivated samples.

strongly depends on the iron loading (% Fe), preparation method, pretreatment conditions, and the temperature and time of reduction.

Before discussing these effects in more detail, some comments will be given concerning the way in which the reducibilities in Table II have been obtained from the Mössbauer spectra. Only *in situ* (in hydrogen) studies have been included in the table. For each of these studies, the lowest temperature (usually room temperature) spectrum has been used to calculate the reducibility from the ratio of the area of the magnetically split metallic iron spectrum to the total absorption area. Thus in the calculations it has been assumed that all the metallic iron behaves ferromagnetically and not superparamagnetically. Recently, however, B. S. Clausen (1979) and B. S. Clausen *et al.* (1979a) have shown that this assumption does not always hold since superparamagnetic metallic iron was observed to be present in a reduced 3% Fe/SiO_2 catalyst at temperatures as low as room temperature. The room temperature spectrum (Fig. 22a) of the catalyst shows only a small intensity of the ferromagnetic six-line α-Fe spectrum. However, application of an external magnetic field of 12.4 kG at 300 K (Fig. 22b) or cooling to 78 K (Fig. 22c) results in a large increase in the intensity of the ferromagnetic six-line component revealing that at room temperature a large fraction of the iron particles were present in the superparamagnetic state. However, the room temperature spectrum (Fig. 22a) shows that the superparamagnetic iron does not give rise to a sharp and well-defined peak at the center of the sextuplet, but rather produces a broad component. This may be the reason why superparamagnetic iron has not been detected in previous studies of this type of sample. The above results imply that some of the reducibilities given in the literature (and in Table II) may be too low.

An understanding of the nonreduced forms of iron for silica-supported catalysts is important since it may help to clarify the nature of the interaction of the iron with the silica support. The fraction of the iron that is not reduced to the metallic state contributes to the central regions of the Mössbauer spectra. This region is quite complex and its interpretation has been the subject of much controversy. (For a discussion, see Berry, 1978). It has been suggested that the nonreduced iron is present as several ferrous species (Gager and Hobson, 1975; Garten, 1976b; Raupp and Delgass, 1979a). However, as discussed above, superparamagnetic iron may give rise to a broad component in the central region and, as such, should be included in the analysis of this region. In a recent study (B. S. Clausen, 1979) where the superparamagnetic component could be accounted for, it was confirmed that two different ferrous species exist. It was, furthermore, observed that the relative abundance of the two species varied with iron loading.

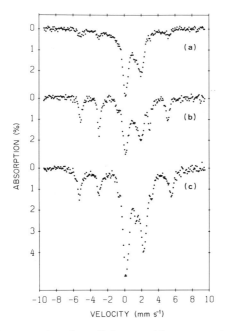

Fig. 22. Mössbauer spectra of small iron particles supported on silica: (a) room temperature, $H = 0$; (b) room temperature, $H = 12.4$ kG; (c) 78 K, $H = 0$ (B. S. Clausen *et al.*, 1979a).

Although, as mentioned above, the reported reducibilities in some instances may be somewhat low, the results given in Table II show that the reducibility strongly depends on the characteristics of the catalyst, in particular the metal loading. In the region of low metal loadings ($< 1\%$) it is generally observed that only a small fraction of the iron can be reduced to the metallic state, even after hydrogen treatment at temperatures several hundred degrees above the point where the bulk iron oxides reduce. These results show that very strong interactions with the silica support may be present. The results in the table show that there is a tendency for the reducibility to increase as the metal loading increases. However, the results also show that at a given loading, different reducibilities may be obtained, depending on the treatment and the structure of the catalyst prior to reduction. It is therefore necessary to discuss some features of oxidized forms of iron supported on silica. The many studies of these systems have been reviewed by Gager and Hobson (1975) and Berry (1978).

Several authors have suggested that the reducibility of supported $\alpha\text{-Fe}_2\text{O}_3$ particles depends on their particle size. For example, Yoshioka *et al.* (1970) proposed that samples exhibiting a magnetically split $\alpha\text{-Fe}_2\text{O}_3$ component in room temperature Mössbauer spectra readily reduce to $\alpha\text{-Fe}$, whereas

α-Fe_2O_3 particles small enough to exhibit superparamagnetic behavior do not reduce below the ferrous state. The above study will be discussed in some detail since similar results and interpretations have been reported by several other authors. Yoshioka *et al.* (1970) found that the room temperature Mössbauer spectra of calcined iron catalysts supported on silica, as well as on γ- and η-Al_2O_3, changed from magnetically split to paramagnetic as the loading decreased. This was taken as evidence for a decrease in the α-Fe_2O_3 particle size with decreasing iron loading. The authors also observed an increase in quadrupole splitting as the loading decreased which, using the results of Kündig *et al.* (1967), also was attributed to a decrease in the α-Fe_2O_3 particle size with decreasing loading. Samples with a very high metal loading showed the presence of antiferromagnetic α-Fe_2O_3 in the room temperature spectra. These samples readily reduced to α-Fe. This result suggests that α-Fe_2O_3 particles greater than about 13.5 nm, the critical diameter for observing superparamagnetism in α-Fe_2O_3 (Kündig *et al.*, 1966), interact weakly with the support, and their reducibility is similar to that of bulk α-Fe_2O_3. Yoshioka *et al.* (1970) also observed that samples exhibiting only paramagnetic behavior at room temperature (samples with low metal loading) did not easily reduce to metallic iron. Neither did these samples show much tendency toward sintering in air at 1053 K. Both results indicate a strong interaction with the support for samples exhibiting paramagnetic behavior. As mentioned above, the paramagnetic spectra were taken as evidence for the presence of small, superparamagnetic α-Fe_2O_3 particles. The interaction between the α-Fe_2O_3 particles and the silica, therefore, seems to increase as the size of the α-Fe_2O_3 particles decreases. Although it is very likely that the size of the α-Fe_2O_3 particles decreases with decreasing metal loading and that the interaction of such particles with the support will become stronger with a decrease in particle size (increased fraction of atoms exposed to the support), this is not the only possible explanation for the above results. It is also possible that the paramagnetic component in the Mössbauer spectra is not only due to superparamagnetic α-Fe_2O_3 particles, but may at least partly be attributed to a significant fraction of iron cations, ion exchanged onto the surface of the support or directly forming a surface or bulk compound or a solid solution with the support. The large quadrupole splittings observed for low metal loadings are consistent with this explanation since ferric iron exchanged at the surface of silica or zeolites (aluminosilicates) (Huang and Anderson, 1975; Garten *et al.*, 1970) shows large quadrupole splittings. The resistance against reduction is also consistent with iron forming surface or bulk compounds with the support since iron silicates are more resistant against reduction than are iron oxides. For example, reduction of Fe_2SiO_4 requires a p_{H_2O}/p_{H_2} ratio about three orders of magnitude lower than that required for reduction of α-Fe_2O_3

(Huang and Anderson, 1975). In agreement with this, these authors observed very low reducibilities for silica samples with iron cations ion exchanged at the surface. At present, a particle-size dependence of the reducibility of supported α-Fe$_2$O$_3$ particles, therefore, does not appear to be firmly established.

The reducibility of a supported iron oxide catalyst and the resulting metallic iron particle size will depend on the degree of interaction with the support, and the extent of oxide particle growth that has occurred prior to the reduction. In this respect, the metal salt used in the impregnation plays an important role. Meisel and Hobert (1968) and Rubaskov et al. (1972) found that silica impregnated with FeCl$_3$ gave large antiferromagnetic α-Fe$_2$O$_3$ crystals after calcination, whereas silica impregnated with Fe(NO$_3$)$_3$ or Fe(CO)$_5$ in absolute ether (Rubaskov et al., 1972) after similar calcination showed a paramagnetic doublet. This result indicates that a much higher dispersion is obtained when using Fe(NO$_3$)$_3$ and Fe(CO)$_5$ than when using FeCl$_3$. The results were interpreted by Rubaskov et al. in terms of the decomposition temperatures of the salts (370, 380, and 770 K for Fe(NO$_3$)$_3$, Fe(CO)$_5$, and FeCl$_3$, respectively). Thus a low decomposition temperature seems favorable for a high dispersion of the oxide.

Raupp and Delgass (1979a) found that dehydration at low temperature also favors a high dispersion of both the oxide and the iron produced by reduction of the oxide (see Table II). The authors observed that as the dispersion increased the fraction of iron reduced to the metallic state decreased. Figure 23 shows room temperature spectra of reduced 10% Fe/SiO$_2$ catalysts as a function of the pretreatment. The spectrum of a dried catalyst calcined in 1% O$_2$/He for 24 hr at 723 K was indicative of large, antiferromagnetic α-Fe$_2$O$_3$ particles. Upon reduction (Fig. 23c) all the iron was reduced to large α-Fe particles. Catalysts calcined at less severe conditions (12 hr at 648 K) showed smaller α-Fe$_2$O$_3$ crystals, and the iron particles formed upon reduction (Fig. 23b) were also smaller than those of the sample shown in Fig. 23a. Correspondingly, the reducibility was now less than 100%. The smallest iron particles and the lowest fraction of reduced iron were obtained for the samples that had first been vacuum dried (Fig. 23a). The spectrum of this sample after the initial vacuum drying showed no magnetically ordered component, indicating that it was composed of very small α-Fe$_2$O$_3$ particles or ferric ions interacting with the silica, as discussed above. A sample, which was first vacuum dried and then calcined in O$_2$ prior to the reduction, gave a spectrum identical to Fig. 23a. These results show that the removal of water or nitrate ions plays an important role in preventing the growth of α-Fe$_2$O$_3$ particles, probably by reducing the surface mobility. As a result, the degree of metal interaction with the support will increase. Suzdalev et al. (1966, 1968) have also observed (see Section III, C, 1) that the mobility of hydrated tin oxide

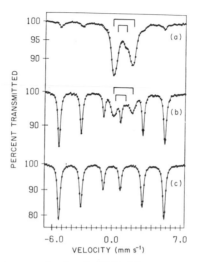

Fig. 23. Mössbauer spectra of reduced 10% Fe/SiO$_2$ catalysts: (a) vacuum-oven dried, reduced in H$_2$ 24 hr at 723 K; (b) calcined in 1% O$_2$/He 12 hr at 648 K, reduced in H$_2$ 8 hr at 723 K; (c) calcined in 1% O$_2$/He 24 hr at 723 K, reduced in H$_2$ 8 hr at 723 K. All spectra were recorded in H$_2$ at room temperature (Raupp and Delgass, 1979a).

species on silica surfaces is greater than that of dehydrated oxide species. The reducibilities reported for Fe/SiO$_2$ catalysts may also be kinetically controlled. Recent Mössbauer experiments (B. S. Clausen, 1979) on a 3% Fe/SiO$_2$ catalyst showed that the reduction at 703 K was initially very fast, followed by a very slow reduction resulting in metallic iron being produced even after three days' reduction. Also the results (see Table II) of Amelse *et al.* (1978) show that the overall reduction process at 628 K is quite slow.

Besides silica, high surface area aluminas have probably been the most extensively used supports for iron-containing catalysts. The behavior of alumina-supported catalysts parallels, in many respects, that of silica-supported catalysts. For example, at low metal loadings (0.05%) iron is not reduced to the metallic state in H$_2$ even at 973 K (Garten and Ollis, 1974). In general, ferric ions will have a strong tendency toward interacting with the alumina surface because of the chemical similarity of the Fe^{3+} and Al^{3+} ions. As the metal loading is increased, metallic iron can be formed upon reduction (e.g., Kölbel and Küspert, 1970; Yoshioka *et al.*, 1970; Hobert and Arnold, 1971). Yoshioka *et al.* (1970) found that the reducibility depends on the type of alumina and increases in the order

$$\gamma - Al_2O_3 > \eta - Al_2O_3 > \alpha - Al_2O_3.$$

Further examples of support interactions in silica- and alumina-supported systems will be given in Section IV, I.

Iron ion-exchanged zeolites are generally very stable against reduction to metallic iron (Garten *et al.*, 1970; Küspert, 1970; Huang and Anderson, 1975). It is found that ferric ions readily reduce to ferrous ions, but that further reduction to the metal does not occur except under conditions where the zeolite structure is destroyed (Küspert, 1970; Yatsimirskii *et al.*, 1976a). In cases where the iron is introduced as a molecular compound into the zeolite, reduction to the metallic state may occur (Wedd *et al.*, 1969). Reduction to the metal may also be achieved by using metallic sodium (Schmidt *et al.*, 1977).

The studies mentioned above have all dealt with supported iron catalysts for which support interactions are very strong and govern the reducibility of such catalysts. In contrast to these systems, one could mention iron supported on carbon or BeO. The support interactions for these systems are in general quite weak, and iron will have a high reducibility to the metallic state even at low temperatures. Furthermore, as a consequence of the weak interactions with the support, the metallic iron particle size, after high temperature reduction, tends to be quite large. For example, Bartholomew and Boudart (1972) found 20–60-nm iron metal particles after reduction of a carbon-supported catalyst at 723 K. Kölbel and Küspert (1970) studied BeO-supported catalysts. After air calcination at 773 K of a sample containing 3 wt% iron, the Mössbauer spectrum was typical of superparamagnetic α-Fe_2O_3 particles, which could subsequently be reduced very easily. Reduction at 423 K converted all the ferric ions to the ferrous state, and at 473 K the degree of reduction to the metallic state was greater than 90%. After reduction at 543 K, the reducibility was found to approach 100%. Küspert (1970) also found high reducibilities under similar conditions for a sample containing 1 wt% iron.

Magnesium oxide has recently received much interest as a support for iron catalysts. Much of this attention started by the finding by Boudart *et al.* (1975a, b) that very small thermostable iron particles can be prepared using this support. The possibility of preparing samples of greatly differing particle size allowed studies of the particle size dependence of chemisorptive and catalytic properties (Dumesic *et al.*, 1975a, b; Boudart *et al.*, 1977; Topsøe *et al.*, 1980). For the production of small metallic iron particles the support interaction using MgO seems to be quite ideal: it is neither strong enough to prevent reduction to the metallic state nor weak enough to bring about rapid sintering of the iron particles. In fact, the reducibilities, which give some indication of the interaction with the support, were found to be different from any of the supported catalysts discussed above. For example, as the iron loading was changed from 1 to 40%, the fraction of iron reduced to the metallic state changed by a factor less than two (from 0.4 to 0.7, see Table III). Mössbauer spectroscopy also showed that the nonreduced iron was present as ferrous ions located in the magnesium oxide.

TABLE III

Reducibility of MgO-Supported Iron Catalysts

wt% Fe	Preparation	Pretreatment	Reduction	Reducibility[a]	Fe⁰ particle size (nm)	References
1	ion exchange[b]	vac., 360 K, 12 hr	700 K, 20 hr[c]	0.39[d]	1.5[e]	Boudart et al. (1975a)
5	ion exchange[b]	vac., 360 K, 12 hr	700 K, 20 hr[c]	~0.38[d]	4.0[e]	Boudart et al. (1975a)
5	coprecipitation	air, 723 K, 24 hr	923 K, 14 hr	~0.5[d]	>13.0	Bussière et al. (1975)
8	ion exchange[b]	vac., 700 K, 20 hr[c]	700 K, 20 hr[c]	0.52[dh]	7.2[e]	Topsøe et al. (1979a)
10	ion exchange[b]	vac., 360 K, 12 hr	700 K, 20 hr[c]	0.49	f	Raupp and Delgass (1979b)
12	ion exchange[b]	vac., 360 K, 12 hr	700 K, 20 hr[c]	0.49[d]	10[e]	Boudart et al. (1975a)
> 5, < 20	impregnation	air, 723 K	~670 K	~0.6	—	Hobert and Arnold (1971)
16	ion exchange[b]	vac., 360 K, 12 hr	700 K, 20 hr[c]	0.49[d]	11[g]	Boudart et al. (1975a)
16	ion exchange[b]	vac., 360 K, 12 hr	700 K, 20 hr[c]	0.62[dh]	8.5[e]	Topsøe et al. (1979a)
16	ion exchange[b]	air, 700 K, 20 hr[c]	700 K, 20 hr[c]	0.64[dh]	12[e]	Topsøe et al. (1979a)
16	coprecipitation	air, 770 K, 24 hr	700 K, 20 hr[c]	0.53[dh]	4.0[e]	Topsøe et al. (1979a)
16	coprecipitation	vac., 360 K, 12 hr	700 K, 20 hr[c]	~0.5[dh]	4.2[e]	Topsøe et al. (1979a)
26	Fe, Mg oxalate	air, 773 K, 24 hr	683 K, 120 hr	~0.55	—	Kölbel and Küspert (1970) and Küspert (1970)
40	ion exchange[b]	vac., 360 K, 12 hr	700 K, 20 hr[c]	0.72	3.0[e]	Boudart et al. (1975a)

[a] Assuming equal recoil-free fraction for Fe⁰ and nonreduced iron.
[b] Ion exchange of magnesium hydroxy carbonate according to Boudart et al. (1975a).
[c] Typical schedule: 370 K, 4 hr; 520 K, 14 hr; 600 K, 4 hr; 650 K, 4 hr; 700 K, 20 hr.
[d] Superparamagnetic iron taken into account.
[e] From chemisorption.
[f] No observable peak by x-ray diffraction on passivated sample.
[g] From x-ray line broadening.
[h] From B. S. Clausen, S. Mørup, and H. Topsøe (unpublished results).

These ions were not distributed randomly, but were present as iron-rich clusters (Boudart *et al.*, 1975a; Topsøe *et al.*, 1979a). Such clusters may be the origin for the special properties of this catalyst system and may explain why reduction to the metal occurs even at low loading. Also, the formation of clusters will tend to give rise to isolated metallic iron particles upon reduction, thereby minimizing agglomeration.

The catalysts studied by Boudart *et al.* (1975a) were prepared by exchanging magnesium cations with iron ions at the surface of magnesium hydroxycarbonate crystals, followed by vacuum drying and temperature programmed decomposition under hydrogen. The genesis and structure of Fe/MgO catalysts have been studied further by Topsøe *et al.* (1979a) using both coprecipitated and "surface-exchanged" magnesium hydroxycarbonate catalyst precursors. They also studied the importance of the activation procedure on the properties of the reduced catalysts. For example, quite different structures were obtained depending on whether the precursors were calcined in air or in vacuum. However, after reduction of these samples, the differences seemed to be small and the catalysts had reducibilities and metallic iron particle sizes similar to those of catalysts prepared by direct reduction of the precursors in hydrogen (Table III). These results are somewhat surprising and quite opposite to the results for, e.g., silica-supported catalysts, for which the reducibilities and metallic iron particle sizes were very sensitive to pretreatment. The similarities between differently prepared magnesium oxide catalysts could be related to the tendency for iron to form stable, Fe^{2+}-rich clusters in magnesium oxide under reduction conditions.

Catalysts prepared from coprecipitated precursors showed a more uniform distribution of iron throughout the support than those prepared from the "surface-exchanged" magnesium hydroxycarbonate (Topsøe *et al.*, 1979a). As a result, the reduced catalysts prepared from coprecipitated precursors consisted of significantly smaller iron particles with a narrower particle-size distribution than those particles prepared from "surface-exchanged" magnesium hydroxycarbonate (Table III). This result was obtained by analysis of the influence of collective magnetic excitations on the Mössbauer spectra.

The reducibilities of magnesium-oxide-supported catalysts prepared by impregnation (Hobert and Arnold, 1971), coprecipitation (Bussiere *et al.*, 1975), and reduction of mixed iron magnesium oxalates (Kölbel and Küspert, 1970) are given in Table III. The results show that the reducibilities of these catalysts are also close to 0.5.

Finally, it could be added that, in general, the reducibility of supported iron catalysts can be enhanced by "alloying" the iron with a noble metal. This will be discussed in Section IV, D.

2. Evidence for Electronic Support Interactions

When catalytic reactions are carried out on small supported metal particles, it is often found that the specific catalytic reaction rate (per metal surface area) differs from that of large particles. There may be many reasons for such changes in catalytic behavior of small metal particles. For example, the intrinsic electronic structure of a metal particle gradually changes as the number of atoms comprising the particle decreases (Kubo, 1962; Baetzold, 1978). Moreover, the distribution in geometric arrangements of the surface atoms is expected to change as the particle size decreases (Van Hardeveld and Hartog, 1969). Changes in the electronic structure caused by the neighboring support may also be expected to become more important as the metal particle size decreases (Schwab et al., 1959). Small platinum particles encaged in Y zeolites were inferred, from their catalytic behavior, to be electron deficient (Dalla Betta and Boudart, 1973). The electron deficiency of the platinum clusters has later been confirmed by various physical measurements (Ioffe et al., 1977; Gallezot et al., 1977; Vedrine et al., 1978; Gallezot et al., 1979). However, as pointed out by Gallezot et al. (1979), it may be difficult to examine the relative importance of the intrinsic size effect compared to the support effect since both induce electron deficient character in the metal. Mössbauer spectroscopy is a sensitive technique for the study of electronic properties of iron, and some results pertaining to intrinsic or support-induced changes in these properties will be discussed below.

The properties of small, metallic iron clusters in zeolites were studied by Schmidt et al. (1975, 1977). Reduction to metallic iron is difficult in zeolites (see Section IV, B, 1) and was achieved by using sodium vapor. The authors showed that a large fraction of the iron was present as metallic iron clusters with extremely narrow particle-size distribution and diameters less than 1.3 nm. A small fraction of the iron was reduced to large iron particles with diameters up to about 50 nm. These particles were located outside the zeolite framework and showed the usual magnetically split six-line spectrum at all temperatures between 4 and 300 K. From magnetic susceptibility and Mössbauer spectroscopy measurements, the authors concluded that the iron clusters located within the zeolites showed superparamagnetic behavior. Even at temperatures down to 4 K, no magnetic hyperfine splitting of the iron cluster component was observed in the Mössbauer spectra. The superparamagnetic component was identified as a quadrupole-split doublet having a splitting of 0.60 mm s^{-1} and a positive isomer shift relative to that of bulk α-Fe. This was taken as indicative of changes in the electronic environment caused by the zeolite support.

It is possible that the electronic properties of the small metallic iron clusters may also be influenced by the presence of metallic sodium since it has been shown that the isomer shift of small iron particles supported on silica was changed by adsorption of metallic potassium (Topsøe and B. S. Clausen, unpublished results). Such changes may, in general, occur upon chemisorption and should be important for the very small clusters studied by Schmidt *et al.* (1975, 1977), having almost all the iron atoms at the surface.

Direct evidence for changes in the electronic properties caused by the contact with a neighboring phase has been obtained in studies of thin films (see Section III, F). For example, for thin iron films the hyperfine field and the isomer shift (and thus the electronic structure) at the iron–substrate interface were observed to depend on the nature of the neighboring solid (see, e.g., Sections III, D, 2 and III, F, 3). From the large number of studies that have been carried out on films of different thickness and different substrates, it seems that these "electronic support interactions" may be quite large and dominate those electronic changes induced by the small thickness of the films (intrinsic size effects).

D. Alloy Catalysts

The recent commercial success of alloy catalysts (e.g., in reforming reactions) has led to an increase in the number of theoretical and experimental studies of these systems. The questions of fundamental importance for the understanding of the catalytic properties of these catalysts are the following: (i) to what extent are the catalytic properties affected by the geometric packing of the alloy components on the surface (ensemble effect), and (ii) to what extent do the different alloy components retain their individual electronic properties on the surface (ligand effect)? In addition, it is well known that the alloy surface composition may be quite different from the alloy bulk composition, and it is, of course, the former that is of primary importance from a catalytic point of view. For well-defined samples (e.g., single crystals, films) the combination of various electron spectroscopies with catalytic measurements allows correlation of the catalytic properties with the surface alloy composition and structure. However, commercial alloy catalysts consist of small particles present at low loadings (≈ 1 wt%) on oxide supports. Before studying the more fundamental questions of alloy catalysis for these materials, it must first be verified that alloy formation has, in fact, taken place during the catalyst preparation. In this respect, Mössbauer spectroscopy is particularly valuable since conventional techniques of structure determination based on x-ray diffraction are difficult to interpret for particles less than several nanometers in size.

An excellent example of the use of Mössbauer spectroscopy to study alloy formation in small particles is due to Sinfelt (1976) in his characterization of a $PtIr/Al_2O_3$ (5 wt% Pt; 5 wt% Ir) bimetallic reforming catalyst. Catalytic, chemisorption, and x-ray diffraction measurements suggested that if the samples were calcined at temperatures less than 523 K and subsequently reduced in hydrogen, then PtIr bimetallic clusters were formed. However, if the samples were calcined at 773 K prior to hydrogen reduction, then phase separation took place with the production of individual particles of Pt and Ir. For patent purposes, it was desirable to obtain additional evidence to support these statements. While it is possible to carry out both [195]Pt and [193]Ir Mössbauer spectroscopy, the former has a broad resonance, and the latter requires use of a source with a short half-life. Moreover, both isotopes require observation of the Mössbauer effect at low temperatures. Therefore, the $PtIr/Al_2O_3$ catalyst was studied by incorporating a small amount (≈ 0.1 wt%) of [57]Fe into the sample, with the objective of using the [57]Fe as a "probe" of its surroundings.

The [57]Fe Mössbauer spectra of the PtIrFe clusters were indicative of superparamagnetism since magnetic splitting (i.e., six-peak pattern) was observed only upon cooling the samples to temperatures below ≈ 25 K. This is consistent with the small size (≈ 3.0 nm) of the metallic clusters. Below this superparamagnetic blocking temperature, the measured magnetic splitting can be used to determine the chemical environment of the [57]Fe. For the sample that has been calcined at 773 K and subsequently reduced, the measured magnetic splitting at 20 K was characteristic of Fe in Pt, while the sample calcined at 523 K prior to hydrogen reduction gave rise to a magnetic splitting at 20 K that was characteristic of neither FePt nor metallic Fe particles. The clear suggestion is that in the latter case PtIrFe clusters are present. Thus the Mössbauer spectroscopy results provide independent support for the earlier proposed behavior of $PtIr/Al_2O_3$ catalysts. For the present discussion, these results show how [57]Fe (or other Mössbauer isotopes) can be used to probe catalysts that otherwise would not be amenable to study using Mössbauer spectroscopy.

More recently, Lam and Garten (1978) have extended the use of [57]Fe, as a probe of bimetallic clusters, to the RuCu system by studying several FeRu, FeCu, and FeRuCu catalysts. It was found that the iron in $FeRu/SiO_2$ catalysts could be reduced from the ferric to the ferrous state upon exposure to hydrogen at room temperature, while this could be accomplished for Fe/SiO_2 samples only at temperatures above 773 K. Furthermore, hydrogen treatment of the $FeRu/SiO_2$ catalysts at 473 K led to the disappearance of the ferrous component in the Mössbauer spectrum, accompanied by the appearance of new peaks attributable to an FeRu alloy. Additional evidence for alloy formation was the observation of reversible oxidation–reduction of the iron during sequential room-temperature, oxygen–hydrogen treatments of FeRu/SiO$_2$. This is quite unlike the behavior of iron alone, as will be discussed later in this section with respect to the FePd alloy system. For the FeRuCu/SiO$_2$ catalyst, however, reduction to the ferrous state in hydrogen took place only at temperatures above 373 K. At still higher temperatures in hydrogen ($\approx$ 773 K), the Mössbauer parameters for the FeRuCu/SiO$_2$ sample were again indicative of an FeRu alloy. Subsequent room-temperature oxygen treatment of the sample showed a smaller extent of iron oxidation than for FeRu/SiO$_2$ catalysts having similar dispersions (as determined by hydrogen chemisorption). Thus the Mössbauer spectra and spectral changes after various treatments were suggestive of the interaction of Fe with both Cu and Ru, indicating indirectly that the Cu and Ru were in close contact with each other. The model proposed by the authors to explain these results was the formation of FeRu alloy clusters covered by "adsorbed" Cu.

If Mössbauer spectroscopy can be used to study one of the primary components of the alloy, then arguments concerning alloy formation and interaction between alloy components become even more direct than for those studies where the Mössbauer isotope must be added as a probe of the alloy. Garten (1976a, b) and Garten and Ollis (1974) have carried out extensive studies of FePd and FePt alloys supported on Al$_2$O$_3$ and SiO$_2$. A detailed discussion of these results, as well as the physical and chemical properties of bimetallic clusters in general, has recently been given by Burton and Garten (1977). The behavior of the FePd/Al$_2$O$_3$ system can be illustrated by considering two such samples: (i) η-Al$_2$O$_3$ with 0.05 wt% Fe and (ii) η-Al$_2$O$_3$ with 0.05 wt% Fe and 2.2 wt% Pd. The room temperature Mössbauer spectrum of the Fe/Al$_2$O$_3$ sample prior to reduction was an Fe^{3+} doublet, and exposure of the sample to hydrogen at room temperature did not change this spectrum. Hydrogen treatment at temperatures between 770 and 970 K, however, led to complete reduction of the Fe^{3+} to Fe^{2+}. Furthermore, subsequent exposure of the reduced sample to oxygen at room temperature did not change the Fe^{2+} Mössbauer spectrum. When the FePd/Al$_2$O$_3$ sample was exposed to hydrogen at room temperature, the Fe^{3+} doublet of the untreated sample was completely transformed into an Fe^{2+} spectral doublet. Then, after this sample was exposed to hydrogen at 770 K, the room temperature Mössbauer spectrum showed two patterns: a doublet due to Fe^{2+}, and a new singlet ($\delta = 0.3$ mm s^{-1} with respect to ^{57}Co in chromium) due to FePd alloy formation. Subsequent treatment of the FePd clusters with oxygen at room temperature led to the formation of Fe^{3+}. Moreover, this room temperature oxidation was completely reversible in that reexposure of sample to hydrogen at room temperature completely converted the Fe^{3+} doublet back into the singlet of the FePd alloy in the Mössbauer spectrum.

Comparison of the effects of the different treatments on the Fe/Al_2O_3 and $FePd/Al_2O_3$ samples clearly shows the Fe–Pd interactions in the latter case. Prior to the high temperature (770 K) reduction of the sample, the Fe^{3+} initially present was reduced by hydrogen at *room temperature* to Fe^{2+}; the high temperature reduction led to a spectral singlet characteristic of an FePd alloy; and, after this reduction the iron in the sample was capable of undergoing reversible oxidation–reduction at *room temperature*. As demonstrated for the Fe/Al_2O_3 sample, the initial room temperature reduction of Fe^{3+} to Fe^{2+} and the subsequent reversible reduction–oxidation of Fe^{3+} are uncharacteristic of the behavior for iron alone.

Similar studies of $FePt/SiO_2$ catalysts have been carried out by Dezsi *et al.* (1978). In particular, the isomer shift for samples reduced in hydrogen at 770 K was characteristic of an FePt alloy, and the partial reversibility of oxygen–hydrogen treatments at room temperature was indicative of Fe–Pt interactions in these samples. More recently, Dezsi *et al.* (1979) have also used Mössbauer spectroscopy to follow the preparation and genesis of FePt and FeRu clusters supported on silica. For example, they found that if the silica was impregnated using a solution with pH greater than ≈ 1, then the iron was hydrolyzed to ferric hydroxide, and highly dispersed FePt or FeRu clusters could not be obtained by subsequent reduction.

After verification of alloy formation in small particles, the more subtle problem of surface alloy composition measurement may be addressed using Mössbauer spectroscopy. Two limiting cases may be imagined, depending on whether the Mössbauer parameters of the "surface" and the "bulk" atoms are or are not sufficiently different to allow the resolution of their separate spectral patterns. In reality, there is undoubtedly a distribution of Mössbauer parameters as one moves from the surface to the bulk. However, for simplicity, the atoms may be considered as effectively being either in a surface layer or in the bulk of the particle. An example of an alloy system for which the "surface" and bulk atoms can apparently be resolved is that of FePt clusters supported on carbon. Bartholomew and Boudart (1973) collected room temperature ^{57}Fe Mössbauer spectra on samples that had been reduced in hydrogen at 670 K, and the resulting spectral doublets had isomer shifts characteristic of FePt alloys. Thus alloy formation was verified. However, the linewidths of these peaks (≈ 1 mm s^{-1}) were substantially larger than those of bulk FePt samples (≈ 0.30 mm s^{-1}), and this led these investigators to propose a fit of the spectrum with two doublets having different quadrupole splittings. Due to the lower symmetry of an atom on the surface compared to an atom in the bulk, the doublet with the larger quadrupole splitting was attributed to the "surface" atoms, while the doublet with the smaller splitting was believed to represent the bulk atoms. The ratio of the surface doublet spectral area to the total spectral area provides a measure of the fraction of the iron atoms that are on the surface, and when this value is combined with an independent measure of the FePt cluster size (e.g., using chemisorption methods), the surface composition of the cluster can be calculated. For example, a 1.5-nm FePt cluster containing an equal number of Pt and Fe atoms was found to have an alloy surface composition equal to its bulk alloy composition after hydrogen reduction and exposure to air at 570 K. However, oxygen treatment of the cluster for 0.1 hr at 570 K led to an increase in the spectral area of the surface doublet at the expense of the bulk doublet in the Mössbauer spectrum. That is, the high temperature oxygen treatment led to an enrichment of Fe in the surface layer, and this is consistent with the fact that the Fe–O bond is stronger than the Pt–O bond. The actual calculation of the surface alloy composition after this treatment showed that the surface mole fraction of iron had increased from ≈ 0.50 (before the treatment) to 0.65. Though it may be argued that the deconvolution of the broad spectral doublet into a "surface" and a bulk doublet is difficult and, furthermore, that the distinction between "surface" and bulk atoms is unclear, the Mössbauer spectroscopy results so obtained are in good agreement with the expected chemical behavior of the FePt clusters. It should be noted that Lam and Garten (1978) have used similar arguments to deconvolute the Mössbauer spectra of FeRu clusters into "surface" and bulk components. The isomer shift for the iron at the surface, however, was markedly different from that of the iron inside the clusters. Such a difference was not found for the FePt system. In addition, the spectral component attributed to the surface by Lam and Garten has been assigned to Fe^{3+} by Dezsi et al. (1979). Further studies are required to better understand this result.

Garten (1976a) has found that unlike the behavior of small FePt and FeRu particles, the surface and bulk contributions to the overall Mössbauer spectrum of reduced FePd clusters cannot be unambiguously resolved. However, room temperature treatment of these clusters with oxygen was found to oxidize the "surface" atoms, while the bulk atoms were not affected by this treatment. The Mössbauer spectrum of reduced FePd clusters was a spectral singlet, the isomer shift of which agreed with that of the value for FePd alloys. Exposure to oxygen at room temperature led to the appearance of an Fe^{3+} spectral doublet and to a corresponding decrease in the area of the FePd spectral singlet. It was also found that as the FePd cluster size increased, the fraction of the iron converted from the FePd alloy into Fe^{3+} decreased. Thus the room temperature oxidation is a "surface" phenomenon. The ambiguity of such a "surface" treatment is that it may well affect a surface layer rather than the surface atoms only. However, this procedure still has qualitative value for alloy surface composition analysis.

The chemistry of PtSn clusters supported on alumina has recently been studied by Bacaud *et al.* (1979). The Mössbauer spectra of calcined $PtSn/Al_2O_3$ samples were characteristic of SnO_2, while those of reduced samples showed the presence of $PtSn_4$. In addition, the spectra of reduced samples also contained components due to stannic and stannous species originating from tin that was not in close contact with platinum. Subsequent vacuum treatment of $PtSn/Al_2O_3$ led to the oxidation of about half of the $PtSn_4$ phase to a stannous species, presumably due to interaction of the metallic particles with OH groups from the support. Chlorination of these samples converted most of the tin to $SnCl_4 \cdot nH_2O$, and rehydrogenation of these materials led, once again, to the formation of the $PtSn_4$ phase.

In addition to the general characterization of alloy samples using Mössbauer spectroscopy as described above, it is also advantageous to attempt to correlate the Mössbauer spectroscopy results with the measured catalytic properties of the alloy clusters. Such a comparison was made by Vannice and Garten (1976) in their study of the Fischer–Tropsch synthesis over the following samples of Fe and Pt supported on γ-Al_2O_3: (i) low loading (0.1 wt%) of Fe on γ-Al_2O_3, (ii) high loading (15 wt%) of Fe on γ-Al_2O_3, and (iii) a series of samples having the same Pt loading (1.75 wt%) but different Fe loadings ranging from 0.1 wt% to 5 wt%. Reduction of these samples in hydrogen at temperatures from 723 to 973 K led to dramatically different Mössbauer spectra as shown in Fig. 24. For the lowest iron loading (i.e., 0.1 wt%), the sample of Fe/Al_2O_3 gave rise to an Fe^{2+} doublet, while the $FePt/Al_2O_3$ sample with the same iron loading had a two-peak Mössbauer spectrum characteristic of FePt alloy. For the remaining samples of the $FePt/Al_2O_3$ series, increasing iron loadings (0.3 wt%–1 wt%) at constant Pt loading led to the appearance of increasing amounts of Fe^{2+} in the Mössbauer spectra; at the highest Fe loading (5 wt%) for the $FePt/Al_2O_3$ series, the presence of the metallic iron six-peak pattern was also noted. Finally, for the heavily loaded (15 wt%) Fe/Al_2O_3 sample, peaks due to Fe^{2+} and metallic iron were present in the Mössbauer spectrum. In short, iron present at low loading cannot be reduced beyond the Fe^{2+} state. The presence of Pt, however, facilitates the reduction of iron to the zero-valent state with the formation of an FePt alloy. For increasing iron loadings, the Pt is unable to facilitate reduction of all the iron, and this leads to increasing amounts of Fe^{2+}. At high iron loadings, those surface sites of γ-Al_2O_3 that stabilize Fe^{2+} upon reduction become saturated with iron, and the iron present in excess of this saturation can be reduced to the metallic state. The results clearly demonstrate

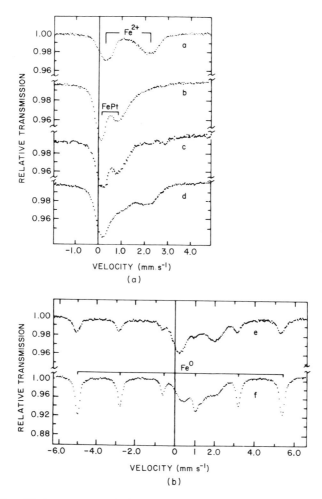

Fig. 24. Mössbauer spectra of alumina-supported PtFe and Fe catalysts reduced in hydrogen; (a) a, 0.1 wt% Fe, reduced 1 hr at 773 K; b, 0.1 wt% Fe, 1.75 wt% Pt, reduced 1 hr at 773 K; c, 0.3 wt% Fe, 1.75 wt% Pt, reduced 2 hr at 973 K; d, 1 wt% Fe, 1.75 wt% Pt, reduced 1 hr at 773 K; (b) e, 5 wt% Fe, 1.75 wt% Pt, reduced 1 hr at 773 K; f, 15 wt% Fe, reduced 1 hr at 723 K. All spectra were recorded at room temperature and under 760 Torr H_2 (Vannice and Garten, 1976).

that the deposition of both Fe and Pt onto the support does not ensure that both of these metals will be present exclusively as an FePt alloy.

After reduction, the samples were exposed to oxygen, followed by hydrogen treatments at room temperature to test the "surface" sensitivity of the Mössbauer spectra. For the Fe/Al_2O_3 sample with low iron loading, the Fe^{2+} was partially oxidized to Fe^{3+} upon exposure to oxygen, and this Fe^{3+} was not converted into Fe^{2+} by subsequent exposure to hydrogen (at room temperature). For $FePt/Al_2O_3$ with the same low iron loading (0.1 wt%), *all* of the iron was oxidized to Fe^{3+} by the room temperature oxygen treatment, and most of this Fe^{3+} was rereduced to the FePt alloy upon exposure to hydrogen. Thus, in a manner similar to the previously discussed FePd system, it may be argued that a large fraction of the iron atoms are near the surface of the FePt particles (i.e., the FePt alloy is present as small clusters). The behavior of the remaining $FePt/Al_2O_3$ samples was similar to that described above in that iron atoms associated with the FePt alloy were able to undergo reversible oxidation (to Fe^{3+}) and reduction (to the FePt alloy) at room temperature. It must, therefore, be argued that the FePt is present as small clusters in these samples as well. The additional Fe^{2+} present after the initial high temperature reduction of these samples, however, was essentially unaffected by the oxygen–hydrogen treatments. For the $FePt/Al_2O_3$ sample with the highest iron loading (5 wt%), the metallic iron present after the initial high temperature reduction was converted into γ-Fe_2O_3 upon exposure to oxygen. Subsequent treatment with hydrogen at room temperature had no observable effect on this six-peak γ-Fe_2O_3 Mössbauer pattern. Finally, the metallic iron present in the reduced Fe/Al_2O_3 sample with heavy iron loading (15 wt%) was not affected by the room temperature oxygen or hydrogen treatments. This metallic iron was evidently in the form of large particles having only a small fraction of the iron atoms near the surface.

The differences seen in the Mössbauer spectra of these samples were used to interpret differences found in the catalytic activity and selectivity of these samples for the Fischer–Tropsch synthesis. The Fe/Al_2O_3 sample with low iron loading (0.1 wt%) was inactive for hydrocarbon formation, while the Fe/Al_2O_3 sample with heavy iron loading gave a product distribution similar to that obtained using a metallic iron catalyst. This is consistent with the presence of Mössbauer spectral peaks due (i) to Fe^{2+} in the reduced Fe/Al_2O_3 sample with low iron loading and (ii) to the metallic iron in the reduced sample with heavy iron loading. For the $FePt/Al_2O_3$ series of catalysts, it was found that the reaction rate per unit surface area (surface area measured by CO adsorption) was approximately equal to the value characteristic of Pt. Yet, pure iron is approximately 30 times more active than pure platinum for the Fischer–Tropsch synthesis. It was only for the $FePt/Al_2O_3$ sample with the heaviest iron loading (5 wt%) that the catalytic activity was greater than that for pure Pt. In fact, the Mössbauer spectrum of this sample showed the presence of metallic iron. The $FePt/Al_2O_3$ catalysts with lower iron loadings were very selective for C_1 and C_2 formation. This again resembles the behavior of Pt, while pure iron produced significant amounts of heavier hydrocarbons.

The catalytic behavior of the FePt/Al$_2$O$_3$ samples is also elucidated by the isomer shift of the Mössbauer spectrum of the FePt alloy. The isomer shift of this alloy is more positive than that of metallic iron, indicating a lower electron density at the iron nucleus in the alloy than in metallic iron. Potassium oxide, a well-known promoter of iron Fischer–Tropsch catalysts, increases the catalytic activity and shifts the product selectivity to higher molecular weights. It is believed that this promotion is due to electron transfer from the alkali metal to iron. Thus, it may be argued that decreasing the electron density on the iron atoms (more positive isomer shift) would decrease the catalytic activity and shift the product distribution to lower molecular weights, and this is in accordance with the observations for the FePt/Al$_2$O$_3$ samples.

The cyclization of *n*-heptane over PtSn/Al$_2$O$_3$ catalysts has recently been discussed by Bacaud *et al.* (1979), with reference to the Mössbauer spectra of these materials. As mentioned earlier in this section, the spectra of reduced catalysts show the presence of PtSn$_4$ and various stannic and stannous species. It was suggested by the authors that the tin alters the catalytic properties of Pt/Al$_2$O$_3$ catalysts by alloying with platinum (PtSn$_4$ component in the spectra) and by changing the acidity of the alumina (stannic and stannous species in the spectra). The former effect should alter the hydrogen–dehydrogenation activity of the catalyst, while the latter should alter the isomerization activity of the catalyst. Indeed, both of these effects should be important in the bifunctional catalysis of *n*-heptane cyclization.

Evans and Swartzendruber (1975) have used Mössbauer spectroscopy to interpret differences in catalytic behavior for a series of related samples. Urushibara-type catalysts were prepared by reacting iron salts with Zn or Al powders. The Fe–Zn and Fe–Al samples have different catalytic activities for hydrogenation reactions, and differences were also observed in the Mössbauer spectra. Specifically, the six-peak spectrum of the Fe–Al sample is identical to that of α-Fe, while the Mössbauer spectrum of the Fe–Zn sample consists of both a magnetically split component and a nonmagnetic component. The magnetic hyperfine field of the former component was significantly smaller than that of α-Fe, and its value is indicative of the presence of ≈ 5 at. % Zn in α-Fe; the nonmagnetic component is probably due to an FeZn alloy with a Zn concentration greater than that present in the magnetic component. The authors proposed that the differences in catalytic behavior of the Fe–Al and Fe–Zn samples were due to the alloying of zinc with iron in the latter case.

Another alloy of catalytic interest is that of Pd–Au since Au atoms are considerably less active than Pd atoms for many catalytic reactions. In this alloy, then, one could investigate the effect of the more "inert" gold on the more active palladium. Furthermore, these two elements form a single

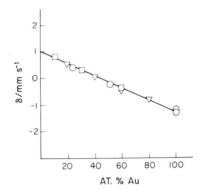

Fig. 25. Isomer shift δ of bulk Au–Pd alloys and supported Au–Pd alloys. O, supported alloys; $\triangledown$ and $\square$, bulk alloys from Longworth (1970) and Roberts *et al.* (1965), respectively. (Zero velocity with respect to [197]Pt in platinum) (Lam and Boudart, 1977. *J. Catal.* **50**, 530).

alloy phase at equilibrium temperatures equal to $\approx$ 573 K, and the metallic radii and heats of vaporization are very similar for Au and Pd. Thus the surface alloy composition under vacuum is expected to be nearly equal to the bulk alloy composition. For these reasons, Lam and Boudart (1977) studied the preparation of PdAu particles supported on silica, and one of the characterization techniques used was [197]Au Mössbauer spectroscopy.

Mössbauer spectra (at 20 K) of Au foils were spectral singlets, and the observed linewidths (2.3 mm s^{-1}) were those expected due to the natural linewidth of the [197]Au resonance (1.9 mm s^{-1}) and to sample "thickness" line broadening. In addition, the Mössbauer spectrum of large Au particles (20–100 nm in diameter) supported on silica was the same as that for the Au foil. Small Au particles (3.0 nm) on silica, however, gave rise to a rather broad (3.4 mm s^{-1}) spectral singlet, interpreted as being due to a distribution of isomer shifts and/or quadrupole splittings of Au atoms near the surface of the small particles.

For a series of PdAu/SiO$_2$ samples with different Pd and Au loadings, broad singlets in the Mössbauer spectra (at 20 K) were again observed. From the relative Pd and Au loadings, one can calculate an "overall alloy composition" for a given sample assuming that all of the metal atoms are present in identical alloy particles. The isomer shift of the spectral singlet for each sample was then compared, as shown in Fig. 25, with measured isomer shifts for a series of bulk PdAu alloys; indeed, the isomer shift of each PdAu/SiO$_2$ sample was indicative of an alloy composition equal to the "overall alloy composition" for that sample. No peaks due to metallic gold were seen in the Mössbauer spectra of the PdAu/SiO$_2$ samples. Thus alloying in these samples seemed to be complete and uniform.

Further evidence for alloy uniformity throughout each sample comes from an analysis of the Mössbauer spectral linewidth. Broadening of the PdAu resonance can be due to three effects: (i) Au nuclei in the alloy have different numbers and arrangements of Pd and Au neighbors, giving rise to a distribution of isomer shifts and quadrupole splittings; (ii) small particles possess a significant number of "surface" atoms, having a distribution of isomer shifts and quadrupole splittings; and (iii) a distribution of alloy compositions exists in the sample, leading to a distribution of isomer shifts and quadrupole splittings. The first effect is seen by Mössbauer spectroscopy in bulk alloys, the second effect was seen for small Au particles supported on silica (as discussed previously), and the third effect would result if the alloying in the $PdAu/SiO_2$ samples were nonuniform. To test for the presence of the third effect, Lam and Boudart (1977) calculated the following parameter $\Delta\Gamma/\Gamma$:

$$\Delta\Gamma/\Gamma \equiv ((\Gamma_{PdAu} - \Gamma_{Au})/\Gamma_{Au})_d,$$

where Γ_{PdAu} and Γ_{Au} are the linewidths for PdAu and Au, respectively, and these linewidths are compared for the same size particle d. The reason for comparison at constant d is to cancel the broadening due to effect (ii) described above. Indeed, the value of $\Delta\Gamma/\Gamma$ was similar for both the bulk and silica-supported PdAu samples, indicating that effect (i) was primarily responsible for the line broadening. Thus, effect (iii) was not important for the $PdAu/SiO_2$ samples, giving support to the statement that the alloying was uniform for these samples. Finally, it should be noted for completeness that the results of both x ray line broadening and CO adsorption studies gave similar estimates for the size of the PdAu particles. In the latter determination of particle size, it was assumed that the surface and bulk alloy compositions were equal, and the agreement between the two particle-size estimates then supports this assumption. This is consistent with the similar metallic radii and similar heats of vaporization for pure Au and Pd.

E. Ammonia Catalysts

1. Classical Ammonia Catalysts

The preparation of the classical promoted iron-based ammonia synthesis catalysts involves two essential steps (see, e.g., Nielsen, 1968): iron oxides are first fused together with small amounts of promoters such as Al_2O_3, CaO, and K_2O to yield a nonporous spinel precursor having negligible surface area. This structure is subsequently activated by reduction in H_2 or H_2/N_2 to yield the active, highly porous metallic structure having a BET surface area in the range of $5-20$ $m^2 g^{-1}$. The size of the metallic iron particles in the reduced catalyst is well above 10 nm, and the classical ammonia catalysts are poorly dispersed. Consequently, Mössbauer spectroscopy is insensitive to surface properties, and most Mössbauer studies have been concerned with the bulk properties of these catalysts.

In order to understand the activity and stability of the ammonia catalyst, it is naturally the structure of the active (reduced) catalyst that is of primary interest. This structure is, however, influenced, to a large extent, by the structure of the unreduced catalyst and the way in which it is reduced. These latter topics will be discussed first.

a. Structure of unreduced catalyst. The unreduced catalyst is made, as mentioned above, by fusion of iron oxides together with promoter oxides. The addition of a promoter like K_2O increases the specific activity (referred to the metallic iron surface area) of the reduced catalyst. Such a promoter is called a chemical promoter. Al_2O_3 is a so-called textural promoter, which is added in order to stabilize a high iron surface area under the reaction conditions. It has been found, for example, by Kuznecov (1963) and Nielsen (1968) that the structure of the unreduced catalyst influences both the reducibility of this phase and the activity of the reduced catalyst.

In the first Mössbauer studies of unreduced ammonia catalysts, Yoshioka et al. (1969) reported several paramagnetic iron species besides the magnetically split magnetite spectrum. Furthermore, the magnetite spectral component was different from that of pure magnetite, indicating that some of the promoters had entered the spinel lattice.

Later, Peev (1976) studied several unreduced industrial-type catalysts with various amounts of promoters and also found in the spectra complex central regions, which were decomposed into three quadrupole-split doublets. He also found indications of a magnetically split component besides that belonging to the magnetite. The new component had a magnetic hyperfine field approximately equal to 405 kG. Although the various iron surroundings were not identified and no correlation with the amount of the various promoters could be established, the study shows that the structure of the unreduced catalyst can be very complex. B. S. Clausen et al. (1976a) also observed paramagnetic iron species in a commercial catalyst. It was established that in their case the paramagnetic compound was wüstite $(Fe_{1-x}O)$.

The structure of an unreduced industrial catalyst is determined by the treatment used in the fusion and the subsequent cooling processes. The multicomponent phase diagram corresponding to the composition of typical industrial catalysts is very complex. Besides taking into account the main promoters (Al_2O_3, K_2O, and CaO), one may also have to consider minor amounts of other substituents like MgO and SiO_2. It may also be necessary to consider the oxygen partial pressure since this can influence the valence of the iron cations. During solidification of the melt, a range of different phases may form as the temperature is lowered, and the resulting catalyst will have a very inhomogeneous structure as observed, for example, by Nielsen (1968) and Jensen *et al.* (1977). The final structure will depend not only on the composition but also on the temperature at which the equilibrium is "frozen" during the cooling process. The study of B. S. Clausen *et al.* (1976a), showing that wüstite may be present in a catalyst, indicates that the "freeze-in temperature" may be quite high since wüstite is not thermodynamically stable below 570°C. Moreover, from the Mössbauer parameters the stoichiometry of the wüstite phase could be determined. Then, from the known temperature dependence of the wüstite stoichiometry, a more precise estimate of the "freeze-in temperature" can be made. In general, the "freeze-in temperature" is expected to depend very much on the cooling rate. This was confirmed by Pernicone *et al.* (1978) who observed quite different x ray phases depending on the cooling rate.

There are many features of the Mössbauer spectra of the unreduced catalyst that have not yet been explained in detail. However, they potentially contain very useful information, which may be unraveled by studies on samples of well-defined composition and thermal history. Such studies may also prove valuable for understanding the reduction properties and the structure of the reduced catalyst.

b. Activation of unreduced catalysts. During the reduction of essentially nonporous oxide, small metallic iron particles are formed. Barański *et al.* (1972, 1979) found that the reduction of a commercial catalyst (KMI) could be described by a core-and-shell model, i.e., a model that is useful for studies of a many solid/liquid and solid/gaseous reactions. According to this model, a partly reduced catalyst consists of an unreduced core surrounded by a porous shell of reduced metal. The rate of reduction depends, among other parameters, on the area of the interface between metal and oxide, and the diffusion rate of gaseous reactants and products through the porous shell.

B. S. Clausen *et al.* (1976a) followed the reduction of commercial catalysts by means of *in situ* Mössbauer spectroscopy. Specifically, they studied the influence of the amount of wüstite in the unreduced catalyst on the reduction kinetics. The presence of wüstite has been reported to increase the activity of the reduced catalyst (Kuznecov, 1963; Nielsen, 1968). For measurements of reduction kinetics, Mössbauer spectroscopy has special advantages compared to the more conventional technique of thermogravimetry since it is possible to follow simultaneously the reduction of several phases (here Fe_3O_4 and $Fe_{1-x}O$).

Figure 26 shows the Mössbauer spectra of a catalyst initially containing about 27 wt% wüstite after different times of reduction at 675 K. The spectrum of the unreduced catalyst (Fig. 26a) shows both a paramagnetic wüstite component and a magnetically split magnetite-like component. The latter component showed broadened lines presumably owing to the presence of Al- and Ca ions in the spinel lattice. From a comparison of the wüstite phase in spectra (a) and (b), it was concluded that the reduction of nonstoichiometric wüstite does not proceed directly to the metal, but goes through a more stoichiometric (less cation deficient) wüstite phase. The results also show that the magnetite phase reduces directly to α-Fe and not via wüstite. Another interesting observation was that the rate of wüstite reduction was much higher than that of the magnetite phase. The wüstite is reduced completely at a time when hardly any reduction of the magnetite has taken place. It was found that for the sample with the high wüstite content, the overall reduction kinetics as obtained from the Mössbauer data, did not follow the core-and-shell model, whereas the reduction kinetics derived from Mössbauer results from a catalyst containing only a small amount of wüstite nicely followed a core-and-shell behavior. The difference was explained by the rapid reduction of the wüstite phase, which is distributed throughout the catalyst structure. This leads to the formation of centers at which the reduction of magnetite can take place in the wüstite-rich catalyst. As a result, the assumptions in the core-and-shell model may not be valid at high wüstite contents, and the reduction kinetics may be better described by the more complicated multiple core-and-shell model by Park and Levenspiel (1975).

The reduction of singly promoted catalysts has been studied recently by Ludwiczek et al. (1978) using x ray diffraction and Mössbauer spectroscopy. For samples containing up to 5 wt% Al_2O_3, the reduction at 400°C was found to proceed directly from the spinel to the metallic iron phase, as was observed for the multiply promoted catalyst discussed above.

c. *Structure of active catalysts.* A structural problem which has attracted much attention is the location of the textural promoters in the reduced ammonia catalyst. Al_2O_3 is such a textural promoter, which is added to stabilize a high metallic surface area. Pure iron, without the addition of Al_2O_3, sinters rapidly even below 400°C, whereas catalysts containing a few weight percent Al_2O_3 almost maintain their original surface area for years under industrial conditions. The way in which alumina stabilizes the catalyst against sintering has been extensively discussed. Early physical and chemisorption studies by Emmett and Brunauer (1937, 1940) revealed that a large fraction of the promoters cover the surface of the α-Fe crystallites, and it has been believed that this covering layer prevents them from sintering. Solbakken et al. (1969) found that the alumina promoter layer covering the iron crystals is only a monolayer

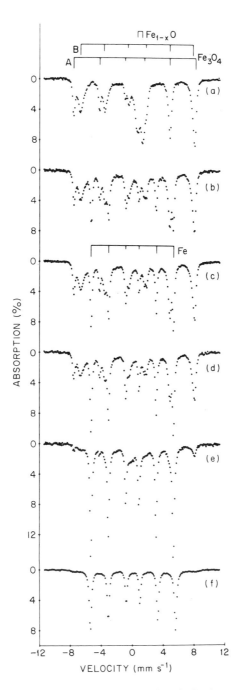

Fig. 26. Mössbauer spectra of an ammonia catalyst obtained at room temperature after different times of reduction in H_2 at 675 K: (a) 0 hr, (b) 0.2 hr, (c) 1 hr, (d) 4 hr, (e) 8 hr, (f) shows the spectrum obtained at 78 K after reduction for several days at 825 K. The bar diagrams indicate the line positions for $Fe_{1-x}O$, Fe_3O_4, and Fe. (Adapted from B. S. Clausen *et al.*, 1976a.)

thick. The remaining alumina ($\sim 50\%$) must be present in some chemical form inside the iron particles. Hosemann et al. (1966) studied an Al_2O_3 promoted catalyst by x ray diffraction and found that the α-Fe lattice contained paracrystalline distortions (i.e., long range disorder). On the basis of these results, the authors suggested an interesting model for the structure of the reduced catalyst according to which $FeAl_2O_4$ groups are inserted in the α-Fe lattice (one group replacing seven iron atoms), thereby giving rise to the paracrystallinity. The stability of the small iron particles toward sintering was also explained by the presence of paracrystalline defects. Although the x-ray diffraction results have been confirmed by Fagherazzi et al. (1972), Pernicone et al. (1973), and Ludwiczek et al. (1978), controversy concerning the size of the inclusions and whether or not these inclusions actually contain iron still exists.

These problems have been investigated using Mössbauer spectroscopy by several authors (Fagherazzi et al., 1972; Topsøe et al., 1973; B. S. Clausen et al., 1976a; B. S. Clausen, 1976; and Ludwiczek et al., 1978). First of all, if the iron atoms are not all reduced to the metallic state, the nonreduced atoms should give rise to a separate component in the spectra. Second, information about the size of the inclusions can be obtained from a detailed analysis of the metallic iron spectrum. The reason for this is that the smaller the inclusions, the larger their dispersion and the higher the number of metallic iron atoms neighboring the clusters. These latter iron atoms will have Mössbauer parameters different from those of pure α-Fe.

Before discussing the results further, it is important to mention that if the experiments are carried out on passivated samples instead of in situ on the reduced catalysts, then several complications may arise in the data interpretation. At first it may not seem important to avoid the passivation since the catalysts are poorly dispersed. However, Topsøe et al. (1973) found that the passivation layer was many atomic layers deep (~ 20) and affected a large fraction of the iron atoms. Furthermore, the passivation layer was observed as a broad, paramagnetic component in the Mössbauer spectrum. The presence of this passivation layer will therefore make it difficult, if not impossible, to firmly establish the presence of other forms of nonreduced iron (e.g., $FeAl_2O_4$). The presence of a passivation layer may also cause stresses in the iron core. Such stresses are undesirable because they may interfere with both the x ray and the Mössbauer results. Van Diepen et al. (1978) have recently studied passivated iron samples and have observed that the passivation layer caused a significant broadening of the α-Fe Mössbauer lines below room temperature. Of the above-mentioned Mössbauer studies on reduced ammonia catalysts, only those by Topsøe et al. (1973), B. S. Clausen et al. (1976a), and B. S. Clausen (1976) were carried out in situ.

Topsøe et al. (1973) considered the proposal by Hosemann et al. (1966) that the aluminum is present inside the iron particles as randomly distributed $FeAl_2O_4$ molecules. The presence of these molecules should both affect the α-Fe spectrum and give rise to an Fe^{2+} component. The symmetry of the iron atoms neighboring the $FeAl_2O_4$ molecules would be distorted and result in larger quadrupole splittings. Even small changes in the quadrupole splitting can be detected by measuring the broadening of the α-Fe spectral lines. The iron atoms neighboring the $FeAl_2O_4$ clusters are also expected to exhibit different magnetic hyperfine fields (see Section III, F).

The catalyst studied by Topsøe et al. (1973), using in situ Mössbauer spectroscopy, was the same Al_2O_3 promoted catalyst (containing 3% Al) as that originally examined by Hosemann et al. (1966). After reduction of the catalyst at 700 K for 23 hr, the metallic iron spectrum did not show excessive line broadening and the observed magnetic fields were equal, within the experimental uncertainty (~ 0.1 kG), to that of an NBS α-Fe standard. It was therefore concluded that only a very small fraction of the iron atoms had nonmagnetic neighbors. From a comparison with other known systems in which iron has nonmagnetic neighbors, it was concluded that less than 2% of the total alumina content could be present as single nonmagnetic molecules or atoms dispersed throughout the α-Fe lattice. From these results the presence of molecules of $FeAl_2O_4$, as proposed by Hosemann et al. (1966), could be excluded. The presence of molecules of Al_2O_3 or an Fe–Al alloy with an aluminum content higher than 0.03% could also be excluded.

The presence of large nonmagnetic clusters containing, for example, 100 $FeAl_2O_4$ molecules in the α-Fe lattice would not be in disagreement with the above Mössbauer result since the presence of such clusters would only affect a very small number of metallic iron atoms. In order to examine whether such clusters contain iron (e.g., $FeAl_2O_4$) or do not contain iron (e.g., Al_2O_3), Topsøe et al. (1973) examined the spectra for the presence of nonreduced forms of iron. Such forms of iron were, as mentioned earlier, observed for passivated samples but not for fully reduced catalysts. It was therefore concluded that the majority of the aluminum in the catalyst must be present as a phase not containing iron. This phase is most likely Al_2O_3 and must, as mentioned before, be present as rather large inclusions.

Ludwiczek et al. (1978) recently investigated a new series of singly promoted catalysts (with Al_2O_3) and found evidence of broadening of the metallic iron lines, as well as the presence of a slight increase in the absorption in the central region of the spectra. These results were taken as evidence for $FeAl_2O_4$ groups in the α-Fe lattice. These results should, however, be taken with reservation since the experiments were performed

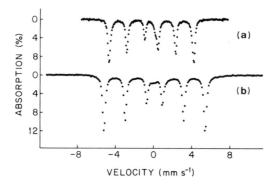

Fig. 27. *In situ* Mössbauer spectra of a reduced multiple promoted ammonia catalyst. The spectra (a) and (b) were obtained at 725 and 300 K, respectively, after several days' reduction at 725 K (B. S. Clausen, 1976).

on passivated samples. At present, therefore, the interpretation of the Mössbauer results is not completely clear. Still, owing to the use of different starting materials, it may be possible that after similar reduction conditions $FeAl_2O_4$ may be present in the samples studied by Ludwiczek *et al.* (1978) and not in those studied by Topsøe *et al.* (1973). Indicative of this is the low reducibility of the starting materials used by Ludwiczek *et al.* (1978).

The *in situ* Mössbauer studies of B. S. Clausen (1976) and B. S. Clausen *et al.* (1976a) give the first clear evidence for the existence of nonreduced forms of iron in reduced ammonia catalysts. These authors studied multiply promoted industrial catalysts (KMI) and found significant amounts of nonreduced iron even after severe reduction. Figure 27 shows the spectra of a catalyst after reduction at 725 K for several days. The spectra were obtained in hydrogen at 725 K (Fig. 27a) and 300 K (Fig. 27b). At room temperature it is difficult to detect the nonreduced iron since it is present as a broad component in the spectrum. However, the detection of nonreduced forms of iron is facilitated by heating the sample to 725 K. At this temperature the broad component has collapsed into an easily detectable paramagnetic peak positioned close to the fourth line in the metallic iron spectrum. From the isomer shift, the nonreduced form of iron was found to be trivalent. This difficult-to-reduce iron-containing phase is not, therefore, $FeAl_2O_4$, but is more likely a calcium-containing ferrite. B. S. Clausen (1976) observed no line broadening of the α-Fe lines in the spectrum obtained in hydrogen at 300 K, indicating that small inclusions are not present in this catalyst. Broadening of the metallic iron lines was, however, observed after cooling a passivated catalyst to 78 K, but it is not yet clear whether this line broadening is caused by stresses originating from the passivation layer or from inclusions in the iron lattice. The former phenomenon is expected to become more pronounced as the temperature is decreased (Van Diepen *et al.*, 1978)

2. Composition and Structure of the Catalyst Surface

A much debated question in ammonia catalyst research concerns the effect of certain surface planes or geometric arrangements of surface atoms on the reaction kinetics.

Dumesic *et al.* (1975a) observed a dependence of the catalytic reaction on the metal surface structure by studying the rate of synthesis as a function of the iron particle size. These results have recently been confirmed by Topsøe *et al.* (1980). Mössbauer spectra of the catalysts under reaction conditions were collected (Dumesic *et al.*, 1975b). The authors found that the reaction rate per metallic iron surface area decreases with decreasing particle size. The isomer shift and internal magnetic field of the metallic iron particles were independent of particle size, thus ruling out particle-size-dependent electronic properties of these metallic iron particles. This suggested that the changes in the catalytic behavior with particle size were caused by changes in surface structure. This conclusion was further supported by Mössbauer studies of the superparamagnetic behavior of the particles (Boudart *et al.*, 1975b) and by chemisorption studies (Topsøe *et al.*, 1980).

For the smallest iron particles, the magnetic anisotropy energy barrier, estimated from the superparamagnetic relaxation of the Mössbauer spectra, was much greater than that expected for magnetocrystalline anisotropy (see Mørup *et al.*, 1980b). From studies of the effect of hydrogen adsorption on the superparamagnetic relaxation, Boudart *et al.* (1977) concluded that the large anisotropy barrier was caused by the presence of magnetic surface anisotropy. This interesting finding enabled the authors to obtain information about surface structure or, more precisely, changes in surface structure.

By measuring the ammonia synthesis reaction kinetics after various treatments, it was shown that the presence of nitrogen during catalyst genesis affected the ultimate catalytic activity. Specifically, it was found that treating H_2-reduced small particles with ammonia at 670 K, followed by rereduction of the catalyst with a $H_2 : N_2$ gas mixture, gave rise to an increase in the catalytic activity compared to the activity measured after H_2 reduction alone. However, when the catalyst in this high-activity state was further treated with H_2 alone at 670 K, the catalytic activity was found to decrease to the value observed before the above ammonia treatment. Subsequent ammonia treatment restored the catalyst to its high-activity state. These effects were observed only for metallic iron particles smaller than about 10 nm in size.

Mössbauer spectra of such small iron particles were obtained under reaction conditions after the various catalyst pretreatments. The magnetically split spectral area versus temperature curves after the various pretreatments are shown in Fig. 28. It is seen that the ammonia treatment, which increases the catalytic activity, decreases the magnetically split spectral area at a given temperature. This was explained by a decrease in the magnetic surface anisotropy energy barrier and a corresponding decrease in the superparamagnetic relaxation time. The important point for surface structure determination is the finding that increased catalytic activity for ammonia synthesis is associated with a decreased magnetic surface anisotropy.

In relating the observed magnetic surface anisotropy energy change to the associated surface structure change, the description of the latter in terms of exposed crystallographic planes was abandoned in view of the many high-index planes (of very small extent) that are undoubtedly present. Instead, the concentrations of the various surface sites on the small particle surface were considered. A surface site C_i is defined as a surface atom with i nearest neighbors. Examples of these sites in the low-index planes of metallic iron are given in Fig. 29. Néel's phenomenological theory of magnetic surface anisotropy (Néel, 1954) was then used to calculate the associated anisotropy energy barriers for these sites (Dumesic, 1974). The surface sites could thereby be arranged in order of decreasing magnetic surface anisotropy: C_6, C_5 or C_7, C_4.

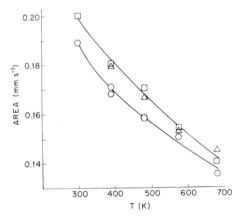

Fig. 28.   Effect of sequential pretreatment on magnetically split spectral area versus temperature of 3% Fe/MgO catalyst, Mössbauer spectra in $H_2:N_2$. Pretreatment sequence: □, H_2 reduction; O, NH_3; △, H_2 reduction; ◯, NH_3 (Dumesic *et al.*, 1975b. *J. Catal.* **37**, 513).

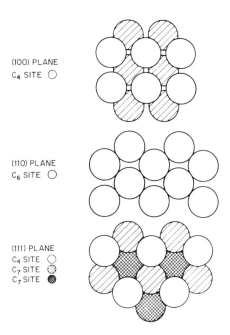

(100) PLANE
C₄ SITE ○

(110) PLANE
C₆ SITE ○

(111) PLANE
C₄ SITE ○
C₇ SITE ◐
C₇ SITE ●

Fig. 29. Surface sites on bcc surface planes (Dumesic *et al.*, 1975b. *J. Catal.* **37**, 513).

The effect of the ammonia treatment, which increased the catalytic activity and decreased the surface anisotropy, is to convert sites from the left to the right in the above sequence.

Clearly, this sequence alone does not provide a unique identification of the surface site(s) that is associated with the increased catalytic activity. Carbon monoxide chemisorption provided the additional information required for this determination. Based on steric considerations, the surface sites can also be arranged in order of decreasing CO chemisorption ability (Dumesic *et al.*, 1975b): C_4, C_5, C_6, C_7.

Experimentally, it was found that the ammonia treatment, which increases the catalytic activity, decreases the CO uptake of the metallic iron surface without sintering the particles (Dumesic *et al.*, 1975b). Thus an increased catalytic activity is accompanied by a conversion of sites on the left into sites on the right in the above sequence of sites. Comparison of the results obtained from Mössbauer spectroscopy with those of CO chemisorption implies that the increased catalytic activity for the ammonia synthesis accompanying the ammonia treatment is associated with an increase in the number of C_7 sites on the metallic iron surface.

Van Hardeveld and Hartog (1969) have calculated that the relative concentration of C_7 sites on octahedral iron crystallites decreases with decreasing particle size. In general, the C_7 site is not common for small particles. The above correlation of increased catalytic activity with increased C_7 site surface concentration thus also explains the observed "structure sensitivity" (particle-size dependence) for this reaction. Finally, this correlation is consistent with results obtained from field emission microscopy of iron (Brill et al., 1967), single crystal reaction studies on tungsten (also a bcc metal) (McAllister and Hansen, 1973), and symmetry considerations (Dumesic and Boudart, 1975).

Another question that has received much interest concerns the possible presence or absence of surface nitrides during ammonia synthesis. Under normal synthesis conditions, the bulk nitride is thermodynamically unstable (see Emmett, 1975); however, the instability of the bulk nitride may not exclude the presence of a surface nitride.

There have been no Mössbauer studies to definitively prove or disprove the presence of surface nitrides under synthesis conditions. Dumesic et al. (1975b) observed surface reconstruction following decomposition of nitrided iron particles supported on MgO. Such a nitrogen-induced surface reconstruction could be taken as evidence for surface nitride formation. Although not directly dealing with the question of surface nitrides, the study of Maksimov et al. (1975c) is of interest in this connection. These authors found by chemical analysis that after ammonia synthesis promoted iron catalysts contained about ten times more nitrogen than the amount that can be dissolved in the bulk. This nitrogen was, therefore, supposed to be associated with defects in the catalyst structure. The Mössbauer spectra of the catalyst contained, in addition to the spectrum of α-Fe, a broad component that was associated with the defect regions. The fact that the Mössbauer measurements were not carried out *in situ* may complicate the interpretation.

In continuation of the above study, Yatsimirskii et al. (1976b) studied nitrided ammonia synthesis catalysts by means of Mössbauer spectroscopy. They found that an increased ability of nitride formation was accompanied by an increased synthesis activity. A high ability toward nitriding was associated with a high concentration of structural defects. The location of the defects was not considered by the authors and is open for discussion. The possibility, however, that the defects may be located at the surface of the catalyst is worth mentioning here since the defects were shown to be associated with nitrogen and, therefore, might resemble a "surface nitride." It is quite probable that surface nitrides may form on real catalyst surfaces since Bozco et al. (1977) have shown that these nitrides can, in fact, form on certain iron single crystal surfaces. It seems worthwhile pursuing this problem further by means of Mössbauer spectroscopy.

3. New Catalyst Systems

It is generally believed that the promoting effect of K_2O in the classical ammonia catalyst is related to electron donation to the iron. If this is correct, one should expect that metallic potassium would be a better promoter due to its much lower ionization potential. Metallic potassium has indeed been found to have a high promoting action for several of the group VIII metals (see for example Ozaki *et al.*, 1971; Aika *et al.*, 1972; Ichikava *et al.*, 1972). Lamellar-type graphite, into which a transition metal and large amounts of metallic potassium had been introduced, also shows high catalytic activity (Ichikava *et al.*, 1972). In a recent study of such iron-containing catalysts, Vol'pin and co-workers (1977) found an interesting relationship between catalytic activity and isomer shift. Figure 30 shows the ammonia synthesis activity and the isomer shift versus the K/Fe atomic ratio. The catalytic activity has a maximum for K/Fe ratios around 5–7. The authors also measured the rate of nitrogen isotope exchange as a function of the K/Fe ratio and found a similar optimum in the exchange rate. The promoting effect of potassium does, therefore, seem to be associated with the activation of the nitrogen molecule. The Mössbauer spectra of these catalyst systems consist of an α-Fe sextet and a paramagnetic singlet. The isomer shift of the α-Fe spectrum did not depend on the K/Fe ratio. However, the isomer shift of the singlet did depend on the K/Fe ratio. This dependence is also shown in Fig. 30, and it is interesting that a minimum in the isomer shift is observed at the K/Fe ratio for which the optimum synthesis and isotope exchange rate occurred. It was, therefore, concluded that the iron species responsible for the singlet is active for ammonia synthesis. The observed initial decrease in the isomer shift with increasing potassium content is most probably due to an increase in the

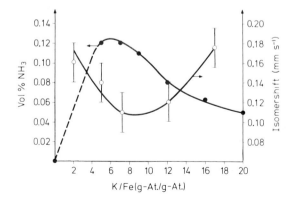

Fig. 30. Ammonia synthesis activity ● and isomer shift of the singlet line ○ versus K/Fe ratio for graphite–Fe–K catalysts. (Adapted from Vol'pin *et al.*, 1977.)

s-electron density at the iron atoms resulting from donation from the potassium. This supports the suggested promoting action of metallic potassium.

F. Hydrocarbon Synthesis Catalysts

When a mixture of CO and H_2 is passed over metals such as Fe, Co, Ni, or Ru, a great variety of organic compounds (alkanes, olefins, alcohols) may be formed through the so-called Fischer–Tropsch-type reactions. With a metal like Ni, the main reaction product is methane, which is formed in the methanation reaction ($CO + 3H_2 \rightarrow CH_4 + H_2O$). However, much higher molecular weight products may form when other catalysts are used. Much of the reawakened interest in hydrocarbon synthesis has turned to developing catalyst systems possessing high selectivity for desirable products such as low molecular weight olefins and alcohols. Traditionally, the Fischer–Tropsch catalysts have been unsupported catalysts (see, e.g., Storch et al., 1951) and the earlier work has concentrated on this type of systems. Mössbauer spectroscopy has been used by Maksimov and co-workers (Maksimov et al., 1972a,b, 1973, 1974a,b) to study the carbide formation on such catalysts.

The trend in much of the recent research work seems to be directed toward supported metal or alloy systems. The primary topics of interest are the determination of specific reaction rates (turnover frequencies), the mechanism(s) of the reactions, and the respective influences of support, promoters, and alloying on the catalyst activity and selectivity (see, e.g., Garten and Ollis, 1974; Vannice, 1976; Vannice and Garten, 1976; Bond and Turnham, 1976; Dalla Betta and Shelef, 1977; Vannice et al., 1978).

Mössbauer spectroscopy was used by Vannice and Garten (1976) and Vannice et al. (1978) to characterize a number of supported Fe–Pt and Fe–Ru bimetallic catalysts. These catalysts were also used in reactions involving CO and H_2. Mössbauer spectroscopy was mainly used to establish alloy formation as discussed in Section IV, D.

Controversy continues as to whether the Fischer–Tropsch synthesis reactions involve carbides or some oxygenated surface complex as the active intermediates. It may well be that both types of intermediates should be considered (Ponec, 1978). Although the catalytic role of carbides remains a controversial subject, they have an important effect on the structure and the properties of the catalyst surface.

Many of the metals used as Fischer–Tropsch catalysts may, in fact, transform into carbides during reaction. This has, for example, been observed for fused and precipitated iron catalysts (Storch et al., 1951;

Sancier *et al.*, 1978). In a series of recent papers, Mössbauer spectroscopy has been employed to study carbide formation in supported catalyst systems. Owing to the applicability of using Mössbauer spectroscopy under *in situ* reaction conditions, it has also been possible to study the catalytic role of carbides.

1. Carbide Formation in Supported Catalysts

Amelse *et al.* (1978) used Mössbauer spectroscopy to investigate carbide formation in supported iron catalysts during Fischer–Tropsch synthesis. The type of carbide formed in a silica-supported catalyst was observed to be different from that encountered in unsupported catalysts. This is another clear example of the strong influence that supports may have. Such effects have been further substantiated in the extensive studies of Raupp and Delgass (1979b, c) who observed significant differences in the carbide phases formed, depending on support, particle size, and alloying with Ni.

The strong influence of metallic iron particle size and the nature of support on the catalyst structure under reaction conditions is clearly seen from the spectra in Fig. 31 (Raupp and Delgass, 1979b). These spectra were obtained after the following treatments: reduction in hydrogen (typically 8 hr at 675–750 K), exposure to a 3.3 : 1 H_2/CO reaction mixture at 523 K for 6 hr, and quenching to room temperature in He. The spectra (a), (b), and (c) in Fig. 31 correspond to samples with iron particle sizes, after reduction, equal to 6.1, 7.4, and 10.1 nm, respectively. Silica was used as a support for these catalysts, whereas the catalyst (d) shown in Fig. 31 was supported on MgO. This latter sample had ~ 4 nm iron particles after reduction.

All spectra of the spent catalysts show that the original metallic iron was transformed into carbides to the extent that none, or very little, of it is left. The smallest iron particles supported on silica were carbided to the hexagonal closepacked ε' carbide ($Fe_{2.2}C$), which has one magnetic hyperfine split sextuplet (Fig. 31a). The large central asymmetric doublet is caused by irreducible ferrous iron in the silica. This carbide was also observed by Amelse *et al.* (1978) after similar carbiding conditions of a silica-supported catalyst. It is interesting that the ε' carbide is formed, although this carbide is the least stable of the known carbides. At least four iron carbides exist and, arranged in order of increasing stability, they are: ε'-$Fe_{2.2}C$ < ε-Fe_2C < χ-Fe_5C_2 (Hägg) < θ-Fe_3C (cementite) (see Chapter 7). As the average particle size was increased to 7.4 nm, Raupp and Delgass (1979b) observed that besides ε' carbide, ε carbide (ε-Fe_2C) was also formed. The Mössbauer spectra of the ε carbide (and the χ

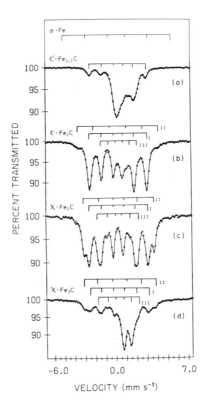

Fig. 31. Mössbauer spectra of 10% Fe/SiO₂ and 10% Fe/MgO catalysts after reaction in 3.3 : 1 H₂ : CO for 6 hr at 523 K. Spectra recorded in He at room temperature. The particle size and support were as follows: (a) 6.1 nm, SiO₂; (b) 7.4 nm, SiO₂; (c) 10.1 nm, SiO₂; (d) ~ 4 nm, MgO. The bar diagrams indicate the line position of metallic iron and different bulk carbides. (Adapted from Raupp and Delgass, 1979b. *J. Catal.* **58**, 348, Part II.)

carbide) are quite complicated since three structurally different sites exist, each giving rise to a magnetic hyperfine split sextuplet. The coexistence of the ε′ and ε carbide phases is most likely a result of an iron particle-size distribution with the smaller particles favoring formation of the ε′ carbide and the larger particles favoring the ε phase.

As shown in Fig. 31c, additional differences in carbiding behavior are observed when the size of the metallic iron particles in the reduced catalysts is further increased. These larger particles are seen to transform into the χ carbide. The influence of the silica on the carbiding behavior of these large iron particles seems to be small since unpromoted fused catalysts usually also form the χ carbide under similar conditions. The silica support, however, does seem to influence the carbiding behavior of the smaller iron particles since these particles were transformed into the normally unstable ε' and ε carbides.

For small iron particles supported on magnesia, the formation of unstable carbides was not observed. This is seen from Fig. 31d, which shows the spectrum of a spent catalyst originally containing ~ 4 nm metallic iron particles. The spectrum shows the presence of χ carbide and irreducible, ferrous ions. This behavior contrasts with that for the silica-supported catalysts, in which an ε carbide was formed for similar iron-particle size. Thus the ε carbide stabilization on silica must primarily be due to a support interaction.

Kölbel and Küspert (1970) and Küspert (1970) studied the carbiding of BeO-supported iron catalysts in CO at 543 K. They found that all the iron transformed into the χ carbide. This finding is not surprising since BeO is a support that only weakly interacts with iron (see Section IV, C).

For catalyst systems in which support interactions are important, the properties of the catalysts will most likely depend on the physical and chemical properties of the support, the drying conditions, the calcination conditions, etc. It is therefore not unexpected that Amelse et al. (1978) observed the ε'-$Fe_{2.2}C$ for somewhat larger iron particles (13 nm) than did Raupp and Delgass (1979b) since different silicas and preparation procedures were used in the two studies.

The carbides discussed above correspond to the steady-state catalyst structure under the reaction conditions, and no structural changes occurred after further exposure of the catalysts to the reaction mixture. However, before reaching the steady state, other carbides may form as intermediates. For example, Raupp and Delgass (1979b) did observe the ε carbide for short carbiding times of both the 10.1-nm iron particles supported on silica and the ~ 4-nm iron particles supported on magnesia.

Raupp and Delgass (1979b) also studied the effect of alloying Fe with Ni on the carbide phases formed after reaction with the $CO-H_2$ mixture. The Mössbauer spectra of the catalyst (5% Fe, 5% Ni/SiO_2) before and after reaction were qualitatively very similar and showed only the presence of spectral components corresponding to the FeNi alloy and some irreducible ferrous iron. It was, therefore, concluded that the formation of bulk Fe or FeNi carbides did not occur. The linewidth of the superparamagnetic alloy peak did, however, significantly narrow after exposure to the reaction mixture. This effect was found to be reversible since the original linewidth was restored upon treatment of the spent catalyst in hydrogen. The results were interpreted in terms of changes in the superparamagnetic relaxation time caused by surface effects and, following arguments similar to those used previously by Dumesic et al. (1975b) to explain the nitrogen-induced reconstruction of iron surfaces (see Section IV, E, 2), the authors concluded that the surface of their spent catalyst has become reconstructed, possibly due to formation of surface carbides.

2. Catalytic Significance of Carbides

The presence of carbides in the catalyst seems to have a strong influence on the catalytic reaction. Both Amelse *et al.* (1978) and Raupp and Delgass (1979c) found that changes in the reaction rate paralleled the extent of bulk carbiding of the supported iron catalysts. The initial metallic iron catalysts showed no activity.

Information regarding the catalytic significance of the carbide formation was obtained by Raupp and Delgass (1979c) by simultaneously measuring the time dependence of the reaction rates, the reaction products, and the *in situ* Mössbauer parameters during carbiding. Some comments on time-dependent Mössbauer studies may be appropriate here. Mössbauer spectroscopy can be used for studies of time-dependent phenomena if the time scale for these changes is greater than the time required to collect the Mössbauer spectrum. There are several ways in which the collection time can be reduced, and many successful kinetic studies have appeared (see, e.g., Vertes, 1975 and Dumesic and Topsøe, 1977). In such studies it is often advantageous to improve the statistics by measuring the absorption of γ rays in certain "regions of interest." Raupp and Delgass (1979c) did this by limiting the measurements to one velocity, namely, that of a metallic iron peak (peak 6), which does not overlap with any carbide peaks. Thus during carbiding they could measure the decrease in the metallic iron spectral component from carbiding by following the corresponding increase in transmission at the position of the metallic iron peak.

Figure 32 compares *in situ* Mössbauer constant velocity measurements of carbiding with conversion of carbon monoxide (x_{CO}) by Fischer–Tropsch synthesis over the 10% Fe/MgO and the 10% Fe/SiO$_2$ (7.4 nm) catalysts. Discussing the silica-supported sample (Fig. 32b) first, it is seen that the initial noncarbided catalyst has no activity. Over the initial two hours of reaction, conversion increases slightly faster than the carbide formation, but past this point, conversion and extent of carbiding increase at almost equal rates. The $\sim$ 4-nm iron particles supported on MgO (Fig. 32a) carbided faster than larger particles supported on silica. The rates of carbiding and reaction are closely matched for the first 30–40 mins, which was the time required for complete carbiding.

The above results show that the carbide formation must play an important role for the catalytic process. It is a somewhat surprising result that there exists a close relationship between the reactivity at the surface and the bulk carbiding reaction. To account for this, Raupp and Delgass

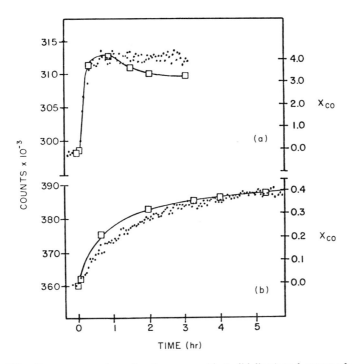

Fig. 32.   Percent conversion of carbon monoxide (solid lines) and extent of carbide formation (dots) versus Fischer–Tropsch synthesis reaction time. (a) 10% Fe/MgO, (b) 10% Fe/SiO₂ (Raupp and Delgass, 1979c. *J. Catal.* **58**, 361, Part III).

(1979c) proposed that the carbiding reaction is initiated by dissociative chemisorption of CO at the surface, giving some form of surface carbon. As long as the metallic iron is not fully carbided, the surface carbon will have a tendency toward diffusing into the iron lattice, leading to the formation of bulk carbides. However, when the iron is fully carbided, and can accommodate no additional carbon from the surface, carbon overlayers may begin to form on the surface.

The above sequence of reactions seems to explain many of the experimental observations. Before the fully carbided state is reached, there may exist a correlation between the compositions of bulk carbon and surface carbon that could account for the observed correlation between the rate of conversion of CO and the extent of bulk carbiding. Also, the observed activity loss for the MgO-supported catalyst (Fig. 32a), after the fully carbided state was reached, may be caused by carbon overlayers building up and partly blocking the surface. As expected from the above reaction sequences, the FeNi alloy may build up surface carbides. Furthermore, the

FeNi alloy catalysts were found to deactivate faster than the iron catalysts. This is not surprising because these catalysts did not bulk carbide, and carbon overlayers may easily form.

In order to gain further insight into the role of carbides in the Fischer–Tropsch reaction, Raupp and Delgass (1979c) also used constant velocity Mössbauer spectroscopy to study the reverse of the carbiding reaction, i.e., the hydrogenation of the carbide. The fully carbided 10% Fe/MgO catalyst was hydrogenated in pure hydrogen at 523 K for 10 hr. From a plot of the intensity of the metallic iron peak versus time, it was observed that the hydrogenation reaction is extremely slow relative to the carbiding reaction. Methane was the only hydrocarbon product detected in the gas phase during hydrogenation, suggesting, as has been reported by Van Barneveld and Ponec (1978), that direct hydrogenation of surface carbides does not contribute to the formation of chain products during Fischer–Tropsch synthesis.

The authors also observed that the rate of methane formation during hydrogenation was always much lower than the methanation rate during Fischer–Tropsch synthesis. This result suggests either that direct hydrogenation of the surface carbides is not responsible for production of a majority of methane during synthesis or that the reaction occurs on a relatively small number of very active sites that are replenished by dissociative chemisorption of CO.

G. Catalyst Studies Using ^{99}Ru

The main limitation in Mössbauer spectroscopy is the relatively restricted number of isotopes that can be studied by this technique. Among such isotopes is, however, ^{99}Ru, which is of great current interest in fields such as automotive emission control, Fischer–Tropsch synthesis, and ammonia synthesis. C. A. Clausen and Good have been the pioneers in the application of Mössbauer spectroscopy to ruthenium catalysts. (For a review see, e.g., C. A. Clausen and Good, 1976, 1977c.)

Ruthenium Mössbauer spectroscopy is best conducted using the ^{99}Ru isotope (Good, 1972), for which the parent nuclide is ^{99}Rh. The half-life of the ^{99}Rh source is approximately 16 days, and the energy of the appropriate γ ray for Mössbauer spectroscopy is quite high (89.36 keV). The high recoil energy requires low temperatures (e.g., liquid helium temperature) to measure the spectrum, and the effect is small. *In situ* studies are, however, conveniently carried out using the following procedure (C. A. Clausen and Good, 1975): (i) samples are placed in small glass cells having thin (e.g., 0.1 mm) glass "windows" for γ-ray transmission, (ii) the catalyst is treated (e.g., calcined, reduced) at a given temperature by flowing gases through the cell and/or evacuating the cell, and (iii) the cell is finally sealed and placed inside a dewar where the low temperature Mössbauer experiments are conducted. For efficient use of the short-lived ^{99}Rh source, a large recoil-free fraction for the source and a relatively short source–detector distance are desired; this may be achieved by the use of a dewar that contains both the source and the glass absorber cell.

The interaction of ruthenium with silica and alumina supports was studied by C. A. Clausen and Good (1975) who prepared supported ruthenium samples (containing approximately 10 wt% Ru) by impregnating SiO_2 and η-Al_2O_3 with an aqueous solution of $RuCl_3 \cdot xH_2O(x = 1-3)$. Some spectra of the silica- and alumina-supported ruthenium samples are shown in Figs. 33 and 34, respectively. After drying at 383 K for 24 hr, the Mössbauer spectra of the silica-impregnated sample (Fig. 33b) was the same as for unsupported $RuCl_3 \cdot 1$-$3H_2O$ (Fig. 33a) (single peak at -0.34 mm s^{-1} with respect to metallic Ru). However, after a similar treatment, the Mössbauer spectrum of the alumina-impregnated sample (Fig. 34a) was an asymmetric quadrupole-split doublet ($\Delta E_Q = 0.45$ mm s^{-1}) with an isomer shift ($\delta = -0.41$ mm s^{-1}) more negative than that of $RuCl_3 \cdot 1$-$3H_2O$. These results show that the structure of the impregnated ruthenium complexes on silica and alumina are different and that the ruthenium complex interacts more strongly with the alumina than with the silica.

Different support interactions on silica and alumina were also found in the Mössbauer spectra of samples that were reduced in hydrogen at 673 K. In the case of ruthenium on silica, no peaks were observed in the Mössbauer spectrum after 2×10^6 counts, and chemical analysis of the sample showed no loss in ruthenium content of the reduced specimen. It must therefore be concluded that the reduced ruthenium species (metallic Ru) is rather weakly bound to the silica support, leading to a very small recoil-free fraction. Hydrogen adsorption measurements on the reduced sample were indicative of metallic ruthenium particles ≈ 9 nm in diameter. If there were no interaction between the support and the ruthenium particles, then a Mössbauer spectrum would be observable only for particles larger than a critical diameter of ≈ 35 nm, because of the recoil energy loss (see Chapter 1 of Volume I of this treatise). The absence of a Mössbauer spectrum for 9-nm metallic ruthenium particles is thus consistent with a weak support interaction for the silica-supported sample. For the alumina-supported ruthenium, however, metallic Ru was observed in the Mössbauer spectrum after reduction of the sample at 673 K (followed by evacuation to 10^{-6} Torr). Hydrogen adsorption measurements and x-ray diffraction indicated that the metallic Ru particles were ≈ 10 nm in

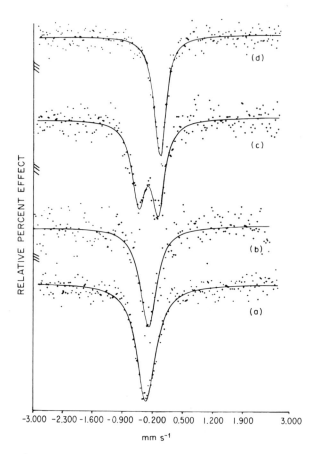

Fig. 33. Mössbauer spectra of (a) $RuCl_3 \cdot 1\text{-}3H_2O$, (b) $RuCl_3 \cdot 1\text{-}3H_2O$ impregnated on a silica support, (c) ruthenium on a silica support after calcination; (d) 24-nm ruthenium metal particles on silica, obtained by reduction of the calcined sample (C. A. Clausen and M. L. Good, 1975. *J. Catal.* **38**, 92).

diameter, i.e., significantly less than the critical diameter of 35 nm. It must, therefore, be concluded that the support interaction between metallic Ru and alumina is fairly strong (compared to silica). It would be interesting to study whether the evacuation step itself (10^{-6} Torr at 400°C) has any effect on the observed recoil-free fraction. This possibility was not investigated by the authors.

After calcination of the impregnated silica-supported sample, however, a Mössbauer spectrum could be observed (Fig. 33c). The isomer shift ($\delta = -0.27$ mm s^{-1} with respect to metallic Ru) and quadrupole splitting ($\Delta E_Q = 0.46$ mm s^{-1}) of the spectrum are indicative of RuO_2. The ab-

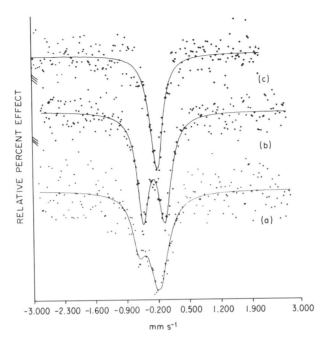

Fig. 34. Mössbauer spectra of (a) $RuCl_3 \cdot 1\text{-}3H_2O$ impregnated on alumina, (b) ruthenium on an alumina support after calcination, (c) 29.5-nm ruthenium metal particles on alumina, obtained by reduction of the calcined sample (C. A. Clausen and M. L. Good, 1975. *J. Catal.* **38**, 92).

sence of other peaks suggests that RuO_2 is the only form of ruthenium present in the sample. Subsequent reduction and evacuation (at 673 K) of the calcined sample led to the formation of large metallic ruthenium particles (≈ 24 nm in diameter), as evidenced by hydrogen adsorption and x-ray diffraction. Although the 24-nm particles are smaller than the critical size of 35 nm, a metallic ruthenium Mössbauer spectrum was observed (Fig. 33d). This suggests that some support interaction with silica in fact still exists. Another possible explanation is that the observed Mössbauer spectrum may be due to a small fraction of metal particles that are larger than the critical size. Similarly, calcination of the alumina-supported sample led to the observation of only RuO_2 in the Mössbauer spectrum (Fig. 34b), and reduction (followed by evacuation) of this sample produced 29.5-nm metallic ruthenium particles with the characteristic metallic ruthenium Mössbauer spectrum (Fig. 34c).

The difference in support interaction for ruthenium catalysts supported on alumina or on silica is qualitatively similar to that observed for supported iron (Section IV, C) and cobalt (Section IV, I) catalysts.

Surface information by Mössbauer spectroscopy was not obtained from the above described samples of ruthenium supported on silica or on alumina because of the large ruthenium particle sizes for these samples. However, C. A. Clausen and Good (1977b) also studied ruthenium exchanged into Na–Y zeolites. The resulting samples had a significant fraction of the ruthenium atoms accessible to gas phase species.

The Na^+ ions in the initial zeolite were exchanged with $[Ru(NH_3)_5N_2]^{2+}$ to approximately 60% completion, and after vacuum drying at 298 K, the Mössbauer spectrum was collected at 4.2 K. A two-peak pattern was observed, the isomer shift of which ($\delta = -0.80$ mm s^{-1} with respect to metallic ruthenium) agreed well with the value for the catalyst starting material $[Ru(NH_3)_5N_2]Cl_2$ ($\delta = -0.76$ mm s^{-1}). The quadrupole splitting for the exchanged zeolite ($\Delta E_Q = 0.56$ mm s^{-1}) was, however, significantly greater than that for $[Ru(NH_3)_5N_2]Cl_2$ ($\Delta E_Q = 0.22$ mm s^{-1}). Thus the oxidation state of the ruthenium did not change during the exchange process (as indicated by the constant isomer shift), but some interaction between the ruthenium complex and the zeolite is evidenced by the change in the quadrupole splitting. Upon exposure of the evacuated sample to air at 298 K, a single peak Mössbauer spectrum was observed with an isomer shift ($\delta = -0.37$ mm s^{-1}) characteristic of Ru^{3+}. That is, exposure to air resulted in the oxidation of Ru^{2+} to Ru^{3+}, with the probable formation of exchanged $[Ru(NH_3)_5OH]^{2+}$ (the isomer shift of $[Ru(NH_3)_5OH]Cl_2$ is $\delta = -0.39$ mm s^{-1}).

To produce small particles of metallic ruthenium, the exchanged zeolite was treated in hydrogen at 673 K, and the Mössbauer spectra of this sample verified that the ruthenium was, in fact, reduced to the metallic state. Furthermore, the metallic particles did not produce detectable x-ray diffraction peaks, which indicates a particle size below ≈ 5.0 nm in diameter. When the reduced sample was exposed to air at 298 K, small but measurable changes in the Mössbauer spectrum were seen (e.g., a decrease in the isomer shift from 0.01 mm s^{-1} for the reduced sample to -0.10 mm s^{-1} for the air-exposed sample). The actual interpretation of these changes is still uncertain, but the sensitivity of the Mössbauer spectrum to this "mild" treatment is consistent with the small size of the ruthenium particles. Indeed, no such observable changes in the Mössbauer spectrum were found for ruthenium supported on silica or alumina.

As a further test of the sensitivity of ruthenium Mössbauer spectroscopy to surface phenomena, the following experiments were conducted. The exchanged zeolite was heated to 673 K under vacuum, and subsequent Mössbauer spectroscopy (which showed a two-peak pattern with $\delta = -0.06$ mm s^{-1}, $\Delta E_Q = 0.43$ mm s^{-1}) indicated an oxidation from (the originally present) Ru^{2+} to Ru^{4+}. When this sample was exposed to air at 298 K, a three-peak Mössbauer spectrum resulted corresponding to a spectral doublet ($\delta = -0.10$ mm s^{-1}, $\Delta E_Q = 0.74$ mm s^{-1}) and a central singlet ($\delta = -0.13$ mm s^{-1}). The isomer shifts of these components show that the air exposure does not lead to any change in the oxidation state of the ruthenium. However, once again the Mössbauer spectrum is sensitive to a fairly mild treatment of the sample. Alternatively, if the exchanged zeolite is treated at 673 K in vacuum and subsequently exposed to NH_3 or NO at room temperature, then broad, single-peak Mössbauer spectra are produced. The isomer shift of the NH_3-treated sample was -0.25 mm s^{-1}, and that of the NO-treated sample was -0.01 mm s^{-1}. Thus, compared to the isomer shift of the vacuum-treated sample ($\delta = -0.06$ mm s^{-1}), NH_3 treatment causes a decrease in the electron density at the ruthenium nucleus and NO treatment causes an increase in the electron density at the nucleus. If the electron transfer is primary to the ruthenium p and d orbitals, then the isomer shift changes are consistent with the electron donating and withdrawing tendencies of NH_3 and NO, respectively.

The above studies by C. A. Clausen and Good illustrate that Mössbauer spectroscopy can be used to study the chemical state of supported ruthenium. This technique was then applied by C. A. Clausen and Good (1977a) in their study of industrial ruthenium-containing catalysts. Ruthenium has a pronounced selectivity for reduction of nitrogen oxides to molecular nitrogen, and attention has been focused on the use of such catalysts for automotive emission control. This process is often carried out in a "dual-bed" catalyst system. The first bed (containing ruthenium) operates under reducing conditions, and its function is to reduce nitrogen oxides. Air is injected into the effluent from the first bed, and the gas mixture is then passed through the second bed where hydrocarbons and carbon monoxide are oxidized to carbon dioxide. During engine warmup, however, the first bed is exposed to net oxidizing conditions, and this leads to possible loss of ruthenium from the first bed due to volatilization of RuO_4.

One method, proposed by Shelef and Gandhi (1972), to minimize the tendency of ruthenium to volatilization is based on the formation of the nonvolatile barium ruthenate ($BaRuO_3$). This stabilization procedure was studied using Mössbauer spectroscopy by C. A. Clausen and Good (1977a) for ruthenium supported on η-Al_2O_3. Samples were prepared with various loadings of ruthenium and barium. After reduction in hydrogen at 673 K, the samples were heated in air to 1173 K and were then cooled to 4.2 K at which temperature Mössbauer spectra were collected. These spectra were composed of three peaks, corresponding to quadrupole-split RuO_2 and single-peak $BaRuO_3$ patterns. It was observed that (i) as the barium content in the sample increased (at constant ruthenium content), a greater fraction of the ruthenium was present as $BaRuO_3$, and (ii) as the ruthenium content in the sample decreased (at constant ruthenium to barium weight ratio), a greater fraction of the ruthenium was present as $BaRuO_3$.

The oxidized samples were then treated at 973 K in a net-reducing simulated auto exhaust (SAE) mixture, and the Mössbauer spectra collected after this treatment were indeed indicative of metallic ruthenium. Furthermore, exposure of the samples to a net-oxidizing SAE mixture at 973 K converted the metallic ruthenium back into RuO_2 and $BaRuO_3$. That is, barium stabilization of the supported ruthenium was successful. Yet, a typical automotive emission control catalyst must withstand oxidation–reduction cycling; therefore, the barium stabilized samples were subjected to as many as 100 cycles between net-reducing and net-oxidizing SAE mixtures at 973 K, ending with a final treatment in the net-oxidizing mixture. From the relative areas of the RuO_2 and $BaRuO_3$ components in the Mössbauer spectra, it could be seen that oxidation–reduction cycling led to a decrease in the relative amount of barium ruthenate. That is, phase separation took place between the ruthenium and the stabilizing agent (barium) during the cyclic treatment. This result explains the experimentally observed loss of ruthenium from barium stabilized catalysts used under normal (cyclic) vehicle operating conditions.

In summary, ^{99}Ru Mössbauer spectroscopy is more difficult to carry out than ^{57}Fe Mössbauer spectroscopy because one must work at low temperatures with short-lived radioactive sources. However, the nuclear parameters of both isotopes allow detailed chemical information to be extracted from the Mössbauer spectrum. This chemical sensitivity combined with the important catalytic properties of ruthenium will undoubtedly lead to continued applications of ^{99}Ru Mössbauer spectroscopy to surface and catalytic phenomena.

H. Partial Oxidation Catalysts

For reactions catalyzed by metals, it is often proposed that the reactants adsorb on the surface, followed by reactions between adsorbed species and desorption of product molecules; that is, the reaction takes place *on* the catalyst surface. For certain types of reactions, however, one can imagine that the catalyst reacts *with* the reactant and product molecules. However, because the catalyst itself remains unchanged by the overall catalytic reaction, all reactions between the catalyst and the reactants and products must lead to closed sequences (e.g., if one reactant oxidizes the catalyst, then another must reduce the catalyst). There are many results, some of

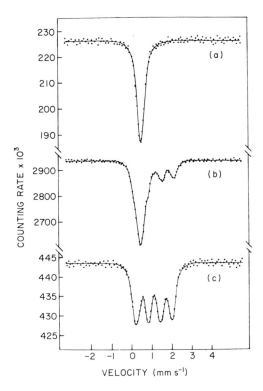

Fig. 35. Mössbauer spectra for a pure $Fe_2(MoO_4)_3$ catalyst that had been kept in a nitrogen–methanol flow for 5 hr at 493 K (a), at 503 K (b), and for 2 hr at 603 K (c) (Carbucicchio and Trifiro, 1976. *J. Catal.* **45**, 77).

them to be discussed in this section, that point to the industrially important partial oxidation catalyst belonging to this latter class of catalysts.

Iron-molybdate-based catalysts are excellent for the partial oxidation of methanol to formaldehyde, and the reaction mechanism over this catalyst is thought to involve a reduction of the catalyst by methanol followed by a reoxidation of the catalyst by oxygen. Such reactions can conveniently be studied using ^{57}Fe Mössbauer spectroscopy as demonstrated by Carbucicchio and Trifiro (1976). The catalysts studied by these authors were pure iron molybdate $[Fe_2(MoO_4)_3]$ and iron molybdate containing 35% excess MoO_3. The Mössbauer spectra of these two samples from 298 to 688 K were identical spectral singlets, and the isomer shifts were indicative of high-spin Fe^{3+}. Thus, before the actual partial oxidation reaction takes place, the excess MoO_3 does not appear to influence the iron-molybdate structure.

In order to study the possible reduction of the catalyst, the samples were treated at various temperatures in a methanol–nitrogen atmosphere and the resulting Mössbauer spectra are shown in Fig. 35. Below 503 K, the Mössbauer spectra of the samples were spectral singlets indicative of unreacted iron molybdate. However, treatment at 503 K for 5 hr led to the appearance of two new spectral doublets in addition to the remaining singlet due to $Fe_2(MoO_4)_3$. Finally, after 2 hr at 603 K in the methanol–nitrogen mixture, the conversion of the spectral singlet into the two spectral doublets was complete. No effect of the excess MoO_3 was seen in the Mössbauer spectra. Comparison of the room temperature isomer shifts and quadrupole splittings for these two doublets ($\delta_1 = 1.23$ mm s^{-1} with respect to ^{57}Co in chromium, $\Delta E_{Q1} = 2.50$ mm s^{-1}; $\delta_2 = 1.25$ mm s^{-1}, $\Delta E_{Q2} = 0.90$ mm s^{-1}) with Mössbauer spectra of known Fe–Mo–O phases suggested that the $Fe_2(MoO_4)_3$ was completely converted into β-$FeMoO_4$. The two doublets are due to Fe^{2+} in two different lattice sites, and the methanol treatment led, in fact, to a net reduction of the entire catalyst sample. The identification of β-$FeMoO_4$ as the reduced phase was confirmed by x-ray diffraction.

After demonstrating that methanol is capable of reducing the catalyst, the authors investigated the second half of the partial oxidation process (i.e., the catalyst reoxidation) by collecting Mössbauer spectra of the β-$FeMoO_4$ after various treatments in oxygen. At temperatures below 543 K, the oxygen was not capable of oxidizing the β-$FeMoO_4$ phase, while treatments in oxygen at temperatures above 573 K lead to Mössbauer spectra containing the spectral singlet characteristic of $Fe_2(MoO_4)_3$. For example, a 20-min treatment at 623 K produced a Mössbauer spectrum with contributions from both β-$FeMoO_4$ and $Fe_2(MoO_4)_3$, and a 7-hr treatment at this temperature produced a complete reoxidation of β-$FeMoO_4$ to $Fe_2(MoO_4)_3$. Again, no effect of the excess MoO_3 was observed in the reoxidation behavior of these samples.

The relevance of these reduction and reoxidation studies to the actual partial oxidation process is ascertained through a discussion of the partial oxidation kinetics. In short, (i) the catalytic activity is appreciable at temperatures above ≈ 523 K; (ii) at temperatures less than 573 K the rate depends on the oxygen partial pressure, while at temperatures above ≈ 573 K the rate is insensitive to the oxygen pressure; (iii) the selectivity to formaldehyde decreases at temperatures above ≈ 573 K; and (iv) while excess MoO_3 does have an effect on the active surface area of the catalyst, the catalytic activity per unit surface area is independent of the excess MoO_3 content. Comparison of these results with the Mössbauer spectroscopy studies indicates: (i) catalytic activity is significant at temperatures for which methanol is capable of reducing the catalyst (i.e., temperatures above ≈ 523 K); (ii) oxygen is capable of reoxidizing the catalyst at temperatures above 573 K, and below this temperature the partial oxidation rate depends on the oxygen partial pressure, while above this temperature it does not; (iii) the selectivity to formaldehyde decreases at temperatures at which oxygen is capable of completely oxidizing the catalyst (above 573 K); and (iv) excess MoO_3 has no effect on the reduction, reoxidation, or catalytic properties of the catalyst. Indeed, the proposed mechanism of partial oxidation involving catalyst reduction and reoxidation is completely consistent with the Mössbauer spectroscopy studies of the Fe–Mo–O system.

A modification of the above $Fe_2(MoO_4)_3$ catalyst, in which Cr^{3+} ions were substituted for part of the Fe^{3+}, was investigated by Popov et al. (1976). Such catalysts, with stoichiometry $Fe_{2-2x}Cr_{2x}(MoO_4)_3$, were found to be more stable for partial methanol oxidation than the unsubstituted iron molybdate, while the catalytic activities for the substituted and unsubstituted samples were the same. Consistent with this latter result, the Mössbauer spectrum of the substituted iron molybdate is a singlet, the isomer shift of which is the same as that for unsubstituted iron molybdate. The primary effect of the chromium appears to be a distortion of the iron–molybdate structure since the spectral linewidth increases with increasing degree of substitution.

While iron molybdate and Cr-substituted iron molybdates are excellent partial methanol oxidation catalysts, a series of Fe–Mo–Bi catalysts has been developed for the selective oxidation and ammoxidation of olefins. Carbucicchio et al. (1976) have used ^{57}Fe Mössbauer spectroscopy to study this Fe–Mo–Bi system in the partial oxidation of butene to butadiene. The Mössbauer spectrum of an oxidized sample containing 14 wt% Bi_2O_3 and 86 wt% $Fe_2(MoO_4)_3$ consisted of a spectral singlet due to Fe^{3+} in $Fe_2(MoO_4)_3$ and a weak shoulder on the low-velocity side of this singlet. This shoulder was taken to be one-half of a quadrupole-split spectrum resulting from Fe^{3+} in a distorted lattice site. Since x-ray diffraction detected both iron molybdate and bismuth molybdate (Bi_2MoO_6), it was concluded that the quadrupole-split doublet was due either to Fe^{3+} in the bismuth molybdate, or to Fe^{3+} in the iron molybdate with distorted surroundings caused by neighboring Bi^{3+} ions. Since Bi–Fe–Mo catalysts, which contain only small amounts of Bi, are selective catalysts for olefin partial oxidation, it was suggested that the Fe^{3+} doublet was due to the effect of Bi^{3+} in the iron–molybdate structure.

The reduction–oxidation nature of this partial oxidation reaction was then investigated by treating the catalyst at various temperatures with 1-butene. For temperatures below 673 K, the Mössbauer spectrum showed that the catalyst remained unchanged upon exposure to 1-butene, while at temperatures above 673 K (e.g., treatment at 683 K for 6 hr) the butene reduced the catalyst to β-$FeMoO_4$ (x-ray diffraction also showed the formation of MoO_2 by this treatment, but this phase would not be detectable in the Mössbauer spectrum). Indeed, the 1-butene reduction of the sample to β-FeMoO$_4$ is very similar to the previously discussed reduction of Fe$_2$(MoO$_4$)$_3$ to β-FeMoO$_4$ by methanol. The major difference between these reactions is the temperature required for complete reduction: 673 K for 1-butene; 503 K for methanol. Subsequent to 1-butene reduction, the sample was treated in oxygen at various temperatures, and it was thereby found that at temperatures higher than $\approx$ 673 K the sample could be completely reoxidized to its initial state (i.e., that state observed prior to the 1-butene reduction).

As for the previously discussed partial methanol oxidation, these Mössbauer spectroscopy results for the Bi–Fe–Mo system must finally be compared with the actual catalytic properties of this system. In short, the catalyst was found active at temperatures above $\approx$ 523 K, but at temperatures above 673 K the selectivity to butadiene was low. The latter observation is in agreement with the result from Mössbauer spectroscopy that oxygen completely oxidizes the catalyst at temperatures above $\approx$ 673 K. In addition, it must be concluded that although β-FeMoO$_4$ can be formed by exposure to 1-butene at temperatures higher than 673 K, this phase is not present under actual reaction conditions (523–673 K). It is this behavior that primarily distinguishes the partial methanol and partial butene oxidation reactions.

In the above study, Carbucicchio *et al.* (1976) investigated iron molybdates with added bismuth. In a similar fashion, the structure of iron-substituted bismuth molybdates was studied by Notermann *et al.* (1975) using ^{57}Fe Mössbauer spectroscopy in combination with x-ray diffraction, infrared spectroscopy, ultraviolet–visible spectroscopy, and x-ray photoelectron spectroscopy. A sample containing an atomic ratio of Bi : Fe : Mo = 3 : 1 : 2 gave a two-peak Mössbauer spectrum at 80 K indicative of Fe^{3+} ($\delta = 0.35$ mm s^{-1} with respect to metallic iron) in a lattice site having less than cubic symmetry ($\Delta E_Q = 1$ mm s^{-1}). Based on the structure of unsubstituted Bi$_2$(MoO$_4$)$_3$, it was suggested that Fe^{3+} was substituted for one Mo^{6+} with the additional Bi^{3+} ion needed for electric neutrality located in one of the vacancies of the structure. When the iron content of the sample was increased to Bi : Fe : Mo = 1 : 1 : 1, the Mössbauer spectral doublet was accompanied by a six-peak pattern at 80 K due to magnetically split Fe^{3+}. For this second Fe^{3+} species, room temperature Mössbauer spectroscopy showed more complex magnetic splitting indicative of, for example, two different crystallographic locations of these ions. Since electric neutrality dictates that two Fe^{3+} are needed to substitute one Mo^{6+}, it was proposed that one Fe^{3+} replaces the Mo^{6+} (giving rise to the spectral doublet) and the second Fe^{3+} enters one of the vacant sites in the structure. Crystallographic ordering (or clustering) of the Fe^{3+} ions in the vacant sites may then lead to magnetic interactions between ions and hence a six-peak Mössbauer spectrum. Furthermore, the quadrupole splitting of the Fe^{3+} doublet suggests that the incorporation of iron into the bismuth molybdate leads to a distortion of the structure, and this was verified by the infrared and ultraviolet–visible spectroscopic results.

A further test of the catalyst reduction–oxidation mechanism of partial oxidation is to collect Mössbauer spectra of the sample under reaction conditions. For this purpose a simple glass cell that serves both as a catalytic reactor and a Mössbuaer spectroscopy cell was constructed by Maksimov *et al.* (1975a). These authors first studied the oxidation of propene to acrolein over a $CoMoO_4$ catalyst doped with ^{57}Fe to a stoichiometry of approximately $Fe_{0.03}Co_{0.97}MoO_4$. The Mössbauer spectrum of this sample at 583 K (in oxygen or air) was essentially a spectral singlet characteristic of Fe^{3+}. However, approximately 4% of the spectral area was present as a weak doublet with an isomer shift indicative of Fe^{2+}, and even prolonged oxygen treatment at 773 K was unable to oxidize this species to Fe^{3+}. When the Mössbauer spectrum was collected with the sample under propene partial oxidation reaction conditions at 583 K, the spectral area of the Fe^{2+} doublet increased at the expense of the Fe^{3+} singlet. Subsequently, when the reactants were pumped out of the cell, the spectral area ratio (measured at 583 K) of the doublet to singlet patterns decreased to that value ($\approx 4\%$) characteristic of the sample prior to the catalytic reaction. Thus a net reduction in the oxidation state of the sample takes place under reaction conditions due to electron transfer from propene to the catalyst. The partial oxidation mechanism involves catalyst reduction (by propene) *followed by* catalyst reoxidation (by oxygen). Therefore, the lifetime of the Fe^{2+} states formed by electron transfer from propene must be longer than $\approx 10^{-8}$ s (one-tenth the lifetime of the nuclear excited state of ^{57}Fe).

For comparison with the work of Carbucicchio and Trifiro (1976) on methanol partial oxidation over iron molybdates, it should be noted that Maksimov *et al.* (1975b) have also studied this system using Mössbauer spectroscopy under reaction conditions. The results of this study are similar to those for the propene partial oxidation, and, therefore, they will not be discussed here at length. The iron is present primarily as Fe^{3+} prior to the reaction. An Fe^{2+} spectral component is observed under reaction conditions (i.e., reduction of the catalyst by methanol), and the Fe^{2+} signal disappears upon termination of the reaction.

Ferrites (MFe_2O_4, where M is a divalent metal ion) are another class of partial oxidation catalysts. In particular, they have high catalytic activity and selectivity for oxidative dehydrogenation reactions, such as the conversion of butene into butadiene. An important consideration in the use of a given ferrite catalyst, however, is its stability against bulk reduction in a hydrocarbon atmosphere. In this respect, Cares and Hightower (1975) used Mössbauer spectroscopy (at room temperature) to study the structure of $CoFe_2O_4$ and $CuFe_2O_4$ catalysts before and after exposure to butene–oxygen mixtures at temperatures above 708 K. The Mössbauer spectrum of the untreated $CoFe_2O_4$ sample consisted of two different six-peak patterns due to Fe^{3+} ions located in the tetrahedral and octahedral sites of the spinel lattice, respectively, In general, there are twice as many available octahedral sites as there are tetrahedral sites. Furthermore, if the M^{2+} ions occupy the tetrahedral sites while the Fe^{3+} ions occupy the octahedral sites, then a spinel is called "normal." Alternatively, if the M^{2+} ions are found in the octahedral sites, while the Fe^{3+} are equally distributed between the tetrahedral and octahedral sites, then the spinel is called "inverse." Finally, in a "random" spinel, the M^{2+} and Fe^{3+} are randomly placed in the tetrahedral and octahedral sites. For the $CoFe_2O_4$ sample, the spectral area ratio S of the octahedral to tetrahedral site six-peak patterns was approximately equal to two, indicating that the sample preparation led to a random spinel. Upon exposure of the sample to a butene–oxygen mixture (during the course of oxidative dehydrogenation), the Mössbauer spectrum remained essentially unchanged from the initial random spinel. Exposure to butene alone produced a decrease in the area ratio S to a value of approximately unity; that is, the spinel was converted from random to inverse.

For the $CuFe_2O_4$ catalyst, the Mössbauer spectrum of the untreated sample consisted primarily of the two six-peak patterns due to Fe^{3+} in the octahedral and tetrahedral sites. The area ratio S was approximately equal to unity indicative of an inverse spinel. In addition, however, a weak spectral doublet was also present in the spectrum. The exact origin of this doublet was unknown, but is could be due (i) to superparamagnetic regions of $CuFe_2O_4$ or (ii) to the formation of a small amount of reduced iron (e.g., as $CuFeO_2$). When the $CuFe_2O_4$ sample was then treated in butene, dramatic changes in the Mössbauer spectrum took place. The six-peak patterns characteristic of the inverse $CuFe_2O_4$ spinel disappeared, the area of the spectral doublet increased, and new peaks due to Fe_3O_4 were formed. Furthermore, a copper mirror on the walls of the reactor was observed after prolonged butene treatments. Thus, unlike the $CoFe_2O_4$ sample, exposure of $CuFe_2O_4$ to butene can lead to extensive reduction of the sample (to metallic copper and Fe_3O_4). If the length of the butene treatment is insufficient to actually distill the metallic copper from the catalyst, then the original $CuFe_2O_4$ spinel can be regenerated by oxygen treatment of the reduced sample at 793 K. Whereas a detailed description of the reduction process for the $CoFe_2O_4$ catalysts is not possible, the different stabilities of these samples against reduction in a hydrocarbon atmosphere are clearly seen from the Mössbauer spectroscopy results. It should be noted here that the doping of ferrites with Cr^{3+} can lead to greater stabilities against reduction. This procedure was studied in detail by Rennard and Kehl (1971) for $MgFe_2O_4$ and $ZnFe_2O_4$ catalysts. They included in their study some preliminary Mössbauer spectroscopy results, which suggested that the introduction of Cr^{3+} into the ferrite decreased the magnetic domain size to the point where superparamagnetism could be observed at room temperature.

The $MgFe_2O_4$ ferrite has more recently been investigated by Gibson and Hightower (1976) with respect to butene oxidative dehydrogenation. Over a period of several hours the catalyst selectivity for butadiene decreased substantially, and this decay could not be reversed by treatment of the sample in oxygen. This irreversible change in the catalyst was then studied using Mössbauer spectroscopy. At room temperature the fresh sample gave rise to patterns in the Mössbauer spectrum characteristic of $MgFe_2O_4$ and a small amount ($\approx 5\%$ of the total spectral area) of α-Fe_2O_3, while the Mössbauer spectrum of the deactivated sample showed only $MgFe_2O_4$. By also collecting Mössbauer spectra at 673 K (above the Curie temperature for $MgFe_2O_4$: 623 K), the overlapping patterns of α-Fe_2O_3 and $MgFe_2O_4$ could be separated for a better estimation of the relative amounts of these phases. In addition, the high temperature Mössbauer spectrum of the deactivated sample indicated the presence of a very small amount of a magnetically ordered, iron-containing phase. While the structure of this phase could not be ascertained from the Mössbauer spectrum (because the pattern was not clearly distinguishable from the background), it was suggested from magnetization measurements that the phase could be a Fe_3O_4-$MgFe_2O_4$ solid solution. In view of these Mössbauer spectroscopy results, the catalyst deactivation is probably due either to (i) the loss of catalytically active α-Fe_2O_3 or (ii) the reduction of $MgFe_2O_4$ to a solid solution of Fe_3O_4-$MgFe_2O_4$ with precipitation of the excess Mg^{2+} ions as MgO. In a more recent publication, Gibson et al. (1977) have chosen the former interpretation.

Misono et al. (1976) found that γ-Fe_2O_3, which has a defect spinel structure, was more selective than α-Fe_2O_3 or Fe_3O_4 in the oxidative dehydrogenation of 1-butene to butadiene. Mössbauer spectroscopy was used for characterizing the bulk structure of the catalysts.

Prasada Rao and Menon (1978) recently investigated silica-supported multicomponent molybdate catalysts before and after they were used for ammoxidation of propylene. Besides Bi–Fe–Mo, these catalysts contained Ni, Co, P, and K. The Mössbauer spectra of the fresh catalyst were typical of $Fe_2(MoO_4)_3$, while after use, the Fe^{3+} had been reduced to Fe^{2+}. The spectrum reported for the used catalyst is similar to that of β-$FeMoO_4$ reported by Carbucicchio and Trifiro (1976) to be present after reacting $Fe_2(MoO_4)_3$ with methanol.

The studies mentioned so far have shown that reduction–oxidation characteristics of a catalyst may be important in determining the catalytic activity toward partial oxidation. Since the reduction and oxidation of the catalysts were found to take place readily throughout the bulk, the surface chemical state may be closely related to that of the bulk. Further evidence for this is seen from a number of studies that have observed correlations between the different Mössbauer parameters, on one hand, and adsorption and catalytic parameters, on the other (Skalkina *et al.*, 1969; Garten *et al.*, 1973; Matsuura, 1976; Kriegsmann *et al.*, 1976). These observations indicate that measurements of bulk Mössbauer parameters may give insight into the reactivity of catalyst surfaces. It does not follow, however, that the surface and bulk properties are identical, but only that they are related to each other in some way.

The correlations observed between the quadrupole splitting and the catalytic activity and selectivity are especially interesting. Kriegsmann *et al.* (1976), for example, studied the ammoxidation of propene on a series of iron–antimony oxide catalysts with different Sb/Fe ratios. These catalysts are interesting since they can be studied by looking at both the ^{57}Fe and the ^{121}Sb resonances. The yield of acrylonitrile is plotted in Fig. 36 against the quadrupole splitting of Fe^{3+} at octahedral sites. This type of volcano-shaped curve was observed by Skalkina *et al.* (1969) for ammoxidation of propylene on mixed iron oxide catalysts. The results suggested that there exists a close relation between the symmetry of the Fe^{3+} ions and the catalytic activity. This was further studied by Kriegsmann *et al.* (1976) by doping $Fe_2Sb_2O_7$ with different cations. The quadrupole splitting was found to increase with increasing ratio of the charge to ionic radius of the doping ion. Again for these catalysts, there was observed a volcano-type relationship between the acrylonitrile yield and the quadrupole splitting. Analysis of the isomer shift gives information about the covalency of the Fe^{3+}–O bond. The authors inferred that the covalency and thereby the

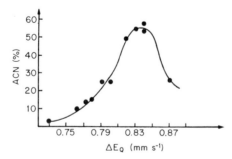

Fig. 36. Acrylonitrile (ACN) yield (mole%) against quadrupole splitting of $^{57}Fe^{3+}$ for different iron–antimony oxide catalysts (Kriegsmann *et al.*, 1976).

Fe^{3+}–O bond strength increased with increasing quadrupole splitting. A combination of the results of Skalkina *et al.* (1969) and those of Kriegsmann *et al.* (1976) gives the following picture for the catalytic behavior: (i) catalysts with very low quadrupole splitting have weak Fe^{3+}–O bonds and low selectivity toward acrylonitrile formation, but will have high activity toward complete oxidation to CO_2 and H_2O; (ii) catalysts with intermediate quadrupole splitting and Fe^{3+}–O bond strength will have high selectivity toward acrylonitrile; and (iii) catalysts with high quadrupole splitting will have both low selectivity toward acrylonitrile formation and low overall oxidation activity. Looking at the ^{121}Sb resonance, Kriegsmann *et al.* (1976) also found a high degree of covalency of the Sb^{5+}–O bond for the most selective catalysts. The increase in covalency of both iron and antimony is suggestive of oxidation states between Fe^{2+} and Fe^{3+} and between Sb^{3+} and Sb^{5+}. In fact, the most selective catalysts showed the simultaneous presence of all the valence states mentioned above. These results point again to reduction–oxidation characteristics of the catalyst being important for the catalytic properties.

The U–Sb–O system forms another group of industrially interesting catalysts. Evans (1976) carried out ^{121}Sb Mössbauer spectroscopy studies on a commercial catalyst and on USb_3O_{10} and $USbO_5$ samples. The last two samples are, indeed, usable acrylonitrile catalysts, but USb_3O_{10} has greater catalytic activity and selectivity. The Mössbauer spectra (at 77 K) of USb_3O_{10} and $USbO_5$ are single-peak patterns, the isomer shifts of which are equal (within the experimental uncertainties) and typical of Sb^{5+}, as shown in Fig. 37. Moreover, since ^{121}Sb isomer shifts are quite sensitive to the chemical bonding of the Sb, it can be said that the covalency of the Sb–O bonds in the two samples are similar. The isomer shift of the U–Sb–O catalysts, however, is more negative than that for Sb^{5+} in many other compounds; since a decrease in isomer shift indicates an increase in the electron density at the nucleus (for ^{121}Sb), it must be concluded that the Sb–O bond has a high covalency in the U–Sb–O samples. Furthermore, while the isomer shifts of the Sb in USb_3O_{10} and $USbO_5$ are the same, the linewidth of the USb_3O_{10} resonance is considerably larger than that for $USbO_5$. This suggests that there are several different crystallographic Sb sites in USb_3O_{10} and a smaller number of different Sb sites (perhaps a single site) in $USbO_5$.

The fact that both USb_3O_{10} and $USbO_5$ are usable acrylonitrile catalysts and that the Sb–O bond has a high covalency in these compounds seems to indicate a relationship between catalytic properties and covalency in the U–Sb–O system. In addition, the catalytic properties of the U–Sb–O system seem to depend on the crystallographic structure (or distortion) of the actual compound since USb_3O_{10} and $USbO_5$ have similar covalency

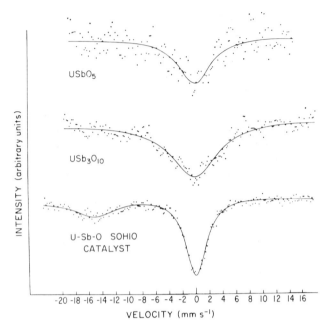

Fig. 37. ^{121}Sb Mössbauer spectra of pure U–Sb–O phases and of a supported catalyst. The points are the experimental data, and the solid lines are the least-mean-squares fit of a single Lorentzian line to each absorption line. The supported catalyst has the composition 60 wt% $USb_{4.6}O_x$, 40 wt% SiO_2 (Evans, 1976. *J. Catal.* **41**, 271).

and different catalytic activities and selectivities. This statement is consistent with the different linewidths of the Mössbauer spectra for these compounds. Finally, Evans also studied a commercial (SOHIO) acrylonitrile catalyst consisting of 65 wt% $USb_{4.6}O_x$ on silica (see Fig. 37). Mössbauer spectroscopy revealed the presence of approximately equal amounts of USb_3O_{10} and Sb_2O_4. (The Sb in Sb_2O_4 is present as equal amounts of Sb^{3+} and Sb^{5+}; the isomer shifts of these two ions are typically -15 and 0 mm s^{-1}, respectively, with respect to a $BaSnO_3$ source).

I. *Hydrodesulfurization (HDS) Catalysts*

In studies of catalysts supported on high surface area materials it is often observed that the support not only ensures a stable, high dispersion of the active phase, but it also influences the form in which the active components are present. Recent Mössbauer studies (Topsøe *et al.*, 1979b; B. S. Clausen *et al.*, 1976b) show that hydrodesulfurization catalysts belong to this latter category of catalysts for which "support effects" play a major role. These studies also revealed a strong influence of the preparation and activation parameters on the final catalyst structure.

Hydrodesulfurization catalysts form an important class of materials. They are used for the removal of sulfur from various fossil hydrocarbon fuel fractions by passing these, together with hydrogen, over the catalyst. In this way the sulfur is removed as H_2S. At the same time other components such as nitrogen and metals may be removed. The application of HDS catalysts in the petrochemical industry contributes to a desirable reduction of atmospheric pollution in utilization of fossil fuels.

The active components in these catalysts are Mo or W promoted with either Co or Ni. Fe is also a promoter, although less active. The components are dispersed on a high surface area support such as γ-alumina, by stepwise impregnation or co-impregnation procedures, usually followed by drying and calcination in air at about 800 K. The active state of the catalyst is reached by sulfiding the calcined catalyst at about 600 K either in an appropriate presulfidation gas mixture (e.g., H_2S/H_2) or by sulfiding directly in the reaction mixture.

Both the calcined and the sulfided states of these quite complex catalysts have been the subject of investigations by many different techniques. It has, however, been difficult to obtain direct structural evidence about the form in which the active elements are present, and many views of the catalyst structure and catalytically active sites have been proposed (see, e.g., Voorhoeve and Stuiver, 1971; Schuit and Gates, 1973; Lo Jacono et al., 1973; Farragher and Cossee, 1973; De Beer and Schuit, 1976; Moné, 1976; Delmon, 1977; Massoth, 1978). In the light of this it is interesting that due to its applicability to studies of highly dispersed systems, Mössbauer spectroscopy has revealed important structural features of HDS catalysts.

Mössbauer spectroscopy has been used in studies of both Co–Mo and Fe–Mo catalysts (Topsøe and Mørup, 1975; B. S. Clausen et al. 1976b; B. S. Clausen 1976, 1979; Topsøe et al., 1979b; Mørup et al., 1979b). Fe–Mo catalysts can be studied directly using [57]Fe transmission Mössbauer spectroscopy. However, since none of the naturally occurring isotopes in the industrially more important Co–Mo catalysts can be used directly for Mössbauer studies, these catalysts were studied by replacing part of the Co by [57]Co or [57]Fe. Interpretation of results of Mössbauer source experiments for this catalyst system has recently been discussed by Mørup et al. (1979b).

1. Calcined State

The thermodynamically stable oxides of Co and Mo under the calcination conditions are Co_3O_4 and MoO_3, respectively; however, when Co and

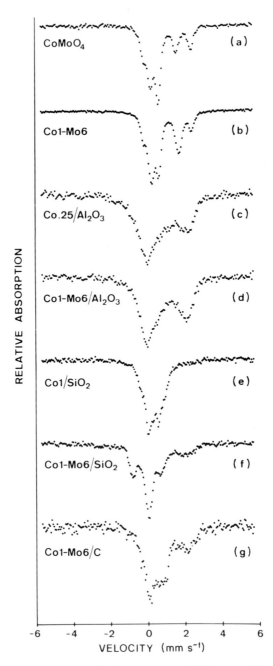

Fig. 38. Room temperature Mössbauer spectra of different model compounds and catalysts in the oxidic state. The metal content is indicated in the figure such that Col-Mo6/Al₂O₃, for example, denotes an alumina-supported sample containing 1% Co and 6% Mo (Topsøe et al., 1979b).

Mo are present jointly, $CoMoO_4$ is expected to form. If the support interaction is weak, then these phases may, in fact, be present after calcination. If, on the other hand, the support interaction is strong, neither of the above phases may be present. This seems to be the situation for Mo supported on alumina, where it is generally believed that a "molybdate" monolayer forms on the alumina surface. Cobalt may also interact strongly with the alumina resulting in the formation of a solid solution of Co in the alumina or a separate cobalt aluminate phase. The influence of the support on the structure of these catalysts has been demonstrated in a recent study by Topsøe *et al.* (1979b) in which Co and Co–Mo catalysts supported on γ-Al_2O_3, SiO_2 (Davison grade 950), and active carbon (900 m^2/g) were studied. For reference, studies were also carried out on unsupported catalysts and model compounds.

Figure 38a shows the spectra of ^{57}Co-doped large crystals of $CoMoO_4$ obtained by B. S. Clausen *et al.* (1977a). The spectrum of an unsupported Co–Mo catalyst (Fig. 38b) is very similar to that of $CoMoO_4$. This is not unexpected since this, as mentioned, is the thermodynamically stable Co–Mo–O compound under calcination conditions. When a γ-alumina support is introduced, a spectrum completely different from that of the unsupported catalyst is observed (Fig. 38d), indicating that the alumina support suppresses the formation of bulk $CoMoO_4$. Indeed, the spectrum of cobalt alone on alumina (Fig. 38c) and the spectrum of cobalt together with molybdenum (Fig. 38d) are quite similar. From this it was concluded that Co in the calcined catalyst is located mainly in the alumina lattice. Some of the small differences observed between the two spectra are probably due to some of the cobalt atoms being associated with the molybdenum monolayer or located in its proximity. Information regarding the support interaction of the Mo was obtained from experiments (B. S. Clausen, 1979) in which the Mo loading was kept constant, while the Co loading was increased beyond a point where all the Co had interacted strongly with the alumina. If the Mo in this case were bound only weakly to the support, one would expect the two weakly bound components to form $CoMoO_4$. The cobalt-containing phase was, however, observed to be Co_3O_4 and not $CoMoO_4$. In this way it could be concluded that Mo must be interacting strongly with the alumina surface. This result is in accordance with the idea of Mo being present as a "molybdate" monolayer on the alumina.

Besides giving information on the support interactions of Mo, the above results also show that the catalyst structure depends very much on preparation parameters such as the metal loading.

The spectra of Co/SiO_2 and $Co–Mo/SiO_2$ samples are shown in Figs. 38e and 38f, respectively. The spectrum in Fig. 38e is quite similar to that of Co_3O_4, except for the small "shoulders" appearing to the right and left

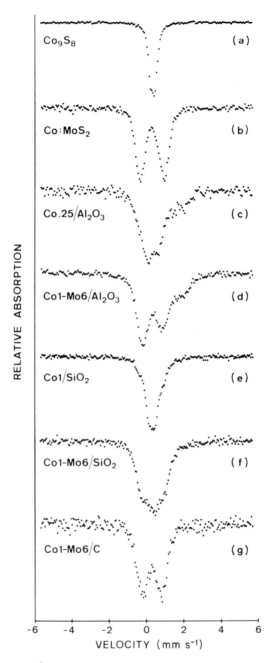

Fig. 39. *In situ* Mössbauer spectra of different model compounds and catalysts in the sulfided state. The spectra were recorded at room temperature in a H_2/H_2S (2% H_2S) mixture. The same nomenclature as in Fig. 38 is used to indicate the metal content (Topsøe *et al.*, 1979b).

of the Co_3O_4 spectral component. This indicates that only a small amount of cobalt has combined with the silica. The spectrum of the calcined $Co-Mo/SiO_2$ catalyst was difficult to interpret in detail. However, formation of a phase containing both Co and Mo (possibly poorly crystallized $CoMoO_4$) was indicated. The behavior of silica-supported catalysts is, therefore, very much in contrast to that of the alumina-supported catalysts. These findings show that both Co and Mo interact less strongly with silica than with alumina. The spectrum of a calcined carbon-supported catalyst (Fig. 38g) indicates the presence of a poorly crystallized cobalt molybdenum-like phase suggesting that a very weak support interaction is also present in this case.

2. Sulfided State

The thermodynamically stable Co and Mo compounds under the sulfiding conditions (623 K, $H_2S/H_2 \sim 0.02$) are Co_9S_8 and MoS_2, respectively. Mössbauer studies of ^{57}Co-doped Co_9S_8 were carried out by B. S. Clausen et al. (1977b). For reference, Fig. 39a shows the spectrum of Co_9S_8 obtained by these authors.

Sulfidation of bulk Co_3O_4 is expected to yield Co_9S_8. It is therefore not unexpected that Co_9S_8 was found in the sulfided Co/SiO_2 (Fig. 39e) catalyst since the calcined sample showed the presence of Co_3O_4. In a calcined Co/Al_2O_3 sample with moderate metal loading, Co is present in the alumina lattice but not as Co_3O_4. However, part of the cobalt was observed to be affected by the sulfiding probably resulting in some Co_9S_8 (Fig. 39c). This result indicates that, although it interacts with the alumina in the calcined state, part of this Co must be located close to the surface of the alumina and is available for sulfiding.

The $Co-Mo/Al_2O_3$ catalyst (Fig. 39d) behaved quite differently, and no Co_9S_8 was formed upon sulfidation. However, a new phase, which has been identified as Co in a MoS_2-like phase ("Co–Mo–S"), was observed (B. S. Clausen et al., 1976b). The chemical state of the Mössbauer atoms in this phase was found to be very sensitive to changes in the gaseous environments, even at room temperature. It was therefore concluded that the Co–Mo–S phase is present in a highly dispersed state (probably two-dimensional).

The above results are most interesting, as some of the recent research on alumina-supported catalysts has been directed toward demonstrating the possible presence of cobalt associated with MoS_2 (e.g., Farragher and Cossee, 1973), or the presence of two distinct phases of Co_9S_8 and MoS_2 exhibiting synergy (e.g., Delmon, 1977). The Mössbauer results are probably the first direct evidence for the presence of a phase containing cobalt, molybdenum, and sulfur. Furthermore, the technique has elucidated several features of this phase.

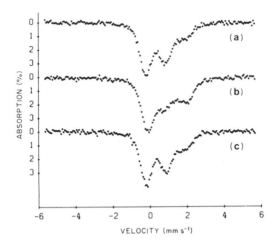

Fig. 40. Room temperature Mössbauer spectra of a ^{57}Co-doped Co–Mo/γ-Al$_2$O$_3$ catalyst: (a) after sulfidation in H$_2$/H$_2$S at 600 K, (b) after reduction in H$_2$ at 600 K, (c) after resulfidation in H$_2$/H$_2$S at 600 K. (Adapted from B. S. Clausen *et al.*, 1976b.)

Since the catalyst structure was found to depend very much on the detailed preparation procedures, the above results should not be taken as generally valid for all types of HDS catalysts. However, the Co–Mo/Al$_2$O$_3$ catalyst discussed here is quite similar to the catalysts used industrially. For such catalysts it is, therefore, important to consider the Co–Mo–S phase. The Co–Mo–S phase was found to be present in these catalysts, but may not be in the major phase in catalysts with much higher Co concentrations (Topsøe *et al.*, 1979b).

In the SiO$_2$-supported Co–Mo catalysts, sulfidation results in the formation of both Co$_9$S$_8$ and a Co–Mo–S-like phase (Fig. 39f), whereas the spectrum of the carbon supported catalyst (Fig. 39g) shows that the cobalt is almost entirely present in the Co–Mo–S-like phase. These results again demonstrate the importance of the support for the type of phases present.

3. Chemical State of Promoter Atoms

The chemical state of the promoter atoms under reaction conditions is of great interest since it may give information on the role of these atoms during hydrodesulfurization.

Unfortunately, this information cannot be obtained directly from the Mössbauer studies of ^{57}Co-doped Co–Mo catalysts because the chemical information obtained concerns the ^{57}Fe atom produced by the decay of ^{57}Co. Since Fe and Co have different chemical and electronic properties, the stable state of ^{57}Fe in the catalysts may be different from that of Co. Moreover, the decay process itself may create unstable but long-living changes in the chemical surroundings of the Mössbauer nucleus (see Section II, D). On the other hand, the chemical state of the promoter atoms in Fe–Mo catalysts is easily determined by use of normal transmission Mössbauer spectroscopy, and comparison of these results with those of the source experiments may, as discussed below, give information on the chemical state of Co in Co–Mo catalysts.

B. S. Clausen *et al.* (1976b) obtained Mössbauer spectra of a ^{57}Co-doped Co – Mo catalyst alternately exposed to sulfiding (H_2/H_2S or H_2/thiophene) and reducing (pure H_2) atmospheres. Some typical spectra are shown in Fig. 40. After sulfiding at 600 K in H_2/H_2S, the spectrum (Fig. 40a) indicated the presence of both Fe^{3+} (low spin) and Fe^{2+} (high spin). After reduction in pure H_2 at 600 K, the room temperature spectrum (Fig. 40b) showed an increase in the Fe^{2+} component and a decrease in the Fe^{3+} component. This change was reversible, as indicated by the spectrum obtained after resulfidation in the H_2/H_2S gas mixture at 600 K (Fig. 40c), and can be expressed by

$$2Fe^{2+} + \square + H_2S \rightleftarrows 2Fe^{3+} + S^{2-}(ads) + H_2 .$$

The spectra obtained with the sample alternatively exposed to H_2/thiophene and H_2 are similar to those shown in Fig. 40. Thus the promoter atoms seem to change their valence reversibly during the catalytical reaction. Sulfur vacancies are generally believed to be the active sites for hydrodesulfurization. Therefore, the present results suggest that the promoting action of cobalt may be associated with this ability to change valence since such valence changes result in creation of sulfur vacancies in the Co–Mo–S surface phase.

As mentioned, these observations concern the chemical state of iron atoms produced by the decay of ^{57}Co in Co–Mo catalysts. However, recent measurements by Mørup *et al.* (1979b) on Fe–Mo catalysts in the sulfided and reduced states yielded spectra with similar Mössbauer parameters. Therefore, Fe and Co atoms were suggested to be present in the same type of sites in the sulfided Fe–Mo/Al_2O_3 and Co–Mo/Al_2O_3 catalysts (i.e., in a MoS_2-like structure). Moreover, since both Fe–Mo and Co–Mo catalysts are active for hydrodesulfurization, it is likely that the promoter atoms behave similarly in the two types of catalysts. The results, therefore, suggest that the valence of Co atoms in Co–Mo catalysts changes reversibly during the catalytic process in the same way as Fe atoms.

The experiments discussed above were all performed in an *in situ* cell allowing measurements to be performed with the catalysts exposed to the appropriate gas mixtures. It was found that exposure to air changed the spectra considerably. In this connection it is noteworthy that many of the measurements reported in literature, using other methods, were not carried out *in situ*. The Mössbauer results, therefore, demonstrate the necessity of carrying out the studies *in situ* and also suggest that many of the results of previous measurements have to be reconsidered.

J. Electroless- and Electrocatalysts

The process of electroless plating of metals onto nonconducting substrates is widely used for manufacture of printed wiring boards in the electronics industry and for decorative plating of plastic parts. The substrate surface must be specially prepared to catalyze deposition of metal from the electroless bath, and two processes are generally used (Goldie, 1968). In the older process the substrate is first "sensitized" in an aqueous solution containing tin chloride and hydrogen chloride and is then "activated" in an acid solution of palladium chloride. The substrate so treated is able to catalyze the deposition of copper from the electroless deposition bath. Qualitatively, the process can be described by the following steps.

The sensitization procedure places Sn^{2+} on the surface. This is followed by an activation step, which involves the reduction of Pd^{2+} to metallic palladium by the oxidation of Sn^{2+} to Sn^{4+}. Electroless metal deposition (for example, of copper) then takes place catalytically on the metallic palladium. This deposition continues even after the palladium has become covered by the deposited metal since the latter itself also serves as a catalyst for the electroless deposition process. Thus the sensitization and activation serve to initiate the autocatalytic deposition in the electroless plating bath.

In the newer electroless metal deposition process, the substrate is treated in a single bath, containing both tin and palladium chloride, and then an "acceleration" bath, before copper deposition. This will be discussed later in this section.

In industrial application of the process selective deposits are often desired (e.g., printed boards). This may be achieved by exposing parts of the sample to ultraviolet light before it is activated in the palladium chloride solution. This treatment destroys the catalytic activity for metal deposition on the UV-exposed parts.

All of the above processes involve a series of complex catalytic and surface chemical reactions, and [119]Sn Mössbauer spectroscopy has provided valuable information for the understanding and improvement of these processes.

The first Mössbauer spectroscopy study of the electroless deposition process was conducted by Cohen *et al.* (1971) and concerned the older process. Kapton (a polyimide) substrates (100 sheets, 2 mil thickness for

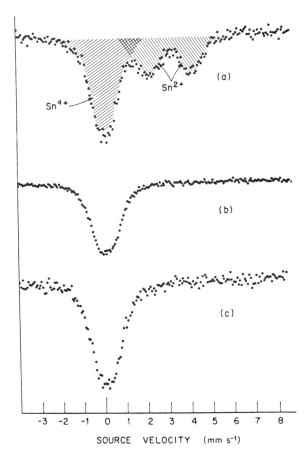

SOURCE   VELOCITY   (mm s⁻¹)

Fig. 41. Mössbauer spectra of tin layers on Kapton foil: (a) sensitized by tin bath, (b) sensitized by tin bath and UV exposed, (c) sensitized by tin bath and then dipped into Pd solution (Cohen *et al.*, 1971).

each sheet) were treated and stacked in the γ-ray beam by these authors. Furthermore, to minimize air oxidation of the surface, the samples were cooled to ≈ 120 K and then evacuated to 10^{-5} Torr after the sensitization or activation steps. The Mössbauer spectrum of the sensitized Kapton shown in Fig. 41a, consisted of a Sn^{2+} doublet and a Sn^{4+} singlet. The Mössbauer parameters of the Sn^{2+} doublet did not correspond to any known oxychloride or hydroxide compounds, and this was interpreted as meaning that the Sn^{4+} and Sn^{2+} ions were both present in one phase. When the sensitized sample was activated in the palladium chloride solution, the intensity of the Sn^{2+} spectral doublet was observed to decrease (Fig. 41c), accompanied by an increase in the spectral area of the Sn^{4+} singlet. This tin oxidation reflects the reduction of palladium according to the reaction scheme: $Sn^{2+} + Pd^{2+} \rightarrow Sn^{4+} + Pd^0$. Thus the Mössbauer spectral changes are consistent with the above qualitative description of the sensitization and activation processes, except for the unexpected presence of Sn^{4+} on the sensitized sample.

The chemical effect of UV irradiation used for selective metal deposition was also studied. The Mössbauer spectrum (Fig. 41b) of a sensitized sample that had been subsequently exposed to UV irradiation showed only the Sn^{4+} singlet. This indicates that the UV radiation has resulted in an oxidation of Sn^{2+} to Sn^{4+}, and without Sn^{2+} the palladium cannot be reduced during the activation procedure.

In the electroless deposition process, it has been found that prolonged exposure of the sensitized sample to the atmosphere (i.e., air and room light) leads to a decrease in the amount of palladium deposited during activation. Mössbauer spectroscopy on sensitized samples before and after exposure to the atmosphere indicated that this treatment led to the expected oxidation of Sn^{2+} to Sn^{4+}. Yet, even prolonged exposure to the atmosphere did not lead to complete oxidation of the sample since the Sn^{2+} spectral doublet was detectable after exposure of the sample to air for one week. Furthermore, the intensity of this Sn^{2+} doublet was not greatly diminished by exposure of the sample to UV radiation or to the palladium chloride activation solution. Thus the Sn^{2+} remaining after prolonged exposure of the sensitized sample to the atmosphere is quite unreactive, which suggests that it is present beneath a coherent "passivation" layer of a Sn^{4+} compound.

The above experiments clearly reveal the general characteristics of the electroless deposition procedure; however, a more complete understanding of the process was obtained from a study of the "structure" of the sensitized surface. Mössbauer spectroscopy was found useful in this respect. The Mössbauer spectrum of a sensitized Kapton sample was essentially the same as the spectrum of the tin in the sensitizing bath itself. This led Cohen and West (1972a) to study, in detail, the properties of the sensitizing solution. For this purpose, the solution was quenched to ≈ 80 K, at which temperature Mössbauer spectra were collected. A Sn^{4+} singlet and a Sn^{2+} doublet were observed in the Mössbauer spectra. It must be noted, however, that the sensitizing solution had a milky color, indicative of the presence of colloidal tin compounds. Therefore, the Mössbauer spectra of the sensitizing bath has contributions from both the tin in the colloidal phase and the tin dissolved in the aqueous solution. In order to separate these contributions, the bath was centrifuged, and Mössbauer spectra were taken for the colloidal particles and for the supernatant (solution). It was thereby found that most of the Sn^{4+} was present in the colloidal phase and that Sn^{2+} was present in both the colloidal phase and in solution. This concentration of Sn^{4+} in the colloidal phase is probably due to the tendency for Sn^{4+} to undergo hydrolysis and polymerization, while the Sn^{2+} (complexed with Cl^-) is quite stable in the acidic sensitizing solution. The continued polymerization of hydrolyzed Sn^{4+} species in the presence of Sn^{2+} then leads to colloidal particles containing mainly Sn^{4+} with some bound (or trapped) Sn^{2+}. The Mössbauer spectral area ratio of the Sn^{4+} to the Sn^{2+} components of the colloidal phase was about 2 : 1.

It was said above that the Mössbauer spectra of the sensitizing bath and of the sensitized Kapton were essentially the same. More correctly, the spectrum of the sensitized Kapton was the same as that for the colloidal phase in the sensitizing bath. Thus the process of sensitizing leads to the adhesion of colloidal tin particles on the surface of the substrate. Moreover, both Sn^{4+} and Sn^{2+} are necessary to achieve sensitization: the Sn^{4+} leads to the necessary colloid formation, and the Sn^{2+} is necessary for the subsequent reduction of palladium in the activation process.

As mentioned earlier, in the newer electroless metal deposition process the sensitizing and activation steps have been modified and combined into a single treatment of the substrate. Sn^{2+} and Pd^{2+} chlorides are mixed together, leading to the production of colloidal particles containing metallic palladium. When the substrate is dipped into this tin–palladium bath, the colloidal particles adhere to the sample surface, and the substrate is ready for the subsequent electroless plating bath. As a first step toward the detailed understanding of the surface chemistry and catalysis involved in this modification of the electroless plating process, Cohen and West (1972b) used ^{119}Sn Mössbauer spectroscopy to study the "structure" of the palladium-containing colloidal particles.

The Mössbauer spectrum (Fig. 42a) at 78 K of the frozen tin palladium bath (containing both the colloidal particles and the complex ions in the aqueous solution) showed three spectral components due to Sn^{4+}, Sn^{2+}, and a SnPd alloy. The latter was identified by comparing its isomer shift with the isomer shift measured for metallurgically prepared Sn_xPd_{1-x} alloys ($x \leqslant 0.15$). Furthermore, after the bath had been centifuged and the colloidal particles were separated from the aqueous solution, the Mössbauer spectrum (Fig. 42b) of the colloidal particles showed only the peaks of Sn^{2+} and the SnPd alloy. Thus the Sn^{4+} species was present only in the aqueous solution. Subsequently, the colloid particle size was varied, and it was found that spectral area ratio of the Sn^{2+} to the SnPd alloy components decreased with increasing particle size (Fig. 42c), indicating that the Sn^{2+} species were present primarily near the surface of the particles. In fact, it was noted that the presence of the Sn^{2+} species is necessary for the preparation of colloidal particles that are stable against

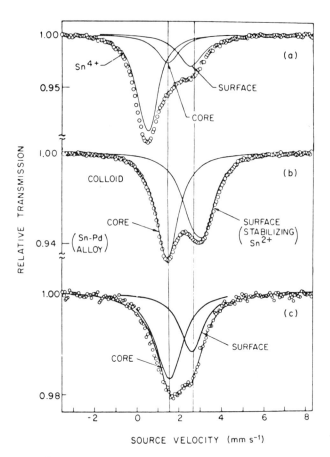

Fig. 42. (a) Mössbauer spectrum of frozen Sn–Pd solution, with decomposition of the spectrum into a constrained Sn^{4+} line and two free Lorentzian lines. (b) spectrum of a centrifugally separated (i.e., particles only) solution, clearly showing two components; (c) similar to (b), but larger particle size, with smaller surface:volume ratio (Cohen and West, 1972b. *Chem. Phys. Lett.* **16,** 128).

coagulation. Thus the model that emerges for the structure of the colloidal particles is the following: the "core" of the particle is a SnPd alloy, and the surface of the particle is covered with a "stabilizing" layer of Sn^{2+} (probably in the form of a tin chloride complex).

The process of colloid formation in the tin palladium bath was studied by Cohen and West (1973). The Sn^{2+} and Pd^{2+} solutions were mixed, and 10 s later the bath was quenched to 78 K, at which temperature a ^{119}Sn Mössbauer spectrum was taken. For solutions with Sn : Pd atomic ratios less than 3, the Mössbauer spectrum was a spectral doublet characteristic of $Sn^{2+} Pd^{2+}$ chloride complexes. However, when the Sn : Pd atomic ratio was larger than 3, additional peaks were seen in the Mössbauer spectrum attributable to the Sn^{2+} chloride complexes. Thus, upon mixing the Sn^{2+} and Pd^{2+} solutions, the first reaction is the formation of $Sn^{2+}Pd^{2+}$ complexes having a Sn : Pd atomic ratio of approximately 3.

A series of Mössbauer spectra were taken for a Sn–Pd bath that had been frozen to 78 K after various "aging" times (where zero time is defined as the moment of mixing of the Sn^{2+} and Pd^{2+} chloride solutions). After 2 h, the Mössbauer spectrum of the initially formed Sn^{2+}–Pd^{2+} complex had been completely replaced by the spectral peaks due to Sn^{4+}, Sn^{2+}, and the SnPd alloy. As discussed above, these peaks are due (i) to colloidal SnPd alloy particles with a Sn^{2+} surface layer and (ii) to Sn^{4+} in aqueous solution. This interpretation of the Mössbauer spectrum was reinforced by exposing the Sn–Pd bath to air. Over a period of 5-day exposure to air, the Sn^{2+} was observed to oxidize to Sn^{4+}, while the Mössbauer spectrum of the SnPd alloy remained unchanged. That is, the treatment with air led to a surface oxidation of the colloidal particles. Moreover, when the Sn^{2+} component disappeared in the Mössbauer spectrum, the colloid particles were observed to coagulate. This is again consistent with the interpretation that Sn^{2+} is present as a stabilizing layer on the colloid particle surface.

Finally, Mössbauer spectra were taken of Kapton substrates that had been dipped into the Sn–Pd bath. Two peaks were seen in this spectrum, and these were due to the SnPd alloy and Sn^{4+}. Undoubtedly, the colloidal particles in the Sn–Pd bath adhered to the Kapton surface, and the Sn^{2+} surface layer on these particles is oxidized to Sn^{4+} by oxygen adsorbed on (or associated with) the Kapton surface.

When this one-step sensitization–activation step in the Sn–Pd bath is used, it is found that an induction period in the electroless plating bath exists, after which electroless metal deposition takes place. Mössbauer spectroscpy was used by Cohen et al. (1976) to study this effect for copper electroless deposition on graphite. Typical spectra are shown in Fig. 43. After the graphite had been dipped into the Sn–Pd bath and subsequently rinsed with water, the Mössbauer spectrum (Fig. 43a) of the treated sample showed peaks due to Sn^{4+} and the SnPd alloy. Actually, the Sn^{4+} peak was much more intense than that of the SnPd alloy. The following interpretation of this spectrum was suggested: (i) after the treatment in the Sn–Pd bath, the graphite surface was covered with both colloid particles and the tin-containing aqueous solution (of low pH) and (ii) rinsing of this surface with water (pH $\sim$ 7) resulted in the precipitation of the Sn^{4+} ions from the adherent tin solution on the graphite surface. It must also be remembered that the Sn^{2+} layer on the SnPd alloy particles had also been oxidized to Sn^{4+} by oxygen adsorbed on the graphite or present in the rinse water. Thus the excess Sn^{4+} on the surface covered the SnPd alloy surface, and the induction period in the electroless plating bath was

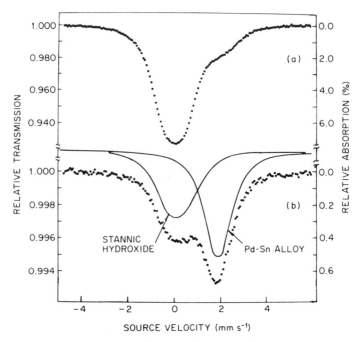

Fig. 43. Mössbauer spectra of graphite samples after processing in: (a) sensitizer and (b) sensitizer followed by accelerator (Cohen *et al.*, 1976).

believed to be due to the time required to remove this Sn^{4+} from the catalytically active SnPd alloy particles.

To test this interpretation, Mössbauer spectra were taken of samples that had been left in the copper electroless plating bath for various periods of time. It was observed that the Sn^{4+} spectral area decreased with time, but surprisingly, the absorption intensity at the SnPd peak position increased with time. This latter result, however, was due to the formation of a CuSn alloy on the surface, and the isomer shift of this alloy is similar to that of the SnPd alloy. Thus the copper electroless plating bath is capable of removing the Sn^{4+} and thereby exposing the catalytically active SnPd alloy particles.

The undesirable induction period and the presence of Sn as an impurity in the copper deposit can be reduced by treating the substrate with a strong base or acid prior to exposure to the electroless plating bath. This strong acid or base is often called an "accelerator" in the electroless metal deposition process. It may be anticipated that the function of this accelerator is to remove the Sn^{4+} from the surface of the SnPd alloy since stannic hydroxide is soluble in both acid and alkali solutions. As expected, the Mössbauer spectra (Fig. 43b) of the treated graphite substrate showed that the accelerator (fluoroboric acid) had, in fact, greatly decreased the intensity of the Sn^{4+} spectral component, relative to that of the SnPd alloy component.

Cohen and Meek (1976a) have suggested that the need for an accelerator can be eliminated if, after the exposure to the Sn–Pd bath, the sample is rinsed with a HCl solution instead of water; in this way, the Sn^{4+} ions (soluble in strong bases or acids) on the surface can be removed by the HCl rinse instead of being precipitated during the water rinse. To test this suggested treatment, these authors obtained a Mössbauer spectrum of graphite substrates after exposure to the Sn–Pd bath and the subsequent HCl rinse. Indeed, this spectrum was essentially the same as that for the colloidal particles in the Sn–Pd bath: only peaks due to the SnPd alloy and the Sn^{2+} layer at the particle surface were observed. Thus the HCl rinse was completely effective in removing the excess Sn^{4+} present on the substrate after treatment in the Sn–Pd bath. Mössbauer spectroscopy also showed that when the HCl-rinsed sample is then rinsed with water, the Sn^{2+} layer on the colloidal particles is oxidized to Sn^{4+}, but this small amount of Sn^{4+} is rapidly removed when the sample is placed in the electroless deposition bath. Cohen and Meek (1976b) have summarized the above Mössbauer spectroscopy studies of the electroless deposition process, and they have compared the results of these studies with information obtained from other physical investigations of the electroless deposition process.

The above series of investigations clearly illustrates that Mössbauer spectral changes can be used to interpret changes made in the operation of the electroless metal depositon process. In a similar manner, it can also be imagined that Mössbauer spectroscopy may be useful in interpreting changes in catalytic activity of electrocatalysts. This was, in fact, successfully done by Appleby et al. (1976) and Appleby and Savy (1976, 1977) for a series of iron phthalocyanines used for the electrochemical reduction of oxygen to water. These authors prepared monomeric and polymeric iron phthalocyanines supported on gold foils or on high surface area carbon. The Mössbauer spectra for these samples, in the presence of oxygen, contained a number of partially overlapping doublets, and the different sample preparations led to significantly different Mössbauer spectra. Since the quadrupole splitting and isomer shift of a spectral doublet provide information about the oxidation state and spin state of the iron ion, these Mössbauer spectral differences (for different sample preparations) were then interpreted in terms of differences in the chemical state of iron. Furthermore, the chemical state of iron in these samples was also studied using optical spectroscopy and x-ray photoelectron spectroscopy. Qualitatively, from the combined results of these spectroscopic techniques, it was concluded that the polymeric samples, in the presence of oxygen, contained primarily high-spin ($S = \frac{5}{2}$) Fe^{3+} and intermediate-spin ($S = \frac{3}{2}$) Fe^{3+}, while the monomeric samples consisted mainly of intermediate-spin ($S = 1$) Fe^{2+}. The catalytic activity for polymeric iron phthalocyanine was much greater than that for the monomeric form of this compound, suggesting that high- and intermediate-spin Fe^{3+} are involved in the oxygen reduction process. The rate-determining step in the process of oxygen reduction is believed to be the cleavage of the O–O bond, and this step would be facilitated by electron transfer to the oxygen molecule (since the added electrons would enter antibonding molecular orbitals). Thus, the presence of Fe^{3+} in the catalytically active samples, in the presence of oxygen, indicates that this electron transfer from iron to oxygen has taken place since in the absence of oxygen, x-ray photoelectron studies showed that the iron in all samples was present primarily as Fe^{2+}. Finally, the origin for the different catalytic activities of the monomeric and polymeric phthalocyanines was suggested as being due to different ligand fields for the iron in these samples. Specifically, it was believed that the ligand field in the monomeric sample was too strong to allow reversible electron transfer between iron (Fe^{2+}) and oxygen.

V. Concluding Remarks

Mössbauer spectroscopy has so many inherent advantages for studies of surface problems and catalysts that the application of the technique will undoubtedly continue to grow. In the field of surface science, the recent move in the direction of studies on well-characterized systems is noteworthy, and the combination of Mössbauer spectroscopy with some of the other UHV surface science tools seems particularly promising. In the field of catalysis, the many important catalyst systems containing Mössbauer isotopes will certainly become further explored. In addition, it is felt that many future studies will be carried out on catalysts that do not contain a Mössbauer isotope, but into which such an isotope can be introduced as a probe. The ease with which *in situ* measurements are carried out should also stimulate many more simultaneous reaction and Mössbauer spectroscopy studies and thereby greatly help to elucidate the nature of the catalytic process. Very few applications of Mössbauer spectroscopy to problems in homogeneous catalysis have appeared. Many such applications do, however, seem possible, and the recent progress in the field of frozen solutions may result in more widespread use of Mössbauer spectroscopy to study this branch of catalysis.

Acknowledgments

The authors are indebted to many people for sending preprints of their work allowing some of the most recent work to be included in the chapter. It is also a great pleasure to acknowledge the many stimulating discussions with and helpful comments from Bjerne S. Clausen, Yuri Maksimov, Helga Stefánsson, Nan Topsøe, and Haldor Topsøe.

References

Aggarwal, K., and Mendiratta, R. G. (1977). *Phys. Rev. B* **16**, 3908.
Aika, K., Hori, H., and Ozaki, A. (1972). *J. Catal.* **27**, 424.
Alvarado, S. F. (1979). Submitted for publication.
Amelse, J. A., Butt, J. B., and Schwartz, L. H. (1978). *J. Phys. Chem.* **82**, 558.
Appleby, A. J., and Savy, M. (1976). National Bureau Standard Special Publication 455 (A. D. Franklin, ed.), p. 241.
Appleby, A. J., and Savy M. (1977). *Symp. Electrochem. Soc., Philadelphia* p. 148.
Appleby, A. J., Fleisch, J., and Savy, M. (1976). *J. Catal.* **44**, 281.
Arnold, D., and Hobert, A. (1968). *Z. Chem.* **8**, 197.
Bacaud, R., Bussiere, P., and Figueras, F. (1979). *J. Phys. Colloq.* **40**, C2-94.
Baetzold, R. C. (1973). *J. Catal.* **29**, 129.
Baetzold, R. C. (1978). *J. Chem. Phys.* **68**, 555.
Barański, A., Bielański, A., and Pattek, A. (1972). *J. Catal.* **26**, 286.
Barański, A., Łagan, M., Pattek, A., Reizer, A., Christiansen, L. J., and Topsøe, H. (1979). *In* "Preparation of Catalysts II" (B. Delmon, P. Grange, P. A. Jacobs, and G. Poncelet, eds.) p. 353. Elsevier, Amsterdam.
Bartholomew, C. H. (1972). Ph.D. Thesis, Stanford Univ.
Bartholomew, C. H., and Boudart, M. (1972). *J. Catal.* **25**, 173.
Bartholomew, C. H., and Boudart, M. (1973). *J. Catal.* **29**, 278.
Bäverstam, U., Ekdahl, T., Bohm, C., Ringström, B., Stefansson, V., and Liljequist, P. (1974a). *Nucl. Instrum. Methods* **115**, 373.
Bäverstam, U., Bohm, C., Ekdahl, T., Liljequist, P., and Ringström, B. (1974b). *In* "Mössbauer Effect Methodology" (I. J. Gruverman, ed.), Vol. 9, p. 259. Plenum Press, New York.
Bergmann, G. (1978). *Phys. Rev. Lett.* **41**, 264.
Bergmann, G. (1979). *Phys. Today* **32** (#4) 25.
Berkowitz, A. E., Lahut, J. A., Jacobs, I. S., Levinson, I. M., and Forester, D. W. (1975). *Phys. Rev. Lett.* **34**, 594.
Berry, F. J. (1978) *Adv. Inorg. Chem. Radio Chem.* **21**, 255.
Bhargava, S. C., Knudsen, J. E., and Mørup, S. (1979). *J. Phys. Chem. Solids* **40**, 45.
Binder, K. and Hohenberg, P. C. (1974). *Phys. Rev. B* **9**, 2194.
Binder, K. and Hohenberg, P. C. (1976). *IEEE Trans. Magn.* **MAG-12**, No. 2, 66.
Bonchev, Z. W., Jordanov, A., and Minkova, A. (1969). *Nucl. Instrum. Methods* **70**, 36.
Bond, G. C., and Turnham, B. D. (1976). *J. Catal.* **45**, 128.
Boudart, M., Delbouille, A., Dumesic, J. A., Khammouma, S., and Topsøe, H. (1975a). *J. Catal.* **37**, 486.
Boudart, M., Topsøe, H., and Dumesic, J. A. (1975b). *In* "The Physical Basis for Heterogeneous Catalysis" (E. Drauglish and R. I. Jaffee, eds.), p. 337. Plenum Press, New York.
Boudart, M., Dumesic, J. A., and Topsøe, H. (1977). *Proc. Nat. Acad. Sci. U.S.* **74**, 806.
Bozso, F., Ertl, G., Grunze, M., and Weiss, M. (1977). *J. Catal.* **49**, 18.
Brill, R., Richter, E. L., and Ruch, F. (1967). *Angew. Chem. Int. Ed. Engl.* **6**, 882.

Bukshpan, S. (1977). *Phys. Lett.* **62A**, 109.

Bukshpan, S., Sonnino, T., and Dash, J. G. (1975). *Surf. Sci.* **52**, 466.

Burton, J. J., and Garten, R. L. (1977). *In* "Advanced Materials in Catalysis" (J. J. Burton and R. L. Garten, eds.), p. 33. Academic Press, New York.

Burton, J. W., and Godwin, R. P. (1967). *Phys. Rev.* **158**, 218.

Bussière, P., Dutartre, R., Martin, G. A., and Mathieu, J. -P. (1975). *C. R. Acad. Sci. Paris Sect. 6* **280C**, 1133.

Carbucicchio, M. (1977). *Nucl. Instrum. Methods* **144**, 225.

Carbucicchio, M., and Trifiro, F. (1976). *J. Catal.* **45**, 77.

Carbucicchio, M., Trifiro, F., and Villa, P. L. (1976). *J. Phys. Colloq.* **37**, C6-253.

Cares, W. R., and Hightower, J. W. (1975). *J. Catal.* **39**, 36.

Chappert, J. (1974). *J. Phys. Colloq.* **35**, C6-71.

Clausen, B. S. (1976). M. S. Theses. L. T. F. II, Technical Univ. of Denmark. Unpublished.

Clausen, B. S. (1979) Ph. D. Dissertation. L. T. F. II, Technical Univ. of Denmark.

Clausen, B. S., Mørup, S., Topsøe, H., Candia, R., Jensen, E. J., Barański, A., and Pattek, A. (1976a). *J. Phys. Colloq.* **37**, C6-245.

Clausen, B. S., Mørup, S., Topsøe, H., and Candia, R. (1976b). *J. Phys. Colloq.* **37**, C6-249.

Clausen, B. S., Topsøe, H., Villadsen, J., and Mørup, S. (1977a). *Proc. Int. Conf. Mössbauer Spectrosc., Bucharest, Romania* (D. Barb and D. Tarina, eds.), Vol. 1, p. 155.

Clausen, B. S., Topsøe, H., Villadsen, J., Mørup, S. and Candia, R. (1977b). *Proc. Int. Conf. Mössbauer Spectrosc., Bucharest, Romania* (D. Barb and D. Tarina, eds.), Vol. 1, p. 177.

Clausen, B. S., Mørup, S., Nielsen, P., Thrane, N., and Topsøe, H. (1979a). *J. Phys. E. Sci. Instrum.* **12**, 439.

Clausen, B. S., Mørup, S., and Topsøe, H. (1979b). *Surf. Sci.* **82**, L589.

Clausen, C. A., and Good, M. L. (1975). *J. Catal.* **38**, 92.

Clausen, C. A., and Good, M. L. (1976). *In* "Mössbauer Effect Methodology" (I. J. Gruverman, ed.), Vol. 10, p. 93. Plenum Press, New York.

Clausen, C. A., and Good, M. L. (1977a). *J. Catal.* **46**, 58.

Clausen, C. A., and Good, M. L. (1977b). *Inorg. Chem.* **16**, 816.

Clausen, C. A., and Good, M. L. (1977c). *In* "Characterization of Metal and Polymer Surfaces" (L. H. Lee, ed.). Vol. 1, p. 65. Academic Press, New York.

Coey, J. M. D. (1971). *Phys. Rev. Lett.* **27**, 1140.

Cohen, R. L., and Meek, R. L. (1976a). *Plat. Surf. Finish.* **63**, 47.

Cohen, R. L., and Meek, R. L. (1976b). *J. Colloid Interface Sci.* **55**, 156.

Cohen, R. L., and West, K. W. (1972a). *J. Electrochem. Soc.* **119**, 433.

Cohen, R. L., and West, K. W. (1972b). *Chem. Phys. Lett.* **16**, 128.

Cohen, R. L., and West, K. W. (1973). *J. Electrochem. Soc.* **120**, 502.

Cohen, R. L., and Wertheim, G. K. (1974). *Methods Exp. Phys.* **11**, 307.

Cohen, R. L., D'Amico, J. F., and West, K. W. (1971). *J. Electrochem. Soc.* **118**, 2042.

Cohen, R. L., Meek, R. L. and West, K. W. (1976). *Plat. Surf. Finish.* **63**, 52.

Collins, R. L. (1968). *In* "Mössbauer Effect Methodology" (I. J. Gruverman, ed.), Vol. 4, p. 129. Plenum Press, New York.

Dalla Betta, R. A., and Boudart, M. (1973). *In Proc. Int. Congr. Catal., 5th* (J. Hightower, ed.), Vol. 2, p. 1329. North-Holland Publ., Amsterdam.

Dalla Betta, R. A., and Shelef, M. (1977). *J. Catal.* **48**, 111.

De Beer, V. H. J., and Schuit, G. C. A. (1976). *In* "Preparation of Catalysts" (B. Delmon, P. A. Jacobs, and G. Poncelet, eds.), p. 343. Elsevier, Amsterdam.

Delgass, W. N. (1976). *In* "Mössbauer Effect Methodology" (I. J. Gruverman, ed.), Vol. 10, p. 1. Plenum Press, New York.

Delgass, W. N., Garten, R. L., and Boudart, M. (1969). *J. Phys. Chem.* **73**, 2970.

Delgass, W. N., Chen, L. Y., and Vogel, G. (1976). *Rev. Sci. Instrum.* **47**, 968.

Delmon, B. (1977). *Am. Chem. Soc. Pet. Div. Reprints* **22**, 503.

Dézsi, I., Nagy, D. L., Eszterle, M., and Guczi, L. (1978). *React. Kinet. Catal. Lett.* **8**, 301.

Dézsi, I., Nagy, D. L., Eszterle, M., and Guczi, L. (1979). *J. Phys. Colloq.* **40**, C2-76.

Dumesic, J. A. (1976). *J. Phys. Colloq.* **37**, C6-233.

Dumesic, J. A., and Boudart, M. (1975). In "The Catalytic Chemistry of Nitrogen Oxide" (R. I. Klimisch and J. G. Larson, eds.), p. 95. Plenum Press, New York.

Dumesic, J. A., and Topsøe, H. (1977). *Adv. Catal.* **26**, 121.

Dumesic, J. A., Topsøe, H., and Boudart, M. (1974). *Proc. Int. Conf. Mössbauer Spectrosc., Krakow, Poland* **1**, 319.

Dumesic, J. A., Topsøe, H., Khammouma, S., and Boudart, M. (1975a). *J. Catal.* **37**, 503.

Dumesic, J. A., Topsøe, H., and Boudart, M. (1975b). *J. Catal.* **37**, 513.

Duncan, S., Owens, A. H., Semper, R. J., and Walker, J. C. (1978). *Hyperfine Interactions* **4**, 886.

Emmett, P. H. (1975). In "The Physical Basis for Heterogeneous Catalysis" (E. Drauglish and R. I. Jaffee, eds.), p. 3. Plenum Press, New York.

Emmett, P. H., and Brunauer, S. J. (1937). *J. Am. Chem. Soc.* **59**, 1553.

Emmett, P. H., and Brunauer, S. J. (1940). *J. Am. Chem. Soc.* **62**, 1732.

Evans, B. J. (1976). *J. Catal.* **41**, 271.

Evans, B. J., and Swartzendruber, L. J. (1975). *AIP Conf. Proc.* **24**, 391.

Fagherazzi, G., Galante, F., Garbassi, F., and Pernicone, N. (1972). *J. Catal.* **26**, 344.

Farragher, A. L., and Cossee, P. (1973). *Proc. Int. Congr. Catal.*, *5th* (J. W. Hightower, ed.), p. 1301. North-Holland Publ., Amsterdam.

Fenger, J. (1969). *Nucl. Instrum. Methods* **69**, 268.

Fenger, J. (1973). *Nucl. Instrum. Methods* **106**, 203.

Forester, D. W. (1973). *Proc. Lunar Sci. Conf. 4th, Suppl. 4 Geochem. Cosmochim. Acta.* **3**, 2697.

Friedt, J. M., and Danon, J. (1972). *Radiochem. Acta.* **17**, 173.

Fulde, P., Luther, A., and Watson, R. E. (1976). *Phys. Rev. B* **8**, 440.

Gager, H. M., and Hobson, M. C. (1975). *Catal. Rev.-Sci. Eng.* **11**, 117.

Gager, H. M., Lefelhocz, J. F., and Hobson, M. C. (1973). *Chem. Phys. Lett.* **23**, 386.

Gallezot, P., Datka, J., Massardier, J., Primet, M., and Imelik, B. (1977). *Proc. Int. Congr. Catal.*, *6th* Vol. 2, p. 696. The Chemical Society, London.

Gallezot, P., Weber, R., Dalla Betta, R. A., and Boudart, M. (1979). *Z. Naturforch A*, **34A**, 40.

Garten, R. L. (1976a). *J. Catal.* **43**, 18.

Garten, R. L. (1976b). In "Mössbauer Effect Methodology" (I. J. Gruverman, ed.), Vol. 10, p. 69. Plenum Press, New York.

Garten, R. L., and Ollis, D. F. (1974). *J. Catal.* **35**, 232.

Garten, R. L., Delgass, W. N., and Boudart, M. (1970). *J. Catal.* **18**, 90.

Garten, R. L., Gallard-Nechtschein, J., and Boudart, M. (1973). *Ind. Eng. Chem. Fundamentals* **12**, 299.

Gibson, M. A., and Hightower, J. W. (1976). *J. Catal.* **41**, 431.

Gibson, M. A., Cares, W. R., and Hightower, J. W. (1977). *Prep. Div. Pet. Chem.* p. 475.

Goldie, W. (1968). In "Metal Coatings of Plastics," Vol. 1. Electrochemical Publ., Middlesex.

Good, C. A. (1972). In "Mössbauer Effect Data Index" (J. G. Stevens and V. E. Stevens, eds.), p. 51. IFI/Plenum Press, New York.

Gradmann, U. (1974). *App. Phys.* **3**, 161.

Gradmann, U., Ullrich, K., Pebler, J., and Schmidt, K. (1977). *J. Magn. Magn. Mater.* **5**, 339.

Graham, M. J., Mitchell, D. F., and Channing, D. A. (1978). *Oxid. Met.* **12**, 247.

Göpel, W. (1978). *Ber. Bunsenges. Phys. Chem.* **82**, 1023.

Haneda, K., and Morrish, A. H. (1977). *Phys. Lett.* **64A**, 259.

Haneda, K., and Morrish, A. H. (1978). *Surf. Sci.* **77**, 584.

Hesse, J., and Rübartsch, A. (1974). *J. Phys. E. Sci. Instrum.* **7**, 526.

Hine, S., Shigematsu, T., Shinjo, T., and Takada, T. (1979). *J. Phys. Colloq.* **40**, C2-84.

Hobert, A., and Arnold, D. (1971). *Proc. Conf. Appl. Mössbauer Effect, Tihany, 1969* p. 325.

Hobson, M. C., and Cambell, A. D. (1967). *J. Catal.* **8**, 294.

Hobson, M. C., and Gager, H. M. (1968). *Proc. Int. Congr. Catal., 4th* p. 28.

Hobson, M. C., and Gager, H. M. (1970). *J. Colloid Interface Sci.* **34**, 357.

Hohenberg, P. C., and Binder, K. (1975). *AIP Conf. Proc.* **24**, 300.

Hoseman, R., Preisinger, A., and Vogel, W. (1966). *Ber. Bunsenges. Phys. Chem.* **70**, 796.

Hrynkiewicz, A. Z., Pustowka, A. J., Sawicha, B. D., and Sawicki, J. A. (1971). *Proc. Conf. Mössbauer Spectrosc., Dresden* p. 631.

Huang, Y. -Y., and Anderson, J. R. (1975). *J. Catal.* **40**, 143.

Ichikava, M., Kondo, T., Kawase, K., Sudo, M., Oniski, T., and Tamaru, K. (1972). *J. Chem. Soc. Chem. Commun.* 176.

Ioffe, M. S., Kuznetsov, B. N., Ryndin, Y., and Yermakov, Y. (1977). *Proc. Int. Congr. Catal., 6th* Vol. 1, p. 131. The Chemical Society, London.

Jensen, E. J., Topsøe, H., Sørensen, O., Krag, F., Candia, R., Clausen, B. S., and Mørup, S. (1977). *Scand. J. Metall.* **6**, 6.

Jones, W., Thomas, J. M., Thorpe, R. K., and Tricker, M. J. (1978). *Appl. Surf. Sci.* **1**, 338.

Keune, W., and Gonser, U. (1971). *Thin Solid Films* **7**, R7.

Keune, W., Lauer, J., and Williamson, D. L. (1974). *J. Phys. Colloq.* **35**, C6-473.

Keune, W., Halbauer, R., Gonser, U., Lauer, J., and Williamson, D. L. (1977a). *J. Appl. Phys.* **48**, 2976.

Keune, W., Halbauer, R., Gonser, U., Lauer, J., and Williamson, D. L. (1977b). *J. Appl. Phys.* **48**, 2976.

Keune, W., Lauer, J., Gonser, U., and Williamson, D. L. (1979). *J. Phys. Colloq.* **40**, C2-69.

Kjeldgård, J., Trumpy, G., Thrane, N., and Mørup, S. (1975). *Proc. Int. Conf. Mössbauer Spectrosc., Cracow, Poland* (A. Z. Hrynkiewicz and J. A. Sawicki, eds.), p. 127.

Knudsen, J. E., and Mørup, S. (1977). *Proc. Int. Conf. Mössbauer Spectrosc.* (D. Barb and D. Tariná, eds.), Vol. 1, p. 205. Bucharest, Romania.

Knudsen, J. E. and Mørup, S. (1980). *J. Phys. Colloq.* **41**, C1-155.

Kölbel, H., and Küspert, B. (1970). *Z. Phys. Chem. Neue Folge* **69**, 313.

Krakowski, R. A., and Miller, R. B. (1972). *Nucl. Instrum. Methods* **100**, 93.

Kriegsmann, H., Öhlmann, G., Scheve, J., and Ulrich, F. -J. (1976). *Proc. Int. Congr. Catal., 6th, London* (G. C. Bond, P. B. Wells, and F. C. Tompkins, eds.), Vol. 2, p. 836. The Chemical Society, London.

Kubo, R. J. (1962). *J. Phys. Soc. Jpn.* **17**, 975.

Kuznecov, L. D. (1963). *Chem. Tech.* **15**, 211.

Kündig, W., Bömmel, H., Constabaris, G., and Lindquist, R. H. (1966). *Phys. Rev.* **142**, 327.

Kündig, W., Ando, K. J., Lindquist, R. H., and Constabaris, G. (1967). *Czech. J. Phys.* **B17**, 467.

Küspert, B. (1970). Thesis, Technical Univ. of Berlin.

Lam, Y. L., and Boudart, M. (1977). *J. Catal.* **50**, 530.

Lam, Y. L., and Garten, R. L. (1978). *Simposio Ibero-Am. Catal., 6th, Rio de Janeiro, Brazil* (in press).

Landolt, M., and Campagna, M. (1978). *Surf. Sci.* **70**, 197.

Lauer, J., Keune, W., and Shinjo, T. (1977). *Physica* **86–88B**, 1409.

Lee, E. L., Bolduc, P. E. and Violet, C. E. (1964). *Phys. Rev. Lett.* **13**, 800.

Lieberman, L. N., Fredkin, D. R., and Shore, H. B. (1969). *Phys. Rev. Lett.* **22**, 539.

Lieberman, L. N., Clinton, J., Edwards, D. M., and Mathon, J. (1970). *Phys. Rev. Lett.* **25**, 232.

Lipka, J., Mørup, S., and Topsøe, H. (1977). *Proc. Int. Conf. Soft Magn. Mater. 3rd, Bratislava, Czechoslovakia* p. 763.

Litster, J. D., and Benedek, G. B. (1963). *J. Appl. Phys.* **34**, 688.

Lo Jacono, M., Cimino, K., and Schuit, G. C. A. (1973). *Gazz. Chim. Ital.* **103**, 1281.

Longworth, G. (1970). *J. Phys. C Metall. Phys. Suppl. No. 1*

Ludwiczek, H., Preisinger, A., Fisher, A., Hoseman, R., Schönfeld, A., and Vogel, W. (1978). *J. Catal.* **51**, 326.

Maksimov, Yu. V., Suzdalev, I. P., Arents, R. A., Goncherov, I. V., Kravtsov, A. V., and Loktev, S. M. (1972a). *Kinet. Katal.* **13**, 1600.

Maksimov, Yu. V., Suzdalev, I. P., and Arents, R. A. (1972b). *Fiz. Tverd. Tela* **14**, 3344.

Maksimov, Yu. V., Suzdalev, I. P., Arents, R. A., Imshevnik, V. K., and Krupyanskii, Yu. F. (1973). *Fiz. Metallov. Metalloved* **36**, 277.

Maksimov, Yu. V., Suzdalev, I. P., Arents, R. A., and Loktev, S. M. (1974a). *Kinet. Katal.* **15**, 1293.

Maksimov, Yu. V., Suzdalev, I. P., Arents, R. A., and Loktev, S. M. (1974b). *Izv. Akad. Nauk. SSSR* **N7**, 1621.

Maksimov, Yu. V., Suzdalev, I. P., Nechitaile, A. I., Gol'danskii, V. I., Krylov, O. V., and Margolis, L. Ya. (1975a). *Chem. Phys. Lett.* **34**, 172.

Maksimov, Yu. V., Suzdalev, I. P., Gol'danskii, V. I., Krylov, O. V., Margolis, L. Ya., and Nechitailo, A. E. (1975b). *Dokl. Akad. Nauk SSSR* **N4**.

Maksimov, Yu. V., Suzdalev, I. P., Arents, R. A., Makarov, E. F., Kushnerev, M. Ya., and Kuznetsov, L. D. (1975c). *Dokl. Akad. Nauk SSSR* **222**, 392.

Mangin, P., Marchal, G., Piecuch, M., and Janot, C. (1976). *J. Phys. E. Sci. Instrum.* **9**, 1101.

Marchal, G., and Janot, C. (1972). *Rev. Phys. Appl.* **7**, 385.

Marshall, S. W., and Wilenzik, R. M. (1966). *Phys. Rev. Lett.* **16**, 219.

Massoth, F. E. (1978). *Adv. Catal.* **27**, 265.

Matsuura, I. (1976). *Proc. Int. Congr. Catal., 6th, London* (G. C. Bond, P. B. Wells, and F. C. Tompkins, eds.), Vol. 2, p. 819. The Chemical Society, London.

McAllister, J., and Hansen, R. S. (1973). *J. Chem. Phys.* **59**, 414.

Meisel, W., and Hobert, H. (1968). *Exp. Tech. Phys.* **16**, 237.

Mercader, R. C., and Cranshaw, T. E. (1975). *J. Phys. F* **5**, L. 124.

Michaelson, H. B. (1977). *J. Appl. Phys.* **48**, 4729.

Minkova, A., and Schunck, J. P. (1975). *C. R. Acad. Sci Bulgaria* **28**, 1171.

Misono, M., Nozawa, Ya., and Yoneda, Yu. (1976). *Proc. Int. Congr. Catal., 6th, London* (G. C. Bond, P. B. Wells, and F. C. Tompkins, eds.), Vol. 1, p. 386. The Chemical Society, London.

Moné, R. (1976). *Int. Conf. Sci. Basis Preparat. Solid Catalysts, Brussels*, 1975 p. 381.

Morrish, A. H., Haneda, K., and Schurer, P. J. (1976). *J. Phys. Colloq.* **37**, C6-301.

Mørup, S., and Topsøe, H. (1976). *Appl. Phys.* **11**, 63.

Mørup, S., and Topsøe, H. (1977). *Proc. Int. Conf. Mössbauer Spectrosc., Bucharest, Romania* (D. Barb and D. Tarina, eds.), Vol. 1, p. 229.

Mørup, S., Knudsen, J. E., Nielsen, M. K., and Trumpy, G. (1976). *J. Chem. Phys.* **65**, 536.

Mørup, S., Clausen, B. S., and Topsøe, H. (1979a). *J. Phys. Colloq.* **40**, C2-78.

Mørup, S., Clausen, B. S., and Topsøe, H. (1979b). *J. Phys. Colloq.* **40**, C2-88.

Mørup, S., Clausen, B. S., and Topsøe, H. (1980a). *J. Phys. Colloq.* **41**, C1-331.

Mørup, S., Dumesic, J. A., and Topsøe, H. (1980b). *In* "Applications of Mössbauer Spectroscopy" (R. L. Cohen, ed.), Vol. 2 p. 1., Academic Press, New York.

Néel, L. (1954). *J. Phys. Radium* **15**, 225.

Nielsen, A. (1968). "An Investigation on Promoted Iron Catalysts for the Synthesis of Ammonia," 3rd ed. Jul. Gjellerups Forlag, Copenhagen.

Notermann, T., Keulka, G. W., Skliarov, A., Maksimov, Yu. V., Margolis, L. Ya., and Krylov, O. V. (1975). *J. Catal.* **39**, 286.

Østergaard, P. (1970). *Nucl. Instrum. Methods* **77**, 328.

Oswald, R., and Ohring, M. (1975). *J. Vac. Sci. Technol.* **12**, 40.

Owens, A. H., Chien, C. L., and Walker, J. C. (1979). *J. Phys. Colloq.* **40**, C2-74.

Ozaki, A., Aika, K., and Hori, H. (1971). *Bull. Chem. Soc. Jpn.* **44**, 3216.

Park, J. Y., and Levenspiel, O. (1975). *Chem. Eng. Sci.* **30**, 1207.

Peev, T. (1976). *Monatsh. Chem.* **107**, 1259.

Pernicone, N., Fagherazzi, G., Galante, F., Garbassi, F., Lazzerin, F., and Mattera, A. (1973). *Proc. Int. Congr. Catal., 5th* (J. W. Hightower, ed.), Vol. 2, p. 1241. North-Holland Publ., Amsterdam.

Pernicone, N., Guacci, U., Cucchetto, M., and Traina, F. (1978). *Symp. Ibero-Am. Catal., Rio de Janeiro, Brazil.*

Petrera, M., Gonser, U., Hasman, U., Keune, W., and Lauer, J. (1976). *J. Phys. Colloq.* **37**, C6-295.

Pipkorn, D. N., Edge, G. K., De Pasquali, G., Drickamer, H. G., and Frauenfelder, H. (1964). *Phys. Rev. A* **135**, 1604.

Ponec, V. (1978). *Catal. Rev. Sci. Eng.* **18**, 151.

Popov, B. I., Sedova, G. N., Kustova, L. M., Plyasova, L. M., Maksimov, Yu. V., and Matveev, A. I. (1976). *React. Kinet. Catal. Lett.* **5**, 43.

Pound, R. V., Benedek, G. B., and Drever, R. (1961). *Phys. Rev. Lett.* **7**, 405.

Prasada Rao, T. S. R., and Menon, P. G. (1978). *J. Catal.* **51**, 64.

Pritchard, A. M., and Dobson, C. M. (1973). *Chem. Phys. Lett.* **23**, 514.

Raupp, G. B., and Delgass, W. N. (1979a). *J. Catal.* **58**, 337.

Raupp, G. B., and Delgass, W. N. (1979b). *J. Catal.* **58**, 348.

Raupp, G. B., and Delgass, W. N. (1979c). *J. Catal.* **58**, 361.

Rees, L. V. C. (1976). *In* "Magnetic Resonance in Colloid and Interface Science" (H. A. Resing and C. G. Wade, eds.), A. C. S. Symp. Series 34. American Chemical Society, Washington, D. C.

Rennard, R. J., and Kehl, W. L. (1971). *J. Catal.* **21**, 282.

Roberts, L. D., Becker, R. L., Obenshain, F. E., and Thomson, J. O. (1965). *Phys. Rev.* **137**, 895.

Roth, S., and Hörl, E. M. (1967). *Phys. Lett.* **A25**, 299.

Rubaskov, A. M., Fabrichny, P. B., Strakhov, B. V., and Babishkin, A. M. (1972). *Russ. J. Phys. Chem.* **46**, 765.

Salomon, D., West, P. J., and Weyer, G. (1977). *Hyperfine Interactions* **5**, 61.

Sancier, K. M., Isakson, W. E., and Wise, H. (1978). *Am. Chem. Soc. Pet. Div. Preprints* **23**, 545.

Sawicki, J. A., Sawicka, B. O., and Stanek, J. (1976). *Nucl. Instrum. Methods* **138**, 565.

Schmidt, F., Gunsser, W., and Knappwost, A. (1975). *Z. Naturforsch.* **30a**, 1627.

Schmidt, F., Gunsser, W., and Adolph, J. (1977). *In* "Molecular Sieves," Vol. II, p. 291. American Chemical Society, Washington, D. C.

Schroeer, D. (1968). *Phys. Lett.* **27A**, 507.

Schroeer, D. (1970). *In* "Mössbauer Effect Methodology" (I. J. Gruverman, ed.), Vol. 5, p. 141. Plenum Press, New York.

Schroeer, D., and Nininger, R. C., Jr. (1967). *Phys. Rev. Lett* **19**, 632.

Schroeer, D., Marzke, R. F., Erickson, D. J., Marshall, S. W., and Wilenzick, R. M. (1970). *Phys. Rev. B* **2**, 4414.

Schuit, G. C. A., and Gates, B. C. (1973). *AIChE J.* **19**, 417.

Schunck, J. P., Friedt, J. M., and Llabador, Y. (1975). *Rev. Phys. Appl.* **10**, 121.

Schwab, G. -M., Block, J., and Schultze, D. (1959). *Angew. Chem.* **71**, 101.

Schwartz, L. H. (1976). *In* "Applications of Mössbauer Spectroscopy" (R. L. Cohen ed.), Vol. 1, p. 37. Academic Press, New York.

Seregin, P. P., Nasredinov, F. S., and Vasilev, L. N. (1978). *Phys. Status Solidi (a)* **45**, 11.

Sette Camara, A., and Keune, W. (1975). *Corros. Sci.* **15**, 441.

Shechter, H. (1977). *Colloq. Int. du CNRS Phases Bidimensionnelles Adsorbées, Marseille, France.*

Shechter, H., Dash, J. G., Mor, M., Ingalls, R., and Bukshpan, S. (1976a). *Phys. Rev. B* **14**, 1876.

Shechter, H., Suzanne, J., and Dash, J. G. (1976b). *Phys. Rev. Lett.* **37**, 706.

Shelef, M., and Gandhi, H. S. (1972). *Ind. Eng. Chem. Prod. Res. Dev.* **11**, 393.

Shinjo, T. (1976). *Trans. IEEE Mag.* **12**, 86.

Shinjo, T. (1979). *J. Phys. Colloq.* **40**, C2-63.

Shinjo, T., Matsuzawa, T., Takada, T., Nasu, S., and Murakami, Y. (1973). *J. Phys. Soc. Jpn.* **35**, 1032.

Shinjo, T., Matsuzawa, T., Mizutani, T., and Takada, T. (1974). *Appl. Phys. Suppl.* 2 Part 2, 729.

Shinjo, T., Hine, S., and Takada, T. (1977). *Proc. Int. Vac. Congr.*, *7th*, and *Int. Conf. Solid Surf.*, *3rd, Vienna* p. 2655.

Shinjo, T., Hine, S., and Takada, T. (1979). *J. Phys. Colloq.* **40**, C2-86.

Sigg, R., and Wicke, E. (1977). *Z. Phys. Chem. Neue Folge* **106**, 65.

Simmons, G. W., and Leidheiser, H. (1976). *In* "Applications of Mössbauer Spectroscopy" (R. L. Cohen, ed.), Vol. 1, p. 85. Academic Press, New York.

Sinfelt, J. H. (1976). US Patent No. 3,953,368.

Skalkina, L. V., Suzdalev, I. P., Kolchin, I. K., and Margolis, L. Ya. (1969). *Kinet. Katal.* **10**, 378.

Solbakken, V., Solbakken, A., and Emmett, P. H. (1969). *J. Catal.* **15**, 90.

Southwell, W. H., Decker, D. L., and Vanfleet, H. B. (1968). *Phys. Rev.* **171**, 354.

Storch, H. H., Golumbic, H., and Anderson, R. B. (1951). *In* "The Fischer-Tropsch and Related Syntheses." Wiley, New York.

Suzdalev, I. P. and Amulyavichus, A. P. (1972). *Zh. Eksp. Teor. Fiz.* **63**, 1758 [*English transl.*: *Sov. Phys. JETP* **36**, 929 (1973)].

Suzdalev, I. P., Gol'danskii, V. I., Makarov, E. F., Plachinda, A. S. and Korytko, L. A. (1966). *Sov. Phys. JETP* **22**, 979.

Suzdalev, I. P., Gen, M. Ya., Gol'danskii, V. I., and Makarov, E. F. (1967). *Sov. Phys. JETP* **24**, 79.

Suzdalev, I. P., Plachinda, A. S., and Makarov, E. F. (1968). *Sov. Phys. JETP* **26**, 897.

Swartzendruber, L. J., and Bennett, L. H. (1972). *Scripta Met.* **6**, 737.

Tachibana, T., Toshie, O., Yoshioka, T., Koezuka, J., and Ikoma, H. (1969). *Bull. Chem. Soc. Jpn.* **42**, 2180.

Topsøe, H., and Mørup, S. (1975). *Proc. Int. Conf. Mössbauer Spectrosc., Cracow, Poland* (A. Z. Hrynkiewich and J. A. Sawicki, eds.), Vol. 1, p. 305.

Topsøe, H., Dumesic, J. A., and Boudart, M. (1973). *J. Catal.* **28**, 477.

Topsøe, H., Dumesic, J. A., Derouane, E. G., Clausen, B. S., Mørup, S., Villadsen, J., and Topsøe, N. (1979a). *In* "Preparation of Catalysts II" (B. Delmon, P. Grange, P. A. Jacobs, and G. Poncelet, eds.), p. 365. Elsevier, Amsterdam.

Topsøe, H., Clausen, B. S., Burriesci, N., Candia, R., and Mørup, S. (1979b). *In* "Preparation of Catalysts II" (B. Delmon, P. Grange, P. A. Jacobs, and G. Poncelet, eds.), p. 479. Elsevier, Amsterdam.

Topsøe, H., Topsøe, N., Bohlbro, H., and Dumesic, J. A. (1980). *Int. Congr. Catal.*, *7th.* Paper A-15 Tokyo, Japan.

Toriyama, T., Kigawa, M., Fujioka, M., and Hisatake, K. (1974). *Proc. Int. Vacuum Congr.*, *6th, Jpn. J. Appl. Phys. Suppl.* 2 Part 1, 733.

Tricker, M. J. (1977). *Surf. Def. Prop. Solids* **6**, 106.

Tricker, M. J., Thomas, J. M., and Winterbottom, A. P. (1974). *Surf. Sci.* **45**, 601.

Tricker, M. J., Freeman, A. G., Winterbottom, A. P., and Jones, J. M. (1976a). *Nucl. Instrum. Methods* **135**, 117.

Tricker, M. J., Winterbottom, A. P., and Freeman, A. G. (1976b). *J. Chem. Soc. Dalton Trans.* 1289.

Tricker, M. J., Cranshaw, T. E., and Ash, L. (1977). *Nucl. Instrum. Methods* **143**, 307.

Tricker, M. J., Ash, L., and Jones, W. (1979). *Surf. Sci.* **79**, L333.

Van Barneveld, W. A. A., and Ponec, V. (1978). *J. Catal.* **51**, 426.

Van der Kraan, A. M. (1972). Thesis, unpublished.

Van der Kraan, A. M. (1973). *Phys. Status Solidi (a)* **18**, 215.

Van Diepen, A. M., Vledder, H. J., and Langereis, C. (1977). *Appl. Phys.* **15**, 163.

Van Diepen, A. M., Vledder, H. J., and Langereis, C. (1978). *Appl. Phys.* (in press).

Van Hardeveld, R., and Hartog, I. (1969). *Surf. Sci.* **15**, 189.

Vannice, M. A. (1975). *J. Catal.* **37**, 462.

Vannice, M. A. (1976). *Catal. Rev.-Sci. Eng.* **14**, 153.

Vannice, M. A. (1977). *J. Catal.* **50**, 228.

Vannice, M. A., and Garten, R. L. (1976). *J. Mol. Catal.* **1**, 201.

Vannice, M. A., Lam, Y. L., and Garten, R. L. (1978). *Am. Chem. Soc. Pet. Div. Preprints* **23**, 545.

Van Wieringen, J. S. (1968). *Phys. Lett.* **26A**, 370.

Varma, M. N., and Hoffman, R. W. (1972). *J. Vac. Sci. Technol.* **9**, 117.

Vedrine, J. C., Dufaux, M., Naccache, C., and Imelik, B. (1978). *J. Chem. Soc. Faraday I* **74**, 440.

Vertes, A. (1975). *Proc. Conf. Mössbauer Spectrosc., 5th*, 1973 p. 179. Czech. Atomic Energy Commission, Praha.

Viegers, M. P. A., and Trooster, J. M. (1977). Phys. Rev. *B* **15**, 72.

Vol'pin, M. E., Novikov, Yu. N., Postnikov, V. A., Shur, V. B., Bayerl, B., Kaden, L., Wahren, M., Dmitrienko, L. M., Stukan, R. A., and Nefed, A. V. (1977). *Z. Anorg. Allg. Chem.* **428**, 231.

Von Eynatten, G., and Bömmel, H. E. (1977). *Appl. Phys.* **14**, 415.

Voorhoeve, R. J. H., and Stuiver, J. M. C. (1971). *J. Catal.* **23**, 243.

Wedd, R. W. J., Liengme, B. V., Scott, J. C., and Sams, J. R. (1969). *Solid State Commun.* **7**, 1091.

Wertheim, G. K. (1971). *Accounts Chem. Res.* **4**, 373.

Weyer, G. (1976). *In* "Mössbauer Effect Methodology" (I. J. Gruverman, ed.), Vol. 10, p. 301. Plenum Press, New York.

Williamson, D. L., Nasu, S., and Gonser, U. (1976). *Acta Metall.* **24**, 1003.

Window, B. (1971). *J. Phys. E. Sci. Instrum.* **4**, 401.

Yatsimirskii, V. K., Piontkovskaya, M. A., Kuz'menko, L. S., Dubovik, M. A., Matyash, I. V., Ivanitskii, V. P., and Kozlova, T. P. (1976a). *Kinet. Katal.* **17**, 1303.

Yatsimirskii, V. K., Girenkova, N. I., and Maksimov, Yu. V. (1976b). *Teor. Exp. Khim. (Kiev)* **12**, 263.

Yoshioka, T., Koezuka, J., and Toyoshima, I. (1969). *J. Catal.* **14**, 281.

Yoshioka, T., Koezuka, J., and Ikoma, H. (1970). *J. Catal.* **16**, 264.

3

Mössbauer Studies
of Surface-Treated Steels

G. P. Huffman

United States Steel Corporation
Research Laboratory
Monroeville, Pennsylvania

I. Introduction

There has been much interest in recent years in the application of Mössbauer spectroscopy to the study of surface layers. Electron reemission Mössbauer (ERM) spectra, obtained by detecting the low energy (typically ~ 1–20 keV) internal conversion and Auger electrons that are reemitted following resonant absorption, are particularly suitable for the study of thin surface layers (~ 10–10^4Å) (Spijkerman, 1971; Huffman, 1976). This technique has been applied by numerous investigators in studies of the oxidation of iron and steel (Simmons et al., 1973; Huffman and Podgurski, 1976; Graham et al., 1977). Thicker surface layers (~ 5–$50\,\mu$m) on bulk samples can be conveniently studied in the so-called "backscatter" geometry by detecting reemitted x rays and gamma rays to obtain x-ray and gamma-ray reemission Mössbauer spectra (Spijkerman, 1971; Flinn, 1974). Conventional transmission Mössbauer spectroscopy is most useful when the surface layer can be selectively isolated from the remainder of the sample, or when it contains an appreciable percentage of the total Fe^{57} in the sample. An additional technique,

189

APPLICATIONS OF MÖSSBAUER
SPECTROSCOPY, VOL. II

in which emission spectra are obtained from Co^{57} doped surfaces, has already been discussed in some detail by Simmons and Leidheiser (1976) in Chapter 3 of Volume I.

Despite the great interest in Mössbauer surface studies, there have been relatively few investigations related to the commercially important topic of protectively coated steels. This chapter will summarize Mössbauer data relevant to four types of coated steel: tinplate, galvanized and aluminized steel and steels treated to produce beneficial oxide coatings. In view of the somewhat limited amount of research on this topic, this chapter is probably best regarded not as a review chapter, but rather as an attempt to assess the potential of Mössbauer spectroscopy in this area and to indicate fruitful directions for future studies.

II. Sn^{119} Electron Reemission Mössbauer Analysis of Tinplate

Since the Fe^{57} and Sn^{119} nuclei both have excellent Mössbauer properties (Cohen, 1976), it is not surprising that there is an abundance of Mössbauer data available on Fe–Sn intermetallic alloys and compounds (Nikolaev et al., 1963; Yamamoto, 1966; Huffman et al., 1969; Trumpy et al., 1970; Vincze and Aldred, 1974). To our knowledge, however, the only publication dealing specifically with tinplate is a Sn^{119} ERM study conducted in this laboratory (Huffman and Dunmyre, 1978).

The experimental techniques used in this ERM investigation are discussed in detail elsewhere (Huffman and Podgurski, 1976; Huffman and Dunmyre, 1978). Briefly, the sample covers the rear window of a flow proportional counter and is irradiated through the front window by a collimated, Pd-filtered beam of 23.8-keV gamma rays emitted by Sn^{119m} in a $BaSnO_3$ matrix. Following resonant absorption of the incident gamma rays, the excited Sn^{119} nuclei in the absorber undergo internal conversion 84% of the time. As summarized in a paper by Yagnik et al. (1974), 84 20-keV internal conversion and 75 3-keV Auger electrons are reemitted for every 100 gamma rays resonantly absorbed. Approximately half of these electrons are emitted in the backward direction, and those that escape from the sample surface produce ionization pulses in an electron-efficient 96% He–4% CH_4 flow gas. The electron energies are such that the bulk of the ERM signal arises from approximately the top 10,000 Å of the sample, which is the range of interest for tinplate with coatings of up to about 1 lb/bb.[†]

[†] lb/bb is the abbreviation for pounds per base box, the standard industrial unit of measurement for tinplate coating thickness. 1 lb/bb = 22.4 g/m^2 and is equivalent to a metallic Sn thickness of 15,380 Å.

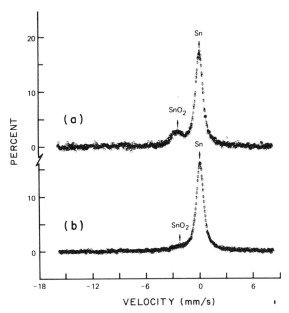

Fig. 1. ERM spectra of a metallic tin foil (a) and of an unmelted 0.5 lb/bb tinplate sample (b).

All samples investigated were commercially prepared tinplate. After plating, but prior to melting, or reflow, the electrodeposited tin coating is gray and nonreflective. The ERM spectra of unmelted tinplate samples are very similar to the ERM spectrum of metallic tin, as illustrated in Fig. 1. The labeled arrows identify the peaks arising from metallic tin (Sn) and stannic oxide (SnO_2). As discussed in a later section, the ERM technique is quite sensitive to oxidized tin, and the oxide layers on these samples are estimated to be very thin (41 Å for the metallic tin foil and 10 Å for the unmelted tinplate sample). Surface oxide layers were present on all tinplate samples investigated.

Melting (reflow) of the electrodeposited tin layer occurs by means of electrical resistance heating of the plated strip to temperatures of approximately 300°C, after which the strip is cooled by passage through a water bath. After melting, the tinplate samples exhibit an additional set of ERM peaks due to the alloy layer ($FeSn_2$). ERM spectra of several samples with tin coatings ranging from 0.1 to 0.7 lb/bb (2.24 to 15.7 g/m²) are shown in Figs. 2 and 3. The spectral components arising from SnO_2, metallic Sn, and $FeSn_2$ are separately indicated in spectrum (a) of Fig. 2. At room temperature $FeSn_2$ is antiferromagnetic, and the average Mössbauer parameters observed for the alloy layer ($H = 24.5 \pm 0.5$ kG, $\delta = -0.07 \pm 0.02$ mm/s,[†] and $\epsilon \sim 0.01 \pm 0.02$ mm/s) are in reasonable

[†]All isomer shifts are measured with respect to metallic Sn at room temperature (β-Sn).

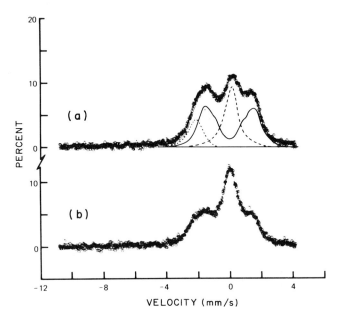

Fig. 2. ERM spectra of tinplate samples with coatings of 0.15 lb/bb (a) and 0.22 lb/bb (b), Sample E. The approximate contributions of the oxide, metal, and alloy layers are indicated separately in spectrum (a) as follows: - - -, SnO_2; – – –, metallic Sn; ——, $FeSn_2$.

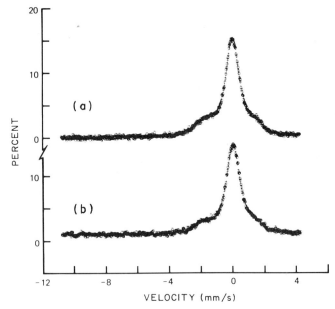

Fig. 3. ERM spectra of tinplate samples with coatings of 0.5 lb/bb, 100 μg Cr/ft^2 (a) and 0.71 lb/bb (b), Sample No. 10.

agreement with those observed for bulk $FeSn_2$ (Nikolaev *et al.*, 1963; Trumpy *et al.*, 1970). However, the average value of the stannic oxide isomer shift[†] for the tinplate samples was found to be -2.27 ± 0.03 mm/s, which is somewhat more positive than the value of -2.50 mm/s observed for SnO_2 (Huffman and Podgurski, 1973). This difference in isomer shift is probably due to the incorporation of some Cr into the SnO_2; the Cr originates from a chromate electrochemical treatment, used to retard oxidation of the tinplate during storage.

The experimental quantities of most interest for present considerations are the percentages of the total ERM spectrum area contributed by the oxide, metallic Sn, and alloy layers. The values of these percentages for all samples investigated are given in Table I. Using a recently developed theory of ERM spectroscopy (Huffman, 1976), these percentages can be processed to determine the separate thickness of the oxide, alloy, and the metallic tin layers. The principal results of this theory, which has also been applied with reasonable success in two studies of oxide layers on metallic iron (Huffman and Podgurski, 1976; 1980; Graham *et al.*, 1977), are summarized below.

The spectral area contributed by the Nth layer is given by

$$A_N = \sum_{j=1}^{M_N} A_{N,j} \tag{1}$$

$$= C \exp(-B_{N-1}) \sum_{j=1}^{M_N} < F_{N,j} > \{\exp(-D_{N-1})H(\beta_{N,j}, Y_{N,j})$$

$$+ 0.89 \exp(-G_{N-1})H(\eta_{N,j}, Y_{N,j})\}, \tag{2}$$

where the Nth layer contributes M_N peaks to the ERM spectrum, and $A_{N,j}$ is the area under the jth peak from layer N. Here C is a constant, and the exponential functions specify the attenuation of the incident gamma rays and outgoing electrons by atomic absorption processes in layers above the Nth layer; specifically,

$$B_{N-1} = \sum_{K=1}^{N-1} \mu_K Z_K, \tag{3}$$

$$D_{N-1} = \sum_{K=1}^{N-1} \rho_K Z_K, \tag{4}$$

and

$$G_{N-1} = \sum_{K=1}^{N-1} \gamma_K Z_K, \tag{5}$$

[†](See footnote on p.191).

TABLE I

ERM Spectrum Area Percentages Contributed by the Oxide (P_{SnO_2}),
Metal (P_{Sn}), and Alloy (P_{FeSn_2}) Layers of Various Tinplate Samples

Sample identification	Tin coating, lb/bb	P_{SnO_2}, (%)	P_{Sn}, (%)	P_{FeSn_2}, (%)
No. 89	0.15[a]	14.8	39.2	46.0
No. 95	0.1	15.2	43.1	41.7
No. 93	0.1	15.2	40.4	44.3
Sample E	0.22[a]	12.4	56.6	31.0
No. 1	0.25	10.3	56.4	33.4
No. 7	0.25	8.9	65.8	25.3
No. 5	0.25	9.7	62.2	28.1
Sample F	0.55[a]	3.5	82.2	14.3
700 Å strip	0.5	2.5	89.0	8.4
100 μg Cr/ft^2	0.5	5.3	82.9	11.8
NCA, good lacquering	0.5	3.1	81.9	15.0
No. 10	0.71[a]	2.8	86.3	10.0

[a] Measured by standard stripping techniques; all other values in column 2 are nominal coating weights.

where μ_K, ρ_K, and γ_K are the attenuation coefficients per unit length in layer K for 23.8-keV gamma rays, 20-keV electrons, and 3-keV electrons, respectively, and Z_K is the thickness of layer K. $\langle F_{N,j} \rangle$ is a more complicated exponential function, which gives the attenuation of the incident gamma rays by Mössbauer absorption. The function H specifies the maximum area contributed by the jth peak of layer N; it has the form

$$H(\beta, Y) = H_0(\beta, Y) - S(\beta, Y), \tag{6}$$

$$H_0(\beta, Y) = (1 - e^{-\beta Y})/(\beta^2 - 1)^{1/2}, \tag{7}$$

where S is a rapidly converging series and is $\ll H_0$ (Huffman, 1976, Appendix).

The parameters $\beta_{N,j}$, $\eta_{N,j}$, and $Y_{N,j}$ are related by simple algebraic expressions to the attenuation coefficients, recoilless fraction, density, and thickness of layer N (Huffman, 1976). Numerical values of all parameters required for ERM analysis of tinplate are derived and tabulated in our original paper (Huffman and Dunmyre, 1978). Since no experimental electron attenuation data were available for the three phases of interest, ρ_N and γ_N were calculated from the empirical expression derived by Cosslett and Thomas (1964):

$$\rho_N = \frac{1.4 \times 10^2 \times d_N}{(E)^{3/2}} (\text{Å})^{-1}, \tag{8}$$

where d_N is the density of phase N and E is the electron energy in electron volts. It should be noted that the validity of using this equation to derive the electron attenuation coefficients required for ERM analysis has not been adequately tested. In particular, the ERM study of magnetite layers on metallic iron by Graham et al. (1977) indicates that the values determined from Eq. (8) for the electrons emitted in the decay of Fe^{57} should be decreased by approximately 50%. Nevertheless, the layer thicknesses determined by ERM analysis of tinplate using the attenuation coefficients of Eq. (8) appeared to be quite reasonable for all samples investigated.

The ERM spectrum area percentages are given by

$$P_N = 100 \, A_N \bigg/ \sum_K A_k = 100 \sum_{j=1}^{M_N} A_{N,j} \bigg/ \sum_K \sum_{l=1}^{M_N} A_{N,l}, \qquad (9)$$

where the sum over K extends over the three layers present, SnO_2, metallic Sn, and $FeSn_2$. Substitution of Eq. (2) into Eq. (9) and insertion of the experimentally determined area percentages (Table I) then gives three nonlinear equations that can be solved for the three layer thicknesses Z_{SnO_2}, Z_{Sn}, and Z_{FeSn_2} (Huffman and Podgurski, 1976; 1980). The results are shown in Table II.

Column 2 gives either the nominal tin coating density or the value determined by electrostripping techniques (McGannon, 1971), and columns 3–5 give the ERM-determined thicknesses of the oxide, metallic Sn, and alloy layers in angstroms. These thicknesses are converted to the conventional units of tin coating density (lb/bb) in columns 6–9. Despite

TABLE II

Layer Thicknesses and Coating Densities Determined by ERM Analysis

Sample identification	W_{tot}, lb/bb	Z_{SnO_2}, Å	Z_{Sn}, Å	Z_{FeSn_2}, Å	W_{SnO_2}, lb/bb	W_{Sn}, lb/bb	W_{FeSn_2}, lb/bb	W_{tot}, lb/bb
No. 89	0.15[a]	35	1153	1586	0.0017	0.075	0.061	0.138
No. 95	0.1	35	1260	1420	0.0017	0.082	0.055	0.139
No. 93	0.1	32	1170	1475	0.0016	0.076	0.057	0.135
Sample E	0.22[a]	30	1975	1420	0.0015	0.128	0.055	0.185
No. 1	0.25	25	1971	1555	0.0012	0.128	0.060	0.189
No. 7	0.25	21	2582	1475	0.0010	0.168	0.057	0.226
No. 5	0.25	23	2316	1477	0.0011	0.151	0.057	0.209
Sample F	0.55[a]	9	5705	3952	0.0004	0.371	0.152	0.523
700 Å strip	0.5	7	6360	2350	0.0003	0.414	0.090	0.504
NCA, 100 μg Cr/sq. ft.	0.5	13	5450	2449	0.0006	0.354	0.094	0.449
NCA, good lacquering	0.5	8	5576	3939	0.0004	0.363	0.151	0.514
No. 10	0.71[a]	7	6950	5651	0.0003	0.452	0.217	0.669

[a] Measured by standard electrostripping technique; all other values in column 2 are nominal coating densitities.

possible inaccuracies in the electron attenuation parameters, the ERM-determined thicknesses and coating densities are seen to be in reasonable agreement with the values in column 2.

III. Mössbauer Studies of Galvanized and Aluminized Steels

Only two published studies have been specifically concerned with galvanized and aluminized steels. Both were conducted by Jones and Denner (1974) and Denner and Jones (1976) and were concerned principally with the kinetics of the hot-dip galvanizing and aluminizing reactions. In both processes, iron reacts with the molten coating metal to form a coatant-rich intermetallic compound that is not magnetically ordered. Reaction rates are therefore readily determined by measuring the nonmagnetic component of Mössbauer transmission spectra obtained from foil samples subsequent to hot-dipping. In the galvanizing study, mild steel foils (0.08% C, 0.3% Mn, 0.1% Si, 0.05% S, 0.04% P) 25-μm thick were dipped into molten zinc at a temperature of 480°C for times ranging from 2 to 60 s. The aluminizing reaction occurs at higher temperature (approximately 700°C) and is considerably more rapid; consequently, the thin iron foils (45 μm; 99.6% Fe) used in this study could be dipped only for relatively short times (< 10 s) without complete conversion to intermetallic phases.

Typical spectra of hot-dipped aluminized iron foils are shown in Fig. 4. Spectra of hot-dipped galvanized samples had a somewhat similar appearance. Although Jones and Denner did not attempt to make conclusive identification of the intermetallic compounds formed, they indicated three Zn-rich phases (gamma, delta, and zeta) as possibilities for the galvanized steel. Graham *et al.* (1980) have investigated the Zn-rich layers on galvanized steel by ERM spectroscopy; this study establishes the presence of a zeta layer, which exhibits a broad single peak, a delta layer, which gives rise to a quadrupole doublet, and a gamma layer, which gives rise to a very asymmetric doublet. In the case of the aluminized samples, the intermetallic spectra appeared rather similar to those obtained by Preston (1972) for Fe_4Al_{13} and $FeAl_6$.

The percentage of iron converted to intermetallic phases α was taken to be proportional to the area under the nonmagnetic peaks near the center of the spectra. This percentage was determined as a function of dipping time t and fitted to a power law of the form

$$\alpha = Kt^n.$$

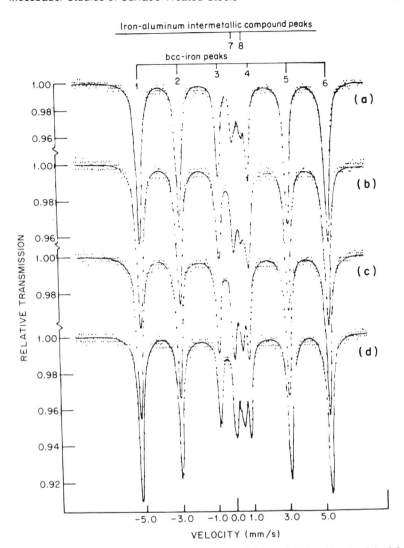

Fig. 4.   Transmission Mössbauer spectra of annealed iron foils hot-dip aluminized for (a) 2 s, (b) 4 s, (c) 7 s, (d) 10 s (Denner and Jones, 1976).

In the hot-dip galvanizing study, the exponent n was determined to be 0.33, while K (evaluated on a weight per unit area basis) had a value of 15.8×10^{-4} g/cm^2/min$^{0.33}$. The aluminizing reaction nearly followed a parabolic rate law, with time exponents of 0.53 at 670°C and 0.48 at 685°C. Assuming an Arrhenius-type temperature dependence, Denner and Jones derived an activation energy of 140 kJ/mole for the rate controlling step in the aluminizing process.

As pointed out by Jones and Denner, reaction rates for hot-dip coating processes can also be determined by metallographic and other techniques. However, Mössbauer spectroscopy appears to be a valuable alternative method, particularly when very short reaction times are of interest.

While not specifically concerned with aluminized steels, a Mössbauer study of solid-state reactions between Fe and Al conducted by Preston (1972) is also of some interest. In this investigation, approximately 250 atomic layers of iron enriched in Fe^{57} was evaporated onto an Al surface in vacuum. Mössbauer spectra obtained immediately after evaporation revealed that approximately $\frac{1}{3}$ of the Fe had reacted with the substrate to form Fe_4Al_{13} and $FeAl_6$. During subsequent annealing treatments at 600°C, it was observed that essentially all of the iron had been incorporated into the Fe_4Al_{13} surface layer after an annealing time of only 5 min. Further annealing at 600°C caused the iron to be liberated from the Fe_4Al_{13} phase and to diffuse into the fcc Al substrate. However, this process was very slow with nearly half of the original Fe_4Al_{13} remaining after 250 hr at 600°C; some typical spectra are shown in Fig. 5. Preston

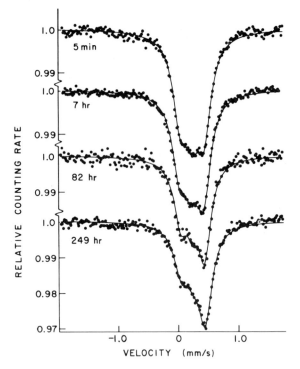

Fig. 5. Room temperature Mössbauer spectra of Fe^{57}-coated aluminum subsequent to annealing at 600°C for the times indicated (Preston, 1972).

speculated that the slowness of the process might be due either to a small interfacial contact area between the Fe_4Al_{13} surface phase and the Al substrate, or to the presence of an oxide film between the Fe_4Al_{13} and the substrate.

IV. Corrosion of Metal-Coated Steels

One of the most interesting aspects of the Sn^{119} ERM study discussed in Section II is the high sensitivity of this technique to very thin surface layers of tin oxide, which arises because the recoilless fraction of SnO_2 is nine times greater than that of metallic Sn ($f_{SnO_2} = 0.45$, $f_{Sn} = 0.05$, at room temperature; Biran *et al.*, 1972). For example, the spectrum area percentages measured for the SnO_2 layers on the metallic Sn and unmelted tinplate samples of Fig. 1 indicate SnO_2 thicknesses of only 41 and 10 Å, respectively, for these two samples. Furthermore, calculation of the expected ERM spectrum area percentages for thin layers of SnO_2 over a thick metallic-Sn substrate using Eqs. (2) and (9) shows that a SnO_2 layer only 270-Å thick will contribute 50% of the spectrum area. A further indication of this sensitivity is found in the ERM study of Bonchev *et al.* (1969), in which strong spectral contributions from $SnBr_2$, $SnBr_4$, and SnO_2 were detected after very short exposures of β-tin of bromine or nitric acid vapors. Electron reemission Mössbauer spectroscopy should therefore be an excellent technique for the investigation of tinplate corrosion. To our knowledge, the only published work in this area is that of Shibuya *et al.* (1978) who utilized Sn^{119} ERM spectroscopy to identify the corrosion products formed by the reaction of metallic tin with various acids. By combining ERM and x-ray diffraction data, the corrosion products formed by 6.7M HNO_3, 5.7M HCl, and 9.0M H_2SO_4 were identified as $SnO_2 \cdot nH_2O$, $Sn_4(OH)_6Cl_2$, and $SnSO_4$, respectively.

Since the principal purpose of most coating processes is to prevent corrosion of the steel substrate, Fe^{57} Mössbauer studies of corrosion products are obviously of interest also. The only published study we are aware of in this area was conducted several years ago in this laboratory (Stoll *et al.*, 1972). It involved Fe^{57} Mössbauer analysis of a black, gelatinous deposit often found in canned seafood, commonly referred to as "sulfide black." Despite the name, no sulfides were identified; instead, the principal constituents observed were vivianite [$Fe_3(PO_4)_2 \cdot 8H_2O$] and ferric oxyhydroxide (a mixture of γ- and β-FeOOH). Mössbauer data obtained over a range of temperatures showed the oxyhydroxide phase to be highly superparamagnetic, corresponding to a mean particle size ~ 20–50 Å; a typical spectrum is shown in Fig. 6. Although Mössbauer analysis of this

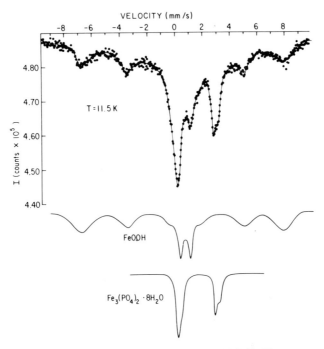

Fig. 6. Mössbauer spectrum of "sulfide black" at 11.5 K. The components of the spectrum arising from oxyhydroxide and vivianite are sketched below the experimental spectrum.

corrosion product proved fairly straightforward, it is worth noting that numerous prior attempts to identify it by other techniques had proved unsuccessful.

There have not yet been any published reports dealing specifically with Mössbauer analysis of the corrosion products of galvanized or aluminized steels. However, there is a significant amount of data available on mixed Fe–Zn and Fe–Al oxides of the type that might be common constituents of such corrosion products, and Graham and Cohen (1976) have reported observation of a spinel of the type $Zn_xFe_{3-x}O_4$ in a corrosion sample removed from a galvanized hot-water pipe. Also worth noting is a study by Montreuill and Evans (1978), in which Mössbauer spectroscopy was used to demonstrate that the Zn-rich gamma phase (Fe_3Zn_{10}) provides the major sacrificial corrosion resistance of amorphous, electrodeposited $Fe_{1-x}Zn_x$ ($0.4 \leqslant x \leqslant 0.6$) alloys in dilute acetic acid solutions. Since a principal concern for all metal-coated steels is the corrosion of the inter-metallic phase and steel substrate underlying a 10–20-μm-thick layer of coating metal, it seems likely that x-ray reemission Mössbauer (XRM) spectroscopy may prove to be a particularly useful technique in this area.

Finally, with regard to other types of coatings, the emission Mössbauer study by Leidheiser *et al.* (1973) of the corrosion of polymer-coated cobalt in NaCl solutions has already been discussed in Chapter 3 of Volume I (Simmons and Leidheiser, 1976).

V. Oxide-Coated Steels

In this section we shall briefly summarize some results of Mössbauer analysis of oxide scales formed on steels subjected to applied oxide coatings. The first example involves some unpublished research (Huffman and Dunmyre, 1971) conducted at this laboratory which involved the analysis of oxide layers formed on slabs of nickel-bearing armor steel (3% Ni), after coating with a slurry containing MgO. The superior properties of this slurry coating, with regard to retardation of oxide scale formation and subsequent descaling behavior, have been documented in a patent by Boggs *et al.* (1974).

Mössbauer spectra of oxide samples removed from the surface of the steel after annealing for 2 and 6 hr at 1300°C in air are shown in Fig. 7.

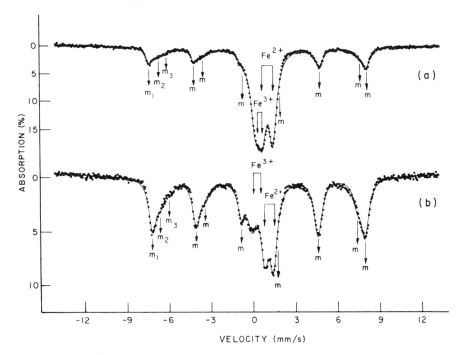

Fig. 7. Mössbauer spectra of Fe–Mg oxides removed from the surface of 3% Ni steel coated with MgO and annealed for 2 hr (a) and 6 hr (b) at 1300°C in air.

The spectra show two overlapping quadrupole doublets arising from Fe^{3+} and Fe^{2+} ions contained in magnesiowustite [$(Fe, Mg)O$], and three sets of magnetic peaks indicated by the arrows labeled m_1, m_2, and m_3. The observed hyperfine fields (470, 431, and 412 kG) indicate that the magnetically ordered constituent of the oxide is an iron-rich magnesioferrite ($Fe_{3-x}Mg_xO_4$). The percentage of the total iron contained in magnesioferrite increases from 26% for the 2-hr anneal to 68% for the 6-hr anneal, demonstrating an increase in the iron content of the oxide scale with increasing annealing time (Levin et al., 1964).

In a more recent study, Vertes et al. (1975) investigated the oxide scales formed on mild steel at 900°C in air before and after immersion in a $Ca(OH)_2$ bath. Wustite, hematite, and magnetite were observed, with wustite being the major component of both scales. However, the relative amount of wustite was considerably greater in the treated than in the untreated scale, suggesting that the $Ca(OH)_2$ treatment inhibits the formation of the higher oxides of iron. Additionally, differences in the wustite spectra of the two scales led the authors to conclude that the wustite phase in the treated scale had a larger vacancy content than that formed on the untreated steel. This difference might also be explained by the formation of some calciowustite on the treated steel.

Finally, although we have limited discussion up to this point to steels to which a coating has been externally applied, it should be noted that there is currently much interest in steels that have a protective oxide coating generated by other means. For example, COR-TEN, a popular structural steel, has superior corrosion resistance due to a protective oxide layer formed during long-term atmospheric exposure; moreover, the oxide layer is sufficiently uniform and attractive that painting is not required. Some spectra obtained in this laboratory, shown in Figs. 8 and 9, demonstrate the usefulness of Mössbauer spectroscopy in the analysis of such oxide layers. The ERM and XRM spectra (Fig. 8) identify the oxide layer as ferric oxyhydroxide; making use of the XRM calibration data of Swanson and Spijkerman (1970), its thickness after three months of atmospheric exposure is estimated to be approximately 13 μm. The low temperature transmission spectra (Fig. 9) of the extracted oxide layer demonstrate its highly superparamagnetic nature and establish that it is a mixture of approximately 67% γ-FeOOH and 33% β-FeOOH. The two oxyhydroxide phases are readily distinguished in the 4.6 K spectrum on the basis of their significantly different magnetic hyperfine fields (Johnson, 1969). A more detailed discussion of the use of Mössbauer spectroscopy to analyze superparamagnetic iron oxides is given in Chapter 1 (Mørup et al., 1980).

Another area of increasing importance involves the development of stainless and other alloy steels that form protective oxide layers in the

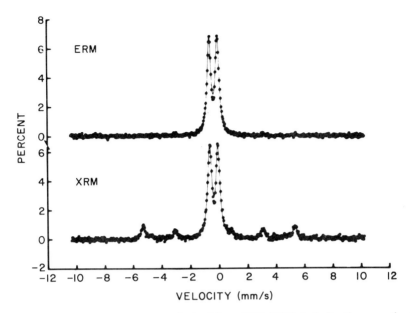

Fig. 8.   ERM and XRM spectra obtained from COR–TEN steel after three months of atmospheric exposure. The percentages of the XRM spectrum area contributed by the oxyhydroxide (74.5%) and by the steel substrate (26.5%) indicate that the average thickness of the oxide layer is 13 μm.

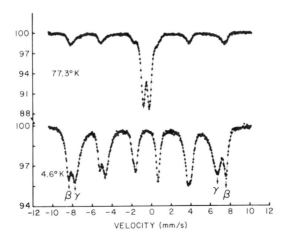

Fig. 9.   Low-temperature transmission spectra of oxide layer extracted from COR–TEN steel after three months of atmospheric exposure, demonstrating its highly superparamagnetic nature. The outermost peaks arising from β- and γ-FeOOH are indicated in the 4.6 K spectrum.

harsh environments associated with coal conversion and other energy-related systems (Foroulis and Smeltzer, 1975). Although there have not been any Mössbauer investigations of the oxidation of any commercial alloys or steels, an interesting study by Channing *et al.* (1977) demonstrates the potential of the Mössbauer technique in this area. In this investigation, Mössbauer spectroscopy was used to determine oxidation rates and oxide scale structures for a series of Fe–Ni alloys oxidized in air at 535 and 635°C. The results are summarized schematically in Fig. 10. Here the block diagrams indicate the relative proportions of different oxide phases formed on alloys of various Ni contents; the phases at the bottom of the diagrams occur at the oxide–alloy interface, and those at the top at the oxide–air interface. The parabolic oxidation rate constants are indicated at the top of Fig. 10, with the oxidation rate constant for the alloy of lowest

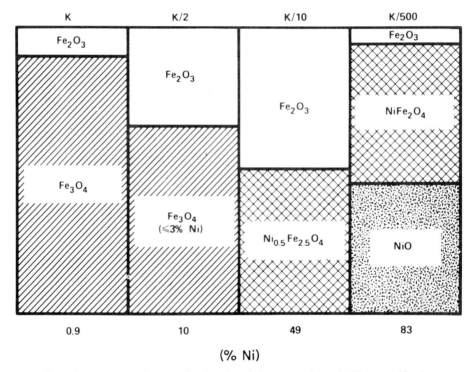

Fig. 10. Schematic diagram showing the relative proportions of different oxide phases observed by Mössbauer spectroscopy in a study of the oxidation of Fe–Ni alloys in air at 535 and 635°C. The oxidation rate constants are indicated across the top of the diagram and the Ni content across the bottom.

Ni content (0.9% Ni) denoted simply as K. As the Ni content of the alloy and oxide scale increases, a dramatic reduction occurs in the oxidation rate. Some possible reasons for this decrease discussed by Channing *et al.* include the much lower diffusion rate of Fe through Ni-containing spinels than through Fe_3O_4, and the slow rate of formation and growth of NiO. In view of the large amount of Mössbauer data available on mixed oxide phases containing two or more elements [see, for example, Sawatzky *et al.* (1969) and Robbins *et al.* (1971)], this type of research would appear to be a very fruitful area for future investigations.

VI. Summary and Conclusions

Mössbauer analysis of surface-coated steels provides much useful information that could not easily be obtained by other techniques. The principal results of the relatively small number of Mössbauer investigations that have been completed in this area are summarized below:

(1) Sn^{119} ERM spectroscopy is a useful new method of determining the thickness of the oxide (SnO_2), metallic Sn, and alloy ($FeSn_2$) layers on tinplate. At present, this method requires some further calibration and is considerably slower than standard techniques. However, it has the advantage of being nondestructive, and there are a number of possible experimental improvements that could make it competitive on a time per sample basis (Huffman and Dunmyre, 1978).

(2) Sn^{119} ERM spectroscopy is particularly sensitive to very thin tin oxide layers. The technique appears to be uniquely well suited for oxidation and corrosion studies of tin since it combines high sensitivity to thin layers with the capability of conclusively identifying the tin corrosion products which are present.

(3) Fe^{57} Mössbauer spectroscopy can be used to identify the products and determine the kinetics of the reactions between steel and molten coating metals and to investigate solid-state reactions of metal-coated steels.

(4) Although there are very few published results, it is clear that Fe^{57} Mössbauer spectroscopy is one of the most powerful methods available for the investigation of the corrosion products of coated steels and protective oxide layers on steel. By combining the various possible Mössbauer modes (ERM, XRM, transmission spectroscopy at various temperatures) it should usually be possible to identify and determine the thicknesses of all iron-containing oxide layers present, even in fairly complex oxide scales. The study of protective oxide layers formed on alloy steels in the corrosive,

high-temperature environment required for coal conversion and other energy-related processes appears to be a very profitable area for future Mössbauer investigations.

References

Biran, A., Yarom, A., Montano, P. A., and Sheckter, H. (1972). *Nucl. Instrum. Methods* **98**, 41.

Boggs, W. E., Linderman, W. A., and Snow, R. B. (1974). Official Patent Gazette, No. 38,827,922, issued August 6.

Bonchev, Zw., Jordanov, A., and Ninkova, A. (1969). *Nucl. Instrum. Methods* **70**, 36.

Channing, D. A., Graham, M. J., and Swallow, G. A. (1977). *J. Mater. Sci.* **12**, 2475.

Cohen, R. L. (1976). *In* "Applications of Mössbauer Spectroscopy" (R. L. Cohen, ed.), Vol. I, pp. 1–32. Academic Press, New York.

Cosslett, V. E., and Thomas, R. N. (1964). *Brit. J. Appl. Phys.* **15**, 883.

Denner, S. G., and Jones, R. D. (1976). *J. Mater. Sci.* **11**, 1778.

Flinn, P. A. (1974). *In* "Mössbauer Effect Methodology"(I. J. Gruverman and C. W. Seidel, eds.), Vol. 9, pp. 245-258. Plenum Press, New York.

Foroulis, Z. A., and Smeltzer, W. A. (eds.) (1975). Proc. of Int. Symposium on "Metal-Gas-Slag Reactions and Processes," Toronto, Canada, Part II, pp. 343–807. The Electrochem. Soc., Inc.

Graham, M. J., and Cohen, M. (1976). *Corrosion* **32**, 42.

Graham, M. J., Mitchell, D. F., and Channing, D. A. (1978), *Oxidat. Met.* **12**, 247.

Graham, M. J., Beaubien, P. E., and Sproule, G. I. (1980). *J. Mat. Sci.* **15**,626.

Huffman, G. P. (1976). *Nucl. Instrum. Methods* **137**, 267.

Huffman, G. P., and Dunmyre, G. R. (1971). Unpublished research.

Huffman, G. P., and Dunmyre, G. R. (1978). *J. Electrochem. Soc.* **125**, 1652.

Huffman, G. P., and Podgurski, H. H. (1973). *Acta Metall.* **21**, 449.

Huffman, G. P., and Podgurski, H. H. (1976). *Oxidat. Met.* **10**, 377.

Huffman, G. P., and Podgurski, H. H. (1980), *Oxidat. Met.* (to be published).

Huffman, G. P., Schwerer, F. C., and Dunmyre, G. R. (1969). *J. Appl. Phys.* **40**, 1487.

Johnson, C. E. (1969). *J. Phys. C. Solid State Phys.* **2**, 1996.

Jones, R. D., and Denner, S. G. (1974). *Scripta Metall.* **8**, 175.

Leidheiser, H., Simmons, G. W., and Kellerman, E. (1973). *J. Electrochem. Soc.* **120**, 1516.

Levin, M., Robbins, C. R., and McMurdie, H. F. (1964), "Phase Diagrams for Ceramists." American Ceramic Society, Columbus, Ohio.

McGannon, H. E. (1971). *In* "The Making, Shaping, and Treating of Steel" (H. E. McGannon, ed.), Chapter 35, pp. 996–1022. U. S. Steel Corp., Pittsburgh, Pennsylvania.

Montreuil, C., and Evans, B. J. (1978). *J. Appl. Phys.* **49**, 1437.

Mørup, S., Dumesic, J. A., and Topsøe, H. (1980). This volume, Chapter 1.

Nikolaev, V. I., Shcherbina, Yu. I., and Yakimov, S. S. (1963). *J. Exp. Theoret. Phys. (USSR)* **45**, 1277.

Preston, R. S. (1972). *Metall. Trans.* **3**, 1831.

Robbins, M., Wertheim, G. K., Sherwood, R. C., and Buchanan, D. N. E. (1971). *J. Phys. Chem. Solids* **32**, 717.

Sawatzky, G. A., van Der Woude, F., and Morrish, A. H. (1969). *Phys. Rev.* **187**, 747.

Shibuya, M., Endo, K., and Sano, H. (1978). *Bull. Chem. Soc. Jpn.* **51**, 1363.

Simmons, G. W., and Leidheiser, H. (1976). *In* "Applications of Mössbauer Spectroscopy" (R. L. Cohen, ed.), Vol. I, pp. 85–125. Academic Press, New York.

Simmons, G. W., Kellerman, E., and Leidheiser, H. (1973). *Corrosion* **29**, 227.

Spijkerman, J. J. (1971). *In* "Mössbauer Effect Methodology" (I. J. Gruverman, ed.), Vol. 7, pp. 85–96. Plenum Press, New York.

Stoll, P. A., Huffman, G. P., Dunmyre, G. R., and Lefkowitz, M. (1972). *J. Food Sci.* **37**, 77.

Swanson, K. R., and Spijkerman, J. J. (1970). *J. Appl. Phys.* **41**, 3155.

Trumpy, G., Both, E., Djega-Mariadassou, C., and Lecocq, P. (1970). *Phys. Rev. B* **2**, 3477.

Vertes, A., Kuzmann, E., Nagy, S., Domonkos, L., and Hegedus, Z. (1975). *Proc. Int. Conf. Mössbauer Spectrosc., Poland-Cracow* (A. Z. Kryniewicz and J. A. Sawicki, eds.), Vol. 1, pp. 317-318; see also *Proc. Int. Conf. Colloid Surf. Sci., Budapest*, 1975 pp. 745-751.

Vincze, I., and Aldred, A. F. (1974). *Phys. Rev. B* **9**, 3845.

Yagnik, C. M., Mazak, R. A., and Collins, R. L. (1974). *Nucl. Instrum. Methods* **114**, 1.

Yamamoto, H. (1966). *J. Phys. Soc. Jpn.* **21**, 1058.

BIOLOGICAL STUDIES

4

Physiological and Medical Applications

D. P. E. Dickson and C. E. Johnson

Department of Physics
University of Liverpool
Liverpool, England

I. Introduction

Physiological and medical applications of Mössbauer spectroscopy come rather more under the category of future possibilities and of potential rather than past achievements. Although all biological Mössbauer

APPLICATIONS OF MÖSSBAUER
SPECTROSCOPY, VOL. II

studies could be regarded as falling into this category, until now relatively few investigations have involved biosystems of a higher complexity than isolated molecules. However, such investigations are now becoming more commonplace, and it is this area that this chapter will explore.

In order to put these uses of Mössbauer spectroscopy into context, it is necessary to consider all levels of investigation, from those on isolated biomolecules to those on complete organisms *in vivo*. It is frequently from the knowledge of relatively simple materials that identification and monitoring of molecules within a more complex system becomes possible.

The problems of sample preparation of biological materials for Mössbauer spectroscopy have been discussed by Lang in Chapter 4 of Volume I. It should be stressed, however, that the preparation of biological samples suitable for Mössbauer spectroscopy is always a problem and that the limitations on what can be achieved arise more frequently as a result of limitations on what is possible in terms of sample preparation than from any lack of ingenuity on the part of the biologists, medics, or Mössbauer spectroscopists. Despite these problems, Mössbauer spectroscopy is now moving into areas where, while the sample preparation problems may be large, the possibilities of assisting in the understanding of biological, physiological, and pathological processes will surely make the effort worthwhile.

Before considering each type of work in more detail, it is worth considering the different areas in which Mössbauer spectroscopy can be applied to the study of systems and processes related to life. Obviously, the divisions are not watertight, but to see how the work can, did, and will develop, it is useful to divide the work into categories.

The work on isolated biomolecules has been on the "simpler" types of molecules isolated from all types of organism. Simple in this context means that while they may have a relative molecular mass of perhaps tens of thousands, they only contain a small number (typically 1, 2, 4, etc.) of atoms of the Mössbauer isotope (most frequently ^{57}Fe). These isolated biomolecules fall into a number of groups and by studying various members of a group a pattern of behavior emerges that is very helpful in understanding the group as a whole. Under this category also comes work on model compounds; these are organic molecules which are much simpler than the biological molecules but which have similar active centers and therefore it is hoped that understanding the model compounds will help in understanding the biomolecules that they emulate. While most of this work has involved the ^{57}Fe Mössbauer isotope, there have also been investigations of iodine compounds using ^{127}I and ^{129}I. The work on isolated biomolecules is now reaching the stage where a complete theoretical understanding of the Mössbauer spectra with computer fitting is possible.

This computer analysis frequently confirms the more qualitative interpretation, which is obtained by inspection of the spectra and gives confidence in using qualitative ideas to interpret the spectra of more complex systems. In essence, our understanding of the isolated biomolecules provides us with a set of fingerprints against which we can compare the spectra from medical or physiological samples.

With the basic information that exists on the simpler molecules, it becomes possible to interpret the Mössbauer spectra of more complex molecules (with typically tens of iron atoms and relative molecular masses of hundreds of thousands) and enzyme systems which may consist of a number of molecules, all of which are required for biological activity. These systems frequently contain a number of species of the Mössbauer isotope giving multicomponent spectra that allow changes in the different components during biological or redox changes to be monitored.

The next level of complexity is that of tissue at which Mössbauer spectroscopy can possibly make the most significant contribution to medical and physiological understanding. To date, there have been relatively few studies in this area. In work on lung samples and blood samples, changes have been observed that result from pathological conditions. Feasibility studies have also shown that Mössbauer spectroscopy could be used for studying bone samples.

Another physiologically relevant application of Mössbauer spectroscopy is to monitor the uptake and use of a Mössbauer isotope by an organism, by, for example, comparing spectra from growth media and from the organism after different periods of growth.

In the vast majority of the investigations carried out so far, the sample (generally the absorber) has consisted of material extracted from an organism and examined under *in vitro* conditions. These are often very different from those of the living organism since low temperatures and high applied magnetic fields can frequently be of great help in the theoretical understanding of the spectra. It can be possible, however, under the right conditions, to obtain spectra from complete organisms, *in vivo*, and this can be very useful in experiments on uptake and metabolism and also in measuring the diffusion rates of various molecules in the organism.

Another rather different use of the Mössbauer effect lies in its sensitivity to small vibrations and movements. This has been used to study the frequency response of auditory mechanisms and the macroscopic motion of complete organisms. In these experiments a Mössbauer nuclide is attached to the system whose motion is to be investigated. This sensitivity also leads to line broadening in the Mössbauer spectra obtained from labeled molecules moving through a membrane. From the measured linewidth, information on this motion can be deduced.

II. Information on Isolated Biomolecules

Many proteins are soluble, and these are generally the easiest biological molecules to isolate. Cells may be broken up by ultrasonic agitation and then filtered and centrifuged to separate the cell walls and membranes from the cytoplasm, which contains the soluble proteins. The different proteins in solution may be separated by chromatography on a column. They may then be washed off and concentrated by ultrafiltration.

The Mössbauer spectra of the isolated proteins (either in frozen solution or when precipitated and/or dried) may be used to characterize the molecule and may enable the molecule to be identified in the cell, where it may occur in the presence of other molecules, either in the cytoplasm or bound to a membrane. The isolated molecules may be studied when frozen in both states of a chemical reaction, e.g., oxidized or reduced, in order to identify their state in the cell. In this way it may be possible to follow the change of a molecule in a metabolic reaction by using the technique of rapid freezing (Bray et al., 1964).

^{57}Fe is the most common Mössbauer isotope. The iron usually occurs in a distinctive environment (or prosthetic group) in several molecules, e.g., the heme group occurs in hemoglogin, myoglobin, cytochromes, etc. The main groups of biological molecules that contain iron at their active centers are shown in Table I. The data on each kind of molecule in the isolated state will be reviewed in the following sections. Other isotopes that may have medical and biochemical applications are ^{127}I and ^{129}I (as in hormones, discussed in Section II, E), ^{67}Zn (in enzymes), ^{133}Cs (substituted for Na^+ and K^+), and ^{151}Eu (since Eu^{2+} can substitute for Ca^{2+} in bones).

TABLE I

Proteins That Contain Iron

Iron Proteins

Heme proteins	Iron-sulfur proteins	Iron transport proteins	Iron storage proteins
hemoglobin	rubredoxins	siderochromes	ferritin
myoglobin	ferredoxins	transferrin	hemosiderin
cytochromes	HiPIPs	conalbumin	gastroferrin
peroxidases	xanthine oxidase	lactoferrin	
catalases		enterobactin	
		mycobactin	

A. Heme Proteins

Heme proteins all contain iron in a fixed and characteristic environment known as the heme group, which is a relatively small planar unit in which the iron is coordinated to four nitrogen ligands in the plane. There may be two further ligands above and below the plane. One of the nonplanar ligands is attached via a nitrogen atom to a chain of amino acids to make up the protein molecule. The heme group is very stable and can exist even when the protein is removed, and it occurs in many biological molecules. The properties of heme compounds have been extensively described in the book by Lemberg and Legge (1949).

The best known examples of heme proteins are hemoglobin, which occurs in blood, and myoglobin, which occurs in muscle and other cells of animals. These proteins are important in the reversible binding of O_2. Hemoglobin, the pigment in red blood cells, was one of the first biological molecules to have its molecular structure determined from x-ray diffraction measurements. In hemoglobin there are four heme groups per molecule; in myoglobin there is only one heme group. The sixth ligand may be varied, and this changes the state of the iron atom. In healthy blood the iron is ferrous. When oxygenated, the iron atom is low spin ($S = 0$) and lies in the heme plane. When deoxygenated, it is high spin ($S = 2$) and lies out of the plane. Abnormal blood may contain ferric iron, which may be high spin ($S = \frac{5}{2}$), as in methemoglobin (where the sixth ligand is a water molecule), or low spin ($S = \frac{1}{2}$), as in hemoglobin cyanide.

A review of work on heme proteins has been given by Lang (1970). Lang and Marshall (1966) made the first systematic measurements of the chemical shifts and quadrupole splittings for each of the four possible spin states of iron. The values found for the common hemoglobin compounds are shown in Table II. The shift indicates the oxidation state; the quadrupole splitting may identify the compound.

At high temperatures (195 or 77 K) the spectrum usually consists of a quadrupole-split doublet, and the spin state may be identified by the shift and splitting. Lang and Marshall (1966) further showed that the magnetic hyperfine splitting observed at low temperatures provides a powerful method of confirming and characterizing the spin states. In oxygenated hemoglobin HbO_2, which occurs in arterial blood, and in hemoglobin carbon monoxide HbCO, the iron is low-spin ferrous and shows no magnetic hyperfine interaction even in the presence of a large external magnetic field. Deoxygenated hemoglobin Hb, which occurs in venous blood, is high-spin ferrous, and the Mössbauer spectrum shows magnetic hyperfine splitting in high fields. In hemoglobin fluoride HiF and methe-

TABLE II

Mössbauer Chemical Shifts and Quadrupole Splittings
of Hemoglobin Derivatives[a]

Oxidation state	Spin	Molecule[b]	Chemical shift[c]	Quadrupole splitting[d]
Fe^{3+}	$S = \frac{5}{2}$	HiF	0.43[e]	0.67[e]
		HiH$_2$O	0.20	2.00
	$S = \frac{1}{2}$	HiCN	0.17	1.39
		HiN$_3$	0.15	2.30
		HiOH	0.18	1.57
Fe^{2+}	$S = 2$	Hb	0.90	2.40
	$S = 0$	HbO$_2$	0.20	1.89
		HbCO	0.18	0.36

[a] From Lang and Marshall (1966).
[b] Hi, ferric hemoglobin; Hb, ferrous hemoglobin.
[c] At 195 K, relative to pure iron metal at room temperature (quoted in mm/s).
[d] At 195 K (quoted in mm/s).
[e] Deduced from 4.2 K values.

moglobin HiH$_2$O, the high-spin ferric atom gives a symmetric six-line hyperfine pattern at low temperatures, corresponding to hyperfine fields of about 50 T.

Diseased blood may contain ferric, usually low-spin, hemoglobin, e.g., hemoglobin cyanide HiCN, azide HiN$_3$, or hydroxide HiOH; in this case the Mössbauer spectrum shows a complex hyperfine splitting at low temperatures. The observation of an EPR spectrum arising from these forms is a possible means of early diagnosis of blood disease. Mössbauer spectra, which take a much longer time to accumulate and are not as sensitive, are less favorable for this kind of use.

The four iron atoms in hemoglobin are associated with different amino-acid chains designated as α and β. Study of these may be valuable in throwing light on the cooperativity of the oxygen uptake, which gets more rapid when more oxygen is bound to the molecule. Mössbauer spectra of dehydrated α and β subunits of deoxyhemoglobin have been measured by Papaefthymiou et al. (1973). The spectra for each subunit show a super-position of a ferrous high-spin ($S = 2$) quadrupole doublet and a ferrous low-spin ($S = 0$) quadrupole doublet. However, the spectrum of anhydrous hemoglobin is not a direct sum of the separate α and β spectra, which is interpreted as showing that the interaction between different subunits in the Hb tetramer ($\alpha_2\beta_2$) results in structural rearrangements near the heme groups that change the relative stability of the low- and high-spin spectra upon dehydration. Since each of the isolated chains shows considerable spin-state mixing, a model previously proposed for anhydrous hemoglobin with the α and β chains in different spin states seems now to be unlikely.

B. Iron–Sulfur Proteins

The iron–sulfur proteins are a group of molecules that are found in many diverse organisms: in plants, animals, and bacteria. They are involved in oxidative electron transfer processes in many different kinds of function, e.g., photosynthesis, nitrogen fixation, digestion, respiration, vision, etc. Their widespread occurrence in nature has only recently been recognized and their importance in biochemistry may well be comparable to that of the better known heme proteins.

Unlike the heme proteins, the iron in the iron–sulfur proteins is not bound in a stable basic structural unit. Instead it is held rather loosely in the chain of amino acids via sulfur atoms. Any attempts to remove the iron chemically are likely to cause the whole molecule to break up. The molecules may contain one, two, three, four or eight iron atoms. The proteins containing three iron atoms have only recently been discovered, and data on them are only just becoming available (Cammack, 1980).

The structural information available on the iron–sulfur proteins indicates that although the various molecules may be different in size, structure, and function, they all contain iron atoms in a similar environment, with the iron atom at the center of four sulfur atoms that form an approximate tetrahedron. The simplest of these molecules are the rubredoxins, since they contain only one iron atom per molecule. Next in order of complexity are the plant-type ferredoxins which contain two iron atoms per molecule. The structure of one of these has only just been determined and hence until now considerable spectroscopic work had been focused on them. More complex still are the bacterial ferredoxins and HiPIPs in which the iron is in cubane units containing four iron atoms and four labile sulfur atoms. For a review of the structure, biochemistry, and function of the iron–sulfur proteins, see Lovenberg (1974, 1977).

Mössbauer effect measurements of the one-iron protein rubredoxin in the oxidized and reduced states have been made in order to find the chemical shifts, quadrupole splittings, and magnetic hyperfine interactions for Fe^{2+} and Fe^{3+} in this tetrahedral sulfur coordination. This effectively calibrates these quantities, i.e., it allows for the effects of covalency of the iron atoms in this environment. This information can then be used in the interpretation of the data on the proteins with two and more iron atoms.

The chemical shift is a very useful parameter for characterizing the formal valences of the iron atoms in these proteins. In Table III the mean chemical shifts in the Mössbauer spectra of the iron–sulfur proteins are listed, together with the assignment of the formal valence states of the iron atoms. However, within the formal valence assignments there are differing degrees of valence electron localization in the different proteins.

In the two-iron centers, the magnetic moments of the two iron atoms in the molecules have been shown to be antiferromagnetically coupled together, and the resulting hyperfine spectrum is very different from that usually observed from a single iron atom. This is a situation unlike that found in inorganic complexes of iron, and it is not easy by any other

TABLE III

Mössbauer Chemical Shifts of the Iron–Sulfur Proteins[a]

Protein	Chemical shift	Formal valences
Fe^{3+} in rubredoxin[b]	0.25	
Fe^{3+} in adrenal ferredoxin[c]	0.26	
Fe^{3+} in spinach ferredoxin[d]	0.26	
Oxidized *Chromatium* HiPIP[e]	0.31	$3\,Fe^{3+}, Fe^{2+}$
Reduced *Chromatium* HiPIP[e]	0.42 ⎱	
Oxidized *B. stearothermophilus* ferredoxin[f]	0.42	
Oxidized *C. pasteurianum* ferredoxin[g]	0.43	$2\,Fe^{3+}, 2\,Fe^{2+}$
Oxidized *Chromatium* ferredoxin[h]	0.41 ⎰	
Super-reduced *Chromatium* HiPIP[i]	0.59	
Reduced *B. stearothermophilus* ferredoxin[f]	0.50, 0.60	
Reduced *C. pasteurianum* ferredoxin[g]	0.57	$1\,Fe^{3+}, 3\,Fe^{2+}$
Reduced *Chromatium* ferredoxin[h]	0.54	
Fe^{2+} in rubredoxin[b]	0.65	
Fe^{2+} in spinach ferredoxin[d]	0.60	

[a] At 77 K, relative to pure iron metal at room temperature (quoted in mm/s).
[b] From Rao *et al.* (1972).
[c] From Cammack *et al.* (1971).
[d] From Rao *et al.* (1971). 195 K value adjusted for second-order Doppler shift.
[e] From Dickson *et al.* (1974).
[f] From Mullinger *et al.* (1975).
[g] From Thompson *et al.* (1974).
[h] From Cammack *et al.* (1977).
[i] From Dickson and Cammack (1974).

method to observe directly the antiferromagnetic coupling between the pairs of iron atoms.

Proteins with a four-iron active center are of two kinds: those with a negative redox potential such as the four-iron ferredoxin from *Bacillus stearothermophilus* and those with a positive redox potential like the HiPIP from *Chromatium*. The iron atoms are again antiferromagnetically coupled in a way similar to that found in two-iron centers, but their Mössbauer spectra are different and therefore may be used to identify these kind of proteins.

Eight-iron ferredoxins such as those from *C. pasteurianum* and *Chromatium* contain two four-iron centers per molecule. Their Mössbauer spectra are similar to those of the four-iron ferredoxins, but there is an additional spin–spin interaction between the two centers, which leads to certain characteristic differences.

In the oxidized state of the four- and eight-iron ferredoxins and the reduced state of HiPIP the iron atoms are equivalent and hence the antiferromagnetic coupling gives rise to a nonmagnetic ground state with a total spin of zero. This gives no magnetic hyperfine interaction, and thus no enhancement of an externally applied field is found in the Mössbauer spectra. This is illustrated in Fig. 1 for the four-iron centers of reduced HiPIP and of three oxidized ferredoxins. Of course, it is not always easy to

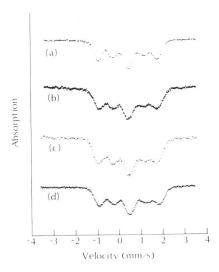

Fig. 1. Mössbauer spectra of proteins with 4Fe–4S centers in the native state, taken at 4.2 K in a perpendicular applied field of 6 T: (a) reduced *Chromatium* HiPIP, (b) oxidized *B. stearothermophilus* ferredoxin (c) oxidized *C. pasteurianum* ferredoxins, and (d) oxidized *Chromatium* ferredoxin. [From Cammack *et al.*, 1977. "Iron–Sulfur Proteins. Vol. III" (W. Lovenberg, ed.). Academic Press, New York.]

distinguish this from a low-spin Fe^{2+} atom, which also has a spin of zero, but the chemical shift should be of help here.

The spectra are also unusual in the four- and eight-iron reduced ferredoxins and oxidized HiPIP in that formally only one of the iron atoms can be reduced, so that the ground state has a total spin of $\frac{1}{2}$. Magnetic hyperfine splitting is now observed, but it is small (compared with uncoupled high-spin Fe^{3+}) and reveals the antiferromagnetic coupling when an external magnetic field is applied.

Xanthine oxidase is another protein that contains eight iron atoms per molecule. The Mössbauer spectra of this (Johnson et al., 1967) are of the two-iron ferredoxin type. This has been interpreted to indicate that xanthine oxidase contains four two-iron centers per molecule.

The Mössbauer effect has been able to contribute to our knowledge of the iron–sulfur proteins by (i) identifying the chemical state of the iron atoms, (ii) by providing information on the localization of the 3d electrons on the iron atoms in the different redox states, (iii) by demonstrating and elucidating the antiferromagnetic coupling between the iron atoms, and (iv) by enabling the different sorts of center to be characterized and identified in new proteins.

C. Iron Transport Proteins

These are molecules that strongly bind iron and that enable iron to be taken up from the environment and incorporated into a cell. They enable the organism to grow and respire. In primitive organisms (bacteria and microbes) these molecules are called siderochromes and have small relative molecular masses of about 1000. In higher organisms (mammals and birds) iron is transported by transferrin, conalbumin, and lactoferrin, which resemble the siderochromes in many ways, but which have much larger relative molecular masses of about 50,000. The soluble transport proteins occur in the extracellular fluid. They generally contain two metal binding sites per molecule. The iron in these proteins is high-spin ferric, and Mössbauer studies have been described in detail by Oosterhuis and Spartalian in Chapter 5 of Volume I.

Most of the work that has been done so far has been on transport proteins from higher organisms, e.g., human and rabbit transferrin and conalbumin from hen egg whites. The ferric ion is coordinated to oxygen in a low-symmetry environment. This is deduced from the ligand field splitting of the kind

$$\mathcal{H} = D\left[S_z^2 - \tfrac{1}{3}S(S + 1)\right] + E\left(S_x^2 - S_y^2\right),$$

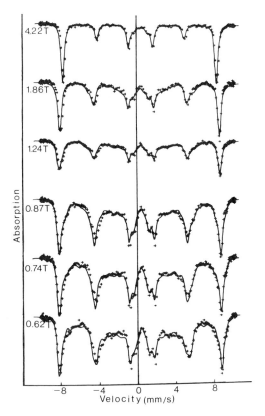

Fig. 2.   Mössbauer spectra of 100% iron-saturated human transferrin at 4.2 K in parallel applied magnetic fields. (From Tsang *et al.*, 1976.)

which is required to explain the spectra. The low symmetry also manifests itself in a characteristic EPR spectrum with $g \sim 4.3$. The Mössbauer spectra show paramagnetic hyperfine splitting at low temperatures when the spin relaxation times are long. This gives a hyperfine field of 50–60 T, with almost symmetrical six-line contributions from each of the three doublets $|m_s = \pm \frac{5}{2}\rangle$, $|m_s = \pm \frac{3}{2}\rangle$, and $|m_s = \pm \frac{1}{2}\rangle$ which are split by the ligand field. Since the iron atoms are paramagnetic, they may be aligned by an external magnetic field. In general, the effect of the field is to sharpen the individual lines in the spectrum and reduce the overall splitting, i.e., to reduce the effective field at the nuclei. This is illustrated in Fig. 2 for human transferrin.

In more primitive organisms the transport proteins may apparently be membrane bound, e.g., in blue–green algae (see Section V, A).

D. Iron Storage Proteins

The iron storage proteins are large molecules that contain appreciable concentrations of iron, up to perhaps 20% by weight. Examples are ferritin, which occurs in human liver and spleen and in plants and fungi, hemosiderin, and gastroferrin. The molecules are approximately spherical with a diameter of typically 12 nm with an inner core of perhaps 7-nm diameter containing the iron. The iron is present in an inorganic form surrounded by a protein skin. In ferritin it consists of ferric oxyhydroxide in the form of finely divided particles, the approximate molecular composition being $(FeOOH)_8 \cdot FeO \cdot PO_4H_2$.

The application of Mössbauer spectroscopy to these proteins has been reviewed by Oosterhuis and Spartalian in Chapter 5 of Volume I. The Mössbauer spectra show the behavior expected of a superparamagnetic (i.e., finely divided antiferromagnetic) material. At low temperatures a simple six-line pattern characteristic of an antiferromagnet is observed, with a hyperfine field of about 50 T. Unlike a bulk antiferromagnet, where the hyperfine field disappears at the Néel temperature T_N, the magnetic

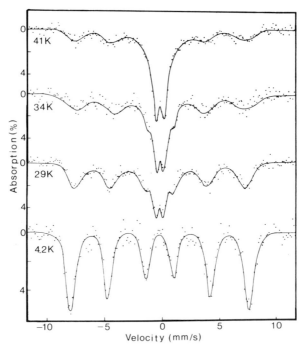

Fig. 3. Mössbauer spectra of horse ferritin at various temperatures. (From Boas and Window, 1966.)

hyperfine pattern transforms to a doublet at a temperature below T_N, where the superparamagnetic spin fluctuations become fast compared with the hyperfine frequency. The temperature at which this occurs and the temperature interval over which it takes place depend on the mean size and the distribution of sizes of the iron particles within the protein. The spectra for horse ferritin are shown in Fig. 3. From spectra of this sort an estimate of the particle size can be made. Spartalian *et al.* (1975) found that the iron clusters in the fungus *Phycomyces blakesleanii* were smaller than those in horse ferritin.

Recent Mössbauer measurements have indicated the presence of a new type of iron storage material in bacteria (Bauminger *et al.*, 1979a, 1980; Dickson and Rottem, 1979).

E. Iodine in Hormones and Other Compounds

Groves *et al.* (1973) have used the ^{129}I Mössbauer effect to study iodine bonding in some iodine-containing hormones. They have made measurements on L-thyroxine (T_4), which is synthesized by the thyroid gland and contains four iodine atoms, and on L-3, 5-di-iodo-tyrosine (DIT), which is believed to be a precursor for T_4 and contains two iodine atoms. Identical Mössbauer spectra were found for the two compounds, showing that the iodine sites in each compound could not be distinguished by Mössbauer spectroscopy.

Oberley *et al.* (1974) have observed significant Mössbauer absorption ($\frac{1}{2}\%$) for these and some similar hormones using the naturally occurring ^{127}I isotope (100% abundant), thus demonstrating the feasibility of using the technique in clinical research. Oberley and Erhardt (1975) have reported ^{127}I measurements on thyroid compounds iodo-tyrosine (MIT) and di-iodo-tyrosine (DIT) and complexes of these compounds T3 and T4. These compounds are important metabolically, and the amounts of them in the blood are intimately related to many of the thyroid diseases. The results obtained are consistent with previous ^{129}I Mössbauer studies.

Haffner *et al.* (1976) have investigated tobacco mosaic virus (TMV) using ^{129}I. The spectra of TMV with both its tyrosine 139 and cysteine 27 amino acids iodinated were compared with the spectra of di-iodo-tyrosine (DIT). The effects of protonation on these compounds was also studied. The Mössbauer parameters of DIT were found to be sensitive to protonation with the DIT in TMV becoming protonated at pH 8. The relative amounts of iodine in tyrosine and cysteine in TMV were found to be stoichiometric. The authors interpret the results as proving the existence of a hydrogen bond involving the hydroxyl group of tyrosine 139, and excluding the formation of a cysteine bridge at cysteine 27 in TMV.

F. Vitamin B₁₂

Vitamin B_{12} and its derivatives (cobalamins) contain cobalt in an environment somewhat similar to that of iron in the heme proteins. If ^{57}Co is incorporated into the molecule, Mössbauer emission spectra may be measured against a single-line absorber. The Mössbauer gamma rays are emitted from an ^{57}Fe nucleus occupying a cobalt site in the molecule.

Nath *et al.* (1968) have demonstrated that the resulting spectra are not drastically affected by multiple charge states associated with the radioactive decay. They found that the Mössbauer parameters of cobalamins are similar to those of iron compounds with similar structures, i.e., vitamin B_{12} and cytochrome c give rise to similar spectra.

In a more recent work Inoue and Nath (1977) have obtained emission Mössbauer spectra from vitamin B_{12} in three redox states with benzimidazole base and with this base replaced by water. The differences in chemical shift and quadrupole splitting observed in the spectra are discussed in terms of the electronic structure and bonding of the ^{57}Fe atom on the cobalt site. These spectra yield fingerprints for the various forms of vitamin B_{12}, which could be useful in identifying the intermediates in enzymatic reactions involving this vitamin.

III. Applcations to Molecular Systems

The molecules discussed in Section II occur in the cells of living organisms, from bacteria to plants and animals. The components of each cell are organized into mitochondria, ribosomes, chloroplasts, membranes, cell walls, etc., depending on the complexity and type of organism. Cells are classified into two broad classes known as prokaryotic (primitive) and eukaryotic (from higher organisms).

A. Enzyme Systems

An enzyme is a protein that acts as a catalyst in a biochemical reaction. It is not used up in the reaction but changes its state, usually from oxidized to reduced or vice versa, and also its conformation as the reaction occurs. Each enzyme is specific to a particular reaction, and this is usually achieved by means of the conformation of the molecule providing easy access for the substrate. In an enzyme system two or more molecules function together, each needing the other for the reaction to occur. It is necessary to study the molecules together in order to elucidate their nature, function, and interrelationship.

The use of the Mössbauer effect in studying enzyme systems has been extensively reviewed by Debrunner in Chapter 6 of Volume I, so we shall mention only a few essential features here. The simplest systems to study are those involving soluble proteins.

The measurements on the putidaredoxin cytochrome P450 system have been described by Debrunner. Cytochrome P450 is a heme protein, which binds molecular oxygen and inserts one atom of it into a specific enzyme, putidaredoxin, which is a two-iron ferredoxin. This particular reaction mechanism occurs in the hydroxylation of camphor in the bacterium *Pseudomonas putida*, and it can be considered as a model for similar enzyme systems that are found in mammalian cells. The reaction was studied by taking Mössbauer spectra of the reaction intermediates, which were frozen at various stages of the cycle.

Another enzyme system containing iron is nitrogenase, which is the enzyme system catalyzing the fixing of atmospheric nitrogen to form ammonia. It consists of two proteins, an iron protein and a molybdenum–iron protein, whose active centers contain many iron atoms. Mössbauer spectroscopy has been used to elucidate the nature of the active centers in nitrogenase by comparison with the spectra of isolated ferredoxins and also by direct analysis (e.g., Smith and Lang, 1974; Münck et al., 1975; Huynh et al., 1979).

Among enzymes that are of great interest for future studies is hydrogenase. Hydrogenase is an enzyme that produces hydrogen in a form that can reduce nitrogen to ammonia. As such, it has a function similar to that of nitrogenase or any other nitrogen fixing enzyme. Hydrogenase is believed to contain ferredoxinlike active centers, though how many and of what kind is not yet known. The Mössbauer spectra could enable these to be identified.

B. Reaction Centers

Photosynthetic organisms fall into two groups: (i) photosynthetic bacteria, which use reduced sulfur compounds or organic substrates as an electron donor and (ii) blue–green algae (prokaryotes), eukaryotic algae, and higher plants, which use water as an electron donor. The basic process is similar in the two groups and involves the absorption of light, followed by the photooxidation of a chlorophyll molecule, which is in a specialized environment known as a reaction center. Two kinds of reaction centers have been recognized: Photosystem II and Photosystem I, which, in sequence, transfer an electron from water to NADP (nicotinamide adenosine diphosphate). Photosystem I uses light of wavelength 720 nm, while Photosystem II can only work with light of wavelength 680 nm or less.

In photosynthetic organisms the light-absorbing pigment (chlorophyll), the electron transport chain, and the ATP (adenosine triphosphate) synthesizing systems are associated with insoluble membrane components of the cell. Hence it is difficult to isolate the component molecules in the reaction centers, and conventional methods of analysis and structure determination cannot readily be applied.

It is not known whether iron is involved in Photosystem II. Photosystem I contains iron thought to be bound in a ferredoxinlike center to which electrons can be transferred from the chlorophyll. It was initially thought that this iron center is the primary electron acceptor, but recent EPR evidence suggests that it is a subsequent acceptor later in the electron

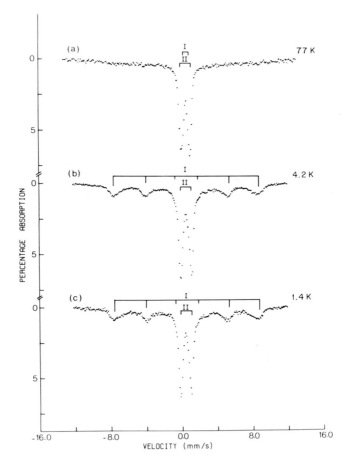

Fig. 4. Mössbauer spectra of *C. fritschii* membranes at (a) 77 K, (b) 4.2 K, and (c) 1.4 K. The stick spectra indicate the two components. (From Evans *et al.*, 1977.)

transport chain. The Mössbauer effect offers the possibility of obtaining data on the iron-containing molecule.

Debrunner *et al.* (1975) have studied reaction centers from *Rhodopseudomonas spheroides*. Both nonreduced and reduced states give a spectrum indicative of high-spin Fe^{2+}. It was suggested that in these bacteria it is unlikely that the iron is the primary electron acceptor of the reaction centers.

Evans *et al.* (1977) have made Mössbauer effect measurements on membrane fragments from the blue–green algae *Chlorogloea fritschii* and *Anacystis nidulans*. These measurements were only performed on oxidized samples. The spectra for *C. fritschii* samples, which are shown in Fig. 4, consist of two components, labeled I and II. Component I is discussed in Section V, A. Component II is shown in detail in Fig. 5. This component consists of a quadrupole-split doublet with a separation $\Delta E_Q = 0.66 \pm 0.02$ mm/s, which is independent of temperature. Measurements in an applied magnetic field (Fig. 5b) show that the hyperfine field at the nuclei is zero, indicating that the iron is in a nonmagnetic state, e.g., low-spin ferrous or a

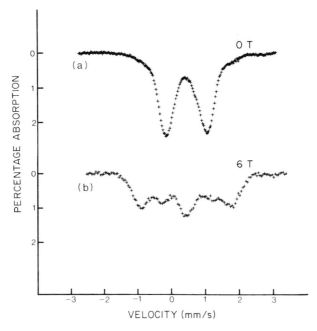

Fig. 5. Mössbauer spectra of *C. fritschii* membranes at 4.2 K in zero field and in a perpendicular applied field of 6 T showing component II. The lower spectrum should be compared with spectra of Fig. 1.

spin-coupled polyatomic center similar to that found in soluble ferredoxins (cf. Fig. 1). The chemical shift ($\delta = 0.44$ mm/s relative to metallic iron) is consistent with the latter. However, it does not exclude the former, although the shift is high for low-spin ferrous (the only well-characterized low-spin ferrous biological compound with a chemical shift greater than 0.25 mm/s is ferrocytochrome c). Both the chemical shift and quadrupole splitting are similar to those of the ferredoxins (e.g., oxidized *B. stearothermophilus* has $\delta = 0.42 \pm 0.01$ mm/s and $\Delta E_Q = 0.98 \pm 0.02$ mm/s—see Table III).

Subsequent work (Evans *et al.*, 1979) has shown the ferredoxinlike material to be associated with photosystem I reaction centers isolated by detergent treatment of the crude membrane fragments. Measurements on chemically reduced samples have yielded spectra very similar to those of reduced four-iron ferredoxins, which confirms that the iron is in a ferredoxinlike center (Evans *et al.*, 1979). This is consistent with the EPR signals with $g \sim 1.94$ which have been obtained from reduced samples.

IV. Applications at the Tissue Level

A. Lung Samples

[57]Fe Mössbauer spectroscopy can be used to monitor the iron content of lung samples. The iron may occur as a result of occupational exposure to iron-containing air, as in the case of welders and hematite workers. Iron buildup can also occur pathologically as in the case of diseases such as hemosiderosis, pneumoconiosis, and bronchitis. Studies in both these areas have been made using Mössbauer spectroscopy, but to date the work has been mainly of a preliminary and investigatory nature.

The study of human subjects can be limited and difficult to interpret because of the lack of a detailed knowledge of their exposure history. It can therefore be useful to investigate lung samples from experimentally exposed animals as well as occupationally exposed human subjects. Rush *et al.* (1975) have made measurements on lyophilized samples of lung tissue taken from rats experimentally exposed to arc weld fume for approximately 4 hr. The spectra from an unexposed control, from the lung of a rat killed seven days after the exposure, and from the weld fume itself are shown in Fig. 6. The spectra show a temperature dependence that is characteristic of a superparamagnetic material and have parameters corresponding to Fe_2O_3, the most likely component of the weld fume. By comparison with the spectra of Fe_2O_3 shown by Kündig *et al.* (1966), the

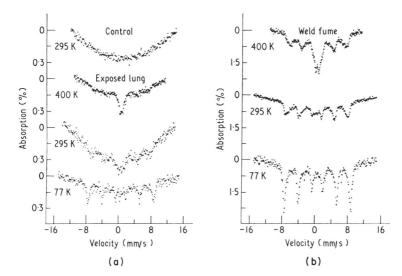

Fig. 6. Mössbauer spectra: (a) from samples of lung taken from a rat which had been exposed to weld fume and from a control and (b) from the weld fume. (From Rush *et al.*, 1975. *Phys. Med. Biol.* **20**, 128. Copyright by The Institute of Physics.)

average particle size is estimated to be approximately 18 nm in the fume and appreciably less in the lung sample.

Johnson (1971) has shown spectra from dried samples of human lung material from both a healthy subject and a victim of hemosiderosis (Fig. 7). The healthy lung shows a weak signal corresponding to the presence of hemoglobin. The diseased lung shows a strong absorption with hyperfine splitting appearing at low temperatures as a result of superparamagnetism. The spectra are consistent with the iron material in the lung being in the form of an iron storage protein like hemosiderin or ferritin. This information had not been obtained previously by other methods. It should be noted that the subject of this work would not have had any occupational exposure to iron, the excess iron in the lung samples being produced metabolically.

In another study Mössbauer spectra were obtained from the lung material of various human subjects (Guest, 1976, 1978). The spectra of the storage iron observed in some of these samples are again similar to the reported spectra of ferritin or hemosiderin. Significant amounts of storage protein are observed in lungs with both high and low dust contents. In some cases the Mössbauer spectra show components resulting both from

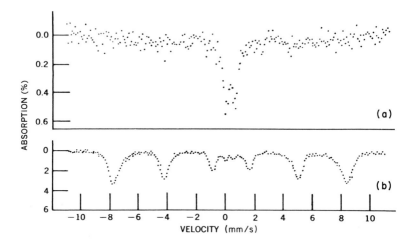

Fig. 7. Mössbauer spectra of human dried lung material at 4.2 K: (a) from a healthy control and (b) from a hemosiderosis victim. (From Johnson, 1971.)

exposure to iron-containing aerosols and a metabolic buildup of iron. In addition to Mössbauer spectroscopy, the samples were analyzed for total dust, blood, total iron, and blood iron. The samples used for Mössbauer analysis contained between 250 and 1000 mg of dried lung.

The absorption due to hemoglobin, the only component in the spectra of normal healthy lung material, was extremely small and could be effectively neglected in the spectra showing large effects. A sample of hemosiderosis lung material showed spectra very similar to those of Johnson (1971), indicating the presence of large numbers of macrophages containing hemosiderin in the interstitium of the lung. A sample of acutely bronchitic lung also showed evidence for large quantities of iron storage protein but with a slightly larger particle size than in the case of the hemosiderosis lung. A sample of pneumoconiotic coal worker's lung again indicated the presence of large amounts of storage protein.

The spectra of lung material from a hematite worker showed two components, one due to the presence of large quantities of hematite and the other due to iron storage protein. Measurements at different temperatures enable the two components to be easily identified. The hematite is the dominant component in these spectra. The spectra from the lung of a foundry worker showed the same two components with the iron storage protein giving a much larger contribution than the hematite. Lung samples from a coal worker with complicated pneumoconiosis also showed the same components. In this lung the PMF lesion was separated from the rest and was analyzed separately. Both parts gave identical spectra, which have a major component resulting from storage protein and an additional small hematite component. This work shows that Mössbauer spectroscopy can be useful in distinguishing between blood iron, storage iron, and inorganic iron and can monitor the relative proportions resulting from pathological and occupational factors.

In addition to the experiments that have been described above, which are all in essence exploratory and suggest that further work in the same direction could be very useful, there are a number of other possibilities. Neutron activation analyses have been made of lung samples, taken at varying intervals after exposure, from rats that were experimentally exposed to arc weld fume for periods of several hours. It was noted that the concentrations of various elements deposited in the lung (of which iron occurs in the highest concentration) decrease with time at a rate that can be described by multicomponent exponentials (Hewitt and Hicks, 1973). Iron is rapidly eliminated over the first few days but much more slowly in the longer term. A controlled series of experiments, using Mössbauer spectroscopy to monitor the chemical and physical state of the iron as a function of time, could be very useful in elucidating the processes involved in its elimination from the lung.

Another possibility would be to monitor changes in the iron content of the lung that occur pathologically during the course of diseases such as hemosiderosis and pneumoconiosis. The relatively large absorptions observed in the experiments described above indicate that adequate measurements could be made on the small samples obtainable from a biopsy.

B. Blood Samples

There have been many Mössbauer studies on the properties of hemoglobin in its various forms. The results are outlined in Section II, A, with a detailed account given in Chapter 4 of Volume I. Relatively few investigations have used blood samples rather than isolated protein, and there has been surprisingly little work with a medical objective. Nevertheless, Mössbauer spectroscopy should be particularly helpful in the study of blood cells in order to understand the changes that occur under pathological conditions.

Kellershohn et al. (1976) have used Mössbauer spectroscopy to study the effects of x radiation and heat treatment on arterial red cells. Before treatment the spectra are those of oxyhemoglobin, which has already been well characterized. The effects of different dosages of x radiation are to produce three additional iron species (Fig. 8), one of which has Mössbauer parameters corresponding very closely to those of deoxyhemoglobin. The other two components (A and B) have parameters corresponding to a high-spin ferric compound in a rhombic environment (confirmed by EPR measurements) and a high-spin ferric heme compound (similar to methemoglobin). The relative proportions of the different iron species are strongly dependent on the dosage. The oxyhemoglobin component gradually diminishes as the dose rises, and it disappears completely at about 12 Mrad, at which dose the spectrum of the high-spin ferric heme compound has also practically disappeared. An earlier study by Kellershohn *et al.* (1975) has shown that the irradiation of red cells made up exclusively of deoxyhemoglobin gave only the high-spin ferric compound, with the deoxyhemoglobin component disappearing completely at doses of around 100 Mrad.

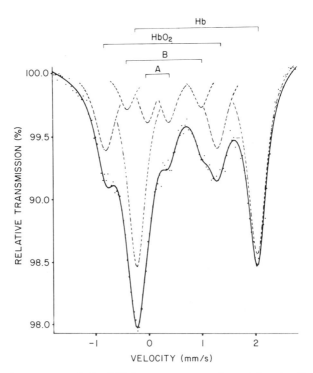

Fig. 8. Components of the Mössbauer spectrum of arterial red cells irradiated at 3.6 Mrad with x rays. HbO$_2$: oxyhemoglobin, Hb: ligand-free deoxy form, A: high-spin ferric compound in a rhombic environment, B: high-spin ferric heme. (From Kellershohn *et al.*, 1976.)

The effect on the red cells of heating for 1 min at 80°C are shown in Fig. 9. Heat acts in the same way as x rays to form the two high-spin ferric compounds but the deoxyhemoglobin component is absent.

The authors discuss the production of the ligand-free deoxyhemoglobin by x radiation in comparison with the well-known phenomenon of oxyhemoglobin photodissociation. Spartalian *et al.* (1976) have shown, by Mössbauer spectroscopy, that the photoproduct resembles ordinary deoxyhemoglobin in most respects but differs in some details. The high-spin ferric compound in a rhombic environment must result from the destruction of the heme group by radiation. The observed iron coordination is found in a number of organic compounds of iron such as ferrichrome A, transferrin, rubredoxin, and iron chelates.

The similarities between the spectra obtained as a result of x radiation and heat treatment emphasize the relationship, that is often observed for many kinds of substances, between radiolysis and thermolysis, which usually cause similar effects. Two important differences should be noted in the present case. First, heat does not dissociate oxyhemoglobin to produce the deoxy form. Second, only radiolysis can produce the high-spin ferric compound on its own.

It has been previously reported by some workers that the action of radiation, heat, or chemicals on oxyhemoglobin can result in the formation

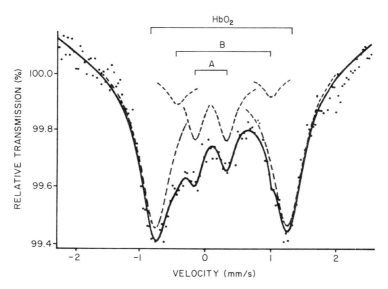

Fig. 9. Mössbauer spectrum of arterial red cells heated for 1 min at 80°C. The components are designated as in Fig. 8. (From Kellershohn *et al.*, 1976.)

of low-spin ferric heme compounds known collectively as hemichromes and low-spin ferrous heme compounds known as hemochromogens. However, Kellershohn et al. (1976) see no evidence for any compounds of these types.

In certain forms of anemia, the blood cells are observed to be the sickle-cell type. Mössbauer measurements on sickle-cell and normal hemoglobin have been carried out by Chow et al. (1971). In spectra obtained in a number of different environments, no differences larger than 0.05 mm/s were observed in the chemical shifts or quadrupole splittings for either oxygenated or deoxygenated samples. This must limit the possible conformational differences between the two molecules. In further work by the same group (Yen et al., 1973), the prominent features of the spectra of anhydrous (lyophilized) samples of both proteins were observed to be essentially identical.

Mössbauer spectra have been obtained from frozen whole blood and red blood cells from patients with different forms of thalassemia and other blood diseases and also from normal healthy adults (Ofer et al., 1976; Bauminger et al., 1979b). All the spectra contain two components that correspond to oxyhemoglobin and deoxyhemoglobin. In the spectra of blood samples from patients with thalassemia, sickle-cell anemia, hemoglobin H disease, and unstable hemoglobin Hammersmith there is also another component, corresponding to the presence of some additional iron compound, with an intensity that is variable among the different samples. A correlation was found between the magnitude of this component and the various types of diseased red blood cells that were examined.

At 82 K the additional component is a quadrupole-split doublet with a chemical shift (relative to iron metal at room temperature) of 0.47 mm/s and a quadrupole splitting of 0.70 mm/s. These parameters are similar to those of synthesized imidazole hemochromes and were first thought to indicate the presence of protein hemochromes in the blood of the thalassemia patients. However, measurements made over a range of temperatures show that the spectra associated with the additional iron-containing material are essentially identical to those obtained from the isolated iron storage proteins ferritin and hemosiderin. These data together with other arguments strongly indicate the presence of a material similar to ferritin or hemosiderin in the diseased blood.

The amounts of iron in the ferritinlike form were determined to be comparable to those in the form of hemoglobin and were particularly large in reticulocytes (young red blood cells). In some cases 1 ml of whole blood contained as much as 0.25 mg of ferritinlike iron in the red blood cells. The diseases in which the ferritinlike iron was observed are those in which there is an abnormality of hemoglobin structure and composition which leads to intracellular denaturation of the hemoglobin, with presumably a deposition of the excess iron in the ferritinlike form. No ferritinlike iron was found in the red blood cells of patients with idiopathic Coombs-positive hemolytic anemia or pernicious anemia. The observed differences in the quantities of ferritinlike iron in the reticulocytes and the mature red blood cells could be an important factor in understanding these diseases, and the authors consider possible reasons for these differences. One possibility is that the mature red blood cells contain the more stable hemoglobin, and there is therefore less likelihood of hemoglobin denaturation and the deposition of the excess iron in the ferritinlike form. The other possibility is that the ferritinlike iron is somehow removed from the red blood cell during its lifespan.

In the work described above Mössbauer spectroscopy has made an important contribution to our knowledge of these blood diseases.

Winterhalter *et al.* (1972) have studied the electronic structure of iron in the diseased form of hemoglobin known as Zurich β (63 His → Arg). They showed by Mössbauer spectroscopy that even in the deoxy form the two heme irons in the abnormal chains are low spin.

C. Bone Samples

Although there is no Mössbauer isotope of calcium, it is desirable to carry out experiments on bone samples using alkaline earth elements. The parent isotope of the Mössbauer nuclide ^{133}Cs is ^{133}Ba. Marshall (1968) has obtained a Mössbauer spectrum with a CsF absorber and a source provided by ^{133}Ba fixed onto bone powder by incubation of the latter in a solution of radioactive ^{133}BaCl$_2$. The spectrum is that of a caesium atom occupying a barium site on the bone, a situation analogous to that in the vitamin B$_{12}$ work described in Section II, F. Unfortunately, the Mössbauer effect in ^{133}Cs can only be observed at liquid helium temperatures, conditions rather far removed from those of physiology, and the relative absorption is rather low.

In addition to the alkaline earths, the rare earths are also "bone seekers." Kellershohn *et al.* (1974, 1979) have investigated both *in vivo* and *in vitro* fixation of rare earths onto bone material using ^{161}Dy Mössbauer spectroscopy.

In order to demonstrate the possibility of using the Mössbauer effect to monitor the metabolic uptake of rare earths onto bone, Kellershohn *et al.* (1979) have injected ^{161}TbCl$_2$ and ^{161}Tb-HEDTA into the muscles of mice. The mice are killed after 24 hr and, following dissection, the skeleton is compressed into a small pellet which is used as the source of ^{161}Dy Mössbauer γ rays, with Dy_2O_3 being used as the absorber. With both source and absorber at room temperature, excellent emission spectra are obtained and interpreted as indicating that the rare earth atom is metabolically fixed onto the bone and is actually incorporated into a solid structure. The spectra obtained in these experiments are the same as those obtained following *in vitro* absorption onto bone powder (Kellershohn *et al.*, 1974).

Another possibility for work on bone samples, which has yet to be explored, is to use the Mössbauer isotope ^{151}Eu since this atom can be isoelectronic with calcium.

V. Applications to Whole or Part Organisms: Uptake and Metabolism

By obtaining Mössbauer spectra of whole or part organisms, it is possible to identify the presence of various chemical species which contain the Mössbauer nuclide. In addition, one can, in principle, identify the location of the various biomolecules within the organism. In its most advanced form, this work can involve obtaining spectra from whole living organisms *in vivo*.

In order to monitor the way in which an organism uses iron (or any other metabolically important Mössbauer isotope), the organism can be grown under varying conditions of nutrient medium, concentration, temperature, etc., and then samples prepared and spectra taken and analysed to give information on the different iron species and their relative proportions.

A. In Vitro Experiments

Moshkovskii *et al.* (1966, 1968) have obtained spectra at 77 K from intact cells of *Bacillus hydrogenomonas* grown in a culture containing ^{57}Fe in the form of a citric acid salt. The spectra consist of the superposition of a single line from a high-spin ferric species and a doublet with parameters characteristic of a high-spin ferrous compound.

In their work on the fungus *Phycomyces blakesleanii*, Spartalian *et al.* (1975) obtained Mössbauer spectra from different parts of the fungus and from the growth medium. They observed that iron in the same chemical state as in the growth medium is present in all parts of the fungus but in diminishing proportions as one looks closer to the top. This can be interpreted to give information on the digestion of the iron as it rises in the fungus, and the role of the various parts in nutrient absorption.

Evans *et al.* (1977) have investigated membrane preparations from the blue–green algae *C. fritschii* and *A. nidulans* by means of Mössbauer spectroscopy. This work was prompted mainly by interest in the photosynthetic processes in these organisms. However, an additional factor was that these and other species of blue–green algae have been shown to accumulate high levels of iron, well in excess of that which could be contained in cytochromes and ferredoxins.

The algae were grown on a medium containing $^{57}FeSO_4$ and then harvested, washed, disrupted, and centrifuged to yield a sample containing the photosynthetic membranes. Mössbauer spectra of *C. fritschii* membranes are shown in Fig. 4. These spectra are composed of two components, I and II, indicated by the stick diagrams, the relative areas of the two components being comparable. The quadrupole-split spectra of component II (see also Fig. 5) are interpreted as resulting from the photosynthetic reaction centers and are discussed in Section III, B.

The other component (I) shows a six-line magnetic hyperfine pattern at 4.2 K, while at 77 K, where the spin–lattice relaxation rate for a ferric ion is fast, it gives a quadrupole-split doublet (Fig. 4a and b). The iron was shown to be in a paramagnetic state from observations of the change in the spectrum in the presence of a large external magnetic field. It was found that the field splits the spectrum and the overall effective field at the nuclei is reduced by large fields, i.e., the outer lines move inward. This is characteristic of paramagnetic iron, i.e., Fe^{3+} which is well separated from other magnetic ions. This is similar to the situation in iron transport proteins (see Section II, D). The results obtained from the *A. nidulans* membranes were in agreement with those from the *C. fritschii* membranes.

In order to investigate the effects on the two components of different growth conditions, and in particular to study the nature of the iron transport protein, Mössbauer spectra were obtained from membranes of *A. nidulans* grown under three different regimes of iron nutrition. The organisms grew well when the iron concentration was reduced. The 10% (relative to the standard conditions) growth showed no significantly reduced mean generation times, while on 1% iron, the mean generation times were reduced by less than a factor of two. Figure 10 shows the 4.2 K spectra of membranes from *A. nidulans* grown under these different conditions. It can be seen that there is a large reduction in the amount of the transport protein in the 10 and 1% growths. In the most iron-deficient growth conditions, the iron transport component virtually disappears and, in addition, the reaction center ferredoxinlike component is also reduced. The authors interpret these measurements as confirming that the component giving rise to the six-line Mössbauer spectrum may be ascribed to a form of iron concerned with the process of iron transport or storage in the membrane. This work is an example of the possibilities of using Mössbauer spectroscopy to monitor the uptake of a certain element by an organism, and to indicate which species of this element is more essential for healthy metabolism. In addition, it makes use of information obtained from Mössbauer spectroscopy of isolated biomolecules to understand the spectrum of a complex system and identify the various compounds present.

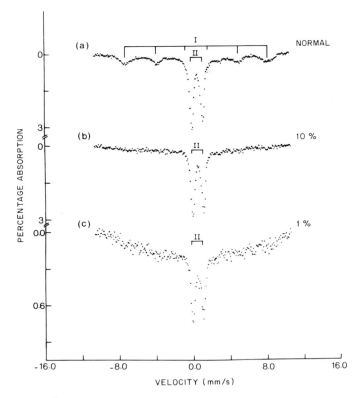

Fig. 10. Mössbauer spectra of *A. nidulans* membranes at 4.2 K after growth with three different concentrations of iron in the medium: (a) normal iron concentration, i.e., 100%, (b) 10% iron concentration, and (c) 1% iron concentration. The lowest spectrum is plotted with a different absorption scale giving an exaggerated background curvature which is a purely geometric factor and should not be confused with absorption. (From Evans *et al.*, 1977.)

B. *In Vivo Experiments*

The vast majority of Mössbauer measurements of biological materials take place under conditions which are far removed from those of a living organism. For example, a temperature of 4.2 K and an applied field of 6 T could hardly be regarded as a normal environment! Although much very useful information can be obtained from measurements made under these sorts of conditions, it would be helpful to obtain spectra under *in vivo* conditions to check the validity of the other measurements. In addition, *in vivo* experiments can enable a particular metabolic process to be monitored as it happens. For example, the observation of diffusion broadening could provide new information on the dynamics of the transport of metal ions through the cell membrane, about which little is known as yet. Obviously, experiments of this nature are extremely difficult but the two reported so far (Giberman *et al.*, 1974; Bauminger *et al.*, 1976), although essentially exploratory, have shown that *in vivo* experiments are both possible and give new and exciting information.

Giberman *et al.* (1974) have obtained Mössbauer spectra from ^{57}Co-enterochelin in *Eschirichia coli* cells, including measurements at 3°C under living conditions. Their experiment involved the chelating agent enterochelin which is associated with the transport of iron into this bacterium. Because the concentration of iron within the bacterium would be very small for use as an absorber, the experiments were carried out using bacteria containing a ^{57}Co-enterochelin complex as a source of Mössbauer γ rays.

E. coli were grown normally, harvested and washed, and a suspension of 3×10^{11} bacterial cells was mixed with ^{57}Co-enterochelin solution. After allowing time for the enterochelin to be absorbed, the bacteria were centrifuged to give a Mössbauer source with a volume of 0.3 mliter and an activity of 0.5 mCi, corresponding to an average concentration of about 3000 ^{57}Co nuclei per bacterium. An enriched stainless steel foil was used for the absorber.

Figure 11 shows the absorption spectrum obtained at 3°C during the first 24 hr after preparation and during the subsequent 72 hr. As can be seen, there is a small but definite absorption of 0.032 ± 0.005%, centered at −0.7 mm/s with a linewidth of about 3 mm/s, obtained in the first 24 hr, but no absorption in the 72 hr after this. The authors ascribe the absorption as resulting from "special sites" in the freshly prepared ("live") bacteria. In order to estimate the number of these special sites, measurements were also carried out on both samples frozen at −190°C. By subtracting the two spectra obtained, a component associated with the

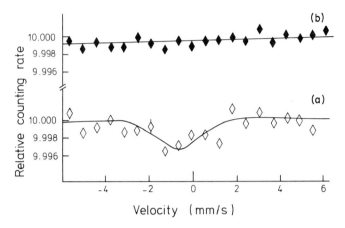

Fig. 11. Mössbauer spectra obtained using a stainless steel absorber and a ^{57}Co-enterochelin source in *E. coli* cells at 3°C: (a) spectrum obtained during the subsequent period of 72 hr, (b) spectrum obtained during the first 24 hr after preparation. (From Giberman *et al.*, 1974.)

special sites could be identified and from this it was estimated that the proportion of special sites was between 2.4 and 5.5% of all the ^{57}Co sites. In addition, it was estimated that at 3°C the recoil-free fraction of the special sites is 0.04.

The authors discuss the origin of these special sites which have a quasi-solid environment and give recoil-free emission of γ rays. One possibility is that the ^{57}Co nuclei at the special sites are in the cell membrane, this is supported by the fact that the proportion of the membrane volume to that of the whole cell in a typical *E. coli* bacterium is a few percent, which is comparable to the estimate of the fraction of special sites. The alternative proposition is that the ^{57}Co is bound to rather large macromolecules situated in the aqueous medium in the cell. This would require a viscosity of at least 10 P with molecules of about 5-nm diameter. The authors feel that this high value is unlikely in a bacterial cell in view of the large water content. For these reasons it appears most likely that the special sites are in the membrane, and on this assumption the broadening of the Mössbauer line at 3°C can be ascribed to the diffusion of the ^{57}Co through the membrane. Assuming continuous diffusion, the experimental value of the linewidth leads to a time of 0.1 ms for the ^{57}Co to diffuse through the cell membrane. On a jump diffusion model the experimental linewidth corresponds to a jump frequency of approximately 2×10^7 Hz.

Although the authors stress their preliminary nature, these measurements do show that spectra can be obtained under *in vivo* conditions and can be interpreted to give information of real biological significance.

A subsequent study involving Mössbauer spectroscopy of whole bacterial cells, including measurements under *in vivo* conditions, has been reported by Bauminger *et al.* (1976). In this work, samples of *E. coli* and *Halobacterium* were grown on an ^{57}Fe-enriched medium and used as the absorber in a Mössbauer spectrometer. One particular motivation in these experiments was to identify any membrane-bound iron sites. In addition to measurements on the whole cell, spectra were also obtained from the membranes of the bacteria and from a ferredoxin isolated from the bacteria.

The samples of whole bacterial cells contained about 10^{12} bacteria/mliter with ^{57}Fe concentrations of up to 1 mg/g of bacteria. In order to compensate for the small Mössbauer absorption in these samples, a high counting rate of about 10^5 counts/s was used.

The native isolated ferredoxin from *Halobacterium* gives a spectrum at 82 K which is similar to that of an oxidized two-iron plant-type ferredoxin, but with the two ferric sites being more different than usual. On reduction, the spectra indicate that one of these sites becomes ferrous. Measurements on whole cells frozen at 82 K show a spectrum of more complex origin than that of the isolated ferredoxin, and indicate that there must be a number of components or line broadening factors. The spectrum of frozen membranes is essentially identical to that of the frozen bacteria which suggests that a large proportion of the iron-containing components are retained within the membrane and are probably membrane bound. These spectra indicate that the ferredoxin represents only a small part of the total iron in the cell.

Spectra were obtained from frozen cells of two mutants of *E. coli* grown with a wide range of iron concentrations both aerobically and anaerobically. The Mössbauer spectra obtained were identical for all growth conditions, and are closely similar to those of the *Halobacterium* cells. These spectra were fitted to two quadrupole-split doublets with large linewidths. A small additional ferrous component was usually observed which was attributed to the presence of reduced material, the amount depending on the treatment of the samples. The intensity of these spectra was dependent on the concentration of ^{57}Fe in the growth medium. The maximum amount of ^{57}Fe that could be introduced was observed from the saturation of the intensity of the spectra. This maximum was strongly dependent on the strain of *E. coli* used, being much less for a cytochrome-free mutant, and much greater in aerobic growth than in anaerobic growth. Much more iron could be incorporated into *E. coli* than *Halobacterium* which is why *in vivo* measurements with the latter species did not produce significant absorptions.

Mössbauer spectra of unfrozen *E. coli* cells were obtained at 3°C and are shown in Fig. 12. These spectra are remarkable in that they exhibit

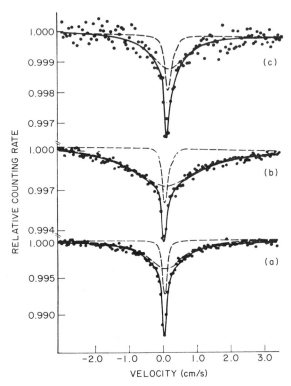

Fig. 12. Mössbauer spectra of *E. coli* cells at 3°C: (a) strain K12 with ∼ 1 mg ^{57}Fe/g of cells, (b) strain K12 with ∼ 500 μg ^{57}Fe/g of cells, and (c) strain H$_7$ with ∼ 60 μg ^{57}Fe/g of cells. The solid lines are computer fits to the spectra using two Lorentzian broadened components indicated by the dashed lines. (From Bauminger *el al.*, 1976.)

appreciable recoil-free effects and are very wide, indicating large motional broadening. The spectra contain two components, one narrow and one wide. The different growth conditions do not affect these basic features although the parameters and the intensity ratio of the components do differ in the various cases. The total intensity of these spectra is an appreciable fraction of the intensity of the spectra of the corresponding frozen samples which suggests that the same iron components must be associated with both sets of spectra.

The spectra were fitted to two Lorentzian lines of different widths and intensities, each corresponding to a broadened version of the original spectrum obtained from the frozen bacteria at low temperatures. This procedure gave linewidths and intensity ratios for the two components. The linewidths for the narrow line are between 1.0 and 1.8 mm/s and for the wide line between 12 and 26 mm/s. The ratio of the intensities of the narrow to the wide line vary between 10 and 31%. The particular values depend on the strain and growth conditions.

Because of the rather remarkable features of these spectra, the authors were concerned that they may be due to some artifact essentially external to the bacterial cells. However, the high degree of reproducibility of the effects under varying conditions of growth and the inability to eliminate them when working under "super-clean" conditions suggested that they must be biologically "real" effects. The main component of the spectra cannot be due to heme iron as there is virtually no difference in the spectral shape for the samples from the heminless mutant.

Obviously a full discussion of these phenomena must await a complete characterization of the material giving rise to the recoil-free effects; recent Mössbauer measurements (Bauminger et al., 1979a, 1980) indicate the presence of considerable quantities of a new type of iron storage material in E. coli grown under these conditions. While expressing the view that further discussion may be premature, the authors suggest some alternative explanations for the large motional broadening observed in the 3°C spectra.

One possibility is that the broadening is produced by Brownian motion of the whole cell. However, rough estimates based on the size of the cells and the viscosity of the whole sample give a value that is too small to explain the wide component, and is also unable to explain the coexistence of two components.

The other possibilities are motion of iron-containing proteins either in the membrane or in solution within the cell. These two effects could be combined to explain both components. Using a jump diffusion model for the motional broadening, and on the assumption that the motion of the protein can be described by the Brownian motion of a particle of a certain diameter (taken to be 4 nm) within a medium of a certain viscosity, values of 7 and 140 P are obtained for the effective viscosities "felt" by the proteins in the two environments.

The experiments described above represent a very exciting development in the application of Mössbauer spectroscopy to biological problems. They also emphasize the need for measurements at as many levels as possible from that of isolated proteins to that of whole organism, if a complete understanding of a biological system is to be achieved.

VI. Measurement of Small Vibrations and Movements

The Mössbauer technique makes the measurement of very small velocities possible. With the ^{57}Fe Mössbauer effect velocities as low as 0.1 mm/s can be measured, representing a displacement of less than 2 nm at 10 kHz. Because of the general convenience of measurement, only ^{57}Fe and ^{119}Sn have been used so far in biological applications of this type. However, with other isotopes having a narrower resonance line even greater sensitivity could, in principle, be achieved.

There are three main areas in which this feature of the Mössbauer effect can be applied. In the first, the study of the auditory mechanism, there has been considerable work on the vibrational mechanism of the basilar membrane in the cochlea of certain animals. There still remains a large amount of work that can be done using the Mössbauer technique which should provide information that could not be obtained by other means. Another possibility is to study the motion of complete organisms under different conditions. There is considerable potential for investigations of this sort, with very little having been done so far. These areas are discussed in Sections A and B. Another application is to measure the rate of movement of a certain molecule containing a Mössbauer nuclide within an organism. This can be derived from the motional broadening observed in the spectral lines. This area also presents considerable possibilities. The work described in Section V, B includes some measurements of this type.

A.   *Applications in Auditory Physiology*

Problems relating to pitch perception and frequency discrimination have been a major area of interest in auditory physiology for many years. The hair cells in the cochlea are generally agreed to be the primary sensors of acoustic energy, their excitation being related to the movement of the cochlear partition. Understanding of cochlear mechanics is therefore of fundamental importance. Measuring the movement of the cochlear partition is particularly difficult as it is transparent, immersed in a fluid environment, and only moves very small distances.

Johnstone and Boyle (1967) have reported measurements of the vibration of the basilar membrane in the first turn of the guinea pig cochlea, *in vivo*, using the Mössbauer effect. These measurements were the first of their kind to be carried out on a living animal and used the relatively low levels of acoustic stimulation made possible by the sensitivity of the technique. The measurements were made with very small (0.3 μg) sources of ^{57}Co in stainless steel surgically mounted on the basilar membrane and the stapes (to monitor the input). An absorber of ^{57}Fe-enriched stainless steel was used so that, with no acoustic stimulation, the source and absorber were stationary and absorption was at a maximum. When the source moved in response to a stimulus of known frequency, the count rate increased by an amount that could be interpreted to give information on the amplitude of the vibration. With sound intensities of between 60 and

95 dB SPL (sound pressure level, i.e., relative to $2 \times 10^{-5} \, \mathrm{N \, m^{-2}}$), there were count rate differences of between 10 and 30%, corresponding to velocities of between 0.2 and 0.6 mm/s. Response curves for both the ratio of the vibration amplitude of the basilar membrane to that of the stapes and the peak amplitude of the vibration of the basilar membrane were obtained at frequencies between 350 Hz and 30 kHz. The data show a maximum in the frequency response curve between 18 and 21 kHz for the 10 animals investigated with a maximum ratio of 50 and a maximum amplitude of vibration of the basilar membrane of 60 nm at 90 dB. The low frequency rise of the response curve is at 13 dB/octave, while the falloff after the maximum is much more rapid at 70 dB/octave. For the range of frequencies and intensities used in these measurements there is no evidence of any nonlinearity in the response. These measurements confirmed the role of the cochlea in mechanical filtering of the acoustic stimulus.

Rhode (1971) has used the same technique, refined to yield both phase and amplitude information, to investigate the response of the basilar membrane of the squirrel monkey. This particular species was chosen because of previous work on auditory-nerve fibers. As in the work of Johnstone and Boyle (1967), the source of ^{57}Fe γ rays (^{57}Co in stainless steel) was mounted surgically on the basilar membrane. In order to investigate any possible effects arising from the source itself, a variety of sources weighing between 0.1 and 0.4 μg were used. These had activities ranging between 25 and 100 μCi. The absorber was an ^{57}Fe-enriched stainless steel foil.

The motion was stimulated using sound in the frequency range between 100 Hz and 25.5 kHz and with sound intensities between 70 and 120 dB SPL. The relative motion of the source and absorber led to an increase in the count rate, and by feeding the counts and the simultaneous stimulus signal into a computer, it was possible to derive information on the amplitude and phase of the vibration of the basilar membrane. Measurements were also made on the bony limbus in the cochlea and the malleus, incus and stapes of the middle ear. Typical data obtained in this investigation are shown in Fig. 13.

An important result of these measurements was that the basilar membrane was observed to vibrate nonlinearly for frequencies which produced the largest deflections at the point of the basilar membrane under observation. The ratio of the displacement of the basilar membrane to that of the malleus was found to increase at a rate of 6 dB/octave until just below the maximum ratio, where it increased by 24 dB/octave. The maximum ratio was observed to be 24 dB for the sound pressure levels used. Above the frequency producing the maximum ratio the dropoff rate is approximately

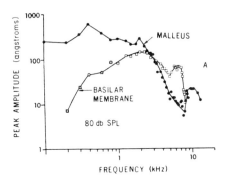

Fig. 13. Frequency dependence of the amplitude of vibration of the malleus and the basilar membrane of the squirrel monkey for a stimulus of 80 dB SPL, as derived from Mössbauer measurements. (From Rhode, 1971.)

100 dB/octave, eventually leveling off at high frequencies. The phase information obtained showed that for frequencies below that producing the maximum ratio the phase difference between the motion of the basilar membrane and that of the malleus is a linear function of frequency. At frequencies near the maximum the phase difference decreases at a faster rate and approaches a constant value at higher frequencies.

The nonlinear behavior of the basilar membrane observed by Rhode (1971) had not been detected in the optical measurements on cadavers by Békésy (1960), in which very high sound levels were used. Johnstone and Boyle (1967) also found linear behaviour in their experiments using the Mössbauer effect technique. In order to investigate whether the observed nonlinearity is a real effect or an artifact of the Mössbauer measurement, Rhode and Robles (1974) have carried out further experiments, including the observation of postmortem changes and the transient response to clicks, in addition to the steady-state response *in vivo*, as investigated previously. The central question is whether the Mössbauer source follows the motion of the basilar membrane in a faithful manner, or somehow interacts with the cochlear mechanism.

The methods used in this investigation (Rhode and Robles, 1974) are essentially those of Rhode (1971). The measurements again resulted in a direct observation of nonlinear vibration of the basilar membrane. The authors claim that these observations are consistent with indirect evidence from physiological experiments which can only be explained by nonlinear cochlear mechanisms.

These Mössbauer data show that the nonlinearity is greatest near the characteristic frequency and spreads to lower frequencies as the sound level is increased. The nonlinearity was observed in both the steady-state and impulse measurements. Measurements with two tones which showed that the amplitude of vibration is not simply additive also indicate nonlinearity of the basilar membrane motion. The linear behavior observed at lower frequencies indicates that the Mössbauer source is not loading the membrane and this is further substantiated by the fact that the size of the source does not affect the membrane motion.

There are a number of postmortem changes, including a rapid decrease in sensitivity. In addition, it was observed that the basilar membrane vibrates linearly after death, in keeping with the measurements of Békésy (1960). The changes that occur on death indicate a physiologically vulnerable acoustic filter.

In order to investigate any possible interaction between the Mössbauer source and the electrical mechanisms of the cochlea, direct electrical stimulation was applied with no effect on the source. However, although this shows that the source is not electrically driven, some other interaction could still be present.

Because of the discrepancy between the various investigations concerning the linearity of the basilar membrane response, Johnstone and Yates (1974) have made a further investigation using guinea pigs. They report that there is very little nonlinearity, which, if it occurs at all, is observed at the foot of the high-frequency slope of the frequence response curve. They suggest that the real differences in linearity observed in the different investigations may be related to species differences, or possibly to the position of measurement on the basilar membrane.

A fuller study of the transient response of the basilar membrane of the squirrel monkey has been made by Robles *et al.* (1976). These workers used acoustic clicks of 150 μs duration repeated many times in order to build up sufficient counts. The responses of the malleus and the basilar membrane were monitored for clicks of varying amplitudes. The basilar membrane was observed to have an oscillatory response with a period close to that of the characteristic frequency. An interesting feature of these results was that the response to the first few clicks was linear, while later responses showed a marked nonlinearity. This behavior is consistent with the nonlinearity reported in the steady-state measurements. The different animals used in these experiments exhibit a wide range of damping. Damage to the cochlea also produces increased damping. In these experiments the delay time of the cochlea was measured to be between 300 and 390 μs, in good agreement with the value derived from the phase information obtained in the steady-state measurements.

B. Motion of Complete Organisms

Bonchev *et al.* (1968) performed an extremely novel series of experiments in which the motion of ants was investigated using Mössbauer spectroscopy. The ants used, *Formica pratensis*, were typically 10 mm long.

The ants had SnO_2 powder glued onto their abdomens, and 50 were packed into a cuvette to form the absorber in a constant velocity ^{119}Sn Mössbauer spectrometer. The influence of temperature and air supply on the width of the Mössbauer line was investigated, the natural linewidth being measured by killing the ants.

The linewidth was observed to increase from under 2 mm/s for dead ants to a value for live ants of about 3 mm/s at 5°C and approximately 5 mm/s at 25°C. It was found that dead ants gave the same linewidth as that of live ants extrapolated to 0°C.

The broadening of the Mössbauer line is affected only by the component of the motion of the separate "absorbers" in the direction of the γ-ray beam. In measurements taken at higher temperatures, nonreproducibility is observed and this is attributed to the ants crawling about and changing the geometry of the experiment. The authors claim that the broadening of the Mössbauer line observed should be regarded as a consequence of breathing and flickering of the abdomen. Ants move quickly at between 30 and 50 mm/s and movements of the whole creatures would be outside the range of the measurement.

Mössbauer spectra for ants at the same temperatures, with and without an inflow of air, showed that there is a slight increase in linewidth when the flow is cut off. This is attributed to an increased frequency of the breathing cycle.

This work suggests many possibilities whereby the laws governing the movement of individual small animals could be tested and changes in the average speed could be readily monitored.

VII. Conclusions

The Mössbauer effect is potentially important for its practical applications, e.g., in medicine. The main difficulties to be overcome, in order for the technique to become widely used, concern the form of the sample, which must be a "solid" and which usually must be cooled in order to get a large effect. For iron, the small amounts generally present in biological tissue and the low abundance of the Mössbauer isotope ^{57}Fe makes the technique insensitive and slow compared with EPR, for example. However, if these difficulties can be overcome, the Mössbauer effect could become a powerful tool for determining the state of iron in a clinical specimen.

It has been shown that measurements on human blood and tissue samples are quite feasible. Whether these will result in any advances in medical diagnosis is not yet clear. However, Mössbauer spectroscopy has been shown in many cases to yield information which had not been obtained previously by any other method. It is quite likely, therefore, that it could become a valuable tool for qualitative and quantitative analysis in medical research, just as it has become a powerful and detailed probe for the study of biological molecules and biochemical reactions.

References

Bauminger, E. R. *et al.* (1976). *J. Physique* **37**, C6-227.

Bauminger, E. R., Cohen, S. G., Dickson, D. P. E., Levy, A., Ofer, S., and Yariv, J. (1979a), *J. Physique* **40**, C2-523.

Bauminger, E. R., Cohen, S. G., Ofer, S., and Rachmilewitz, E. A. (1979b). *Proc. Natl. Acad. Sci. USA* **76**, 939.

Bauminger, E. R., Cohen, S. G., Dickson, D. P. E., Levy, A., Ofer, S., and Yariv, J. (1980). *Biochim. Biophys. Acta* **623**, 237.

Békésy, G. von (1960). *In* "Experiments in Hearing" (E. G. Wever, ed.). McGraw-Hill, New York.

Boas, J. F., and Window, B. (1966). *Aust. J. Phys.* **19**, 573.

Bonchev, T., Vassilev, T., Sapundzhiev, T., and Evtimov, K. (1968). *Nature (London)* **217**, 96.

Bray, R. C., Palmer, G., and Beinert, H. (1964). *J. Biol. Chem.* **239**, 2667.

Cammack, R., Rao, K. K., Hall, D. O., and Johnson, C. E. (1971). *Biochem. Biophys. Res. Commun.* **125**, 849.

Cammack, R., Dickson, D. P. E., and Johnson, C. E. (1977). *In* "Iron–Sulfur Proteins" (W. Lovenberg, ed.) Vol. III, p. 319. Academic Press, New York.

Cammack, R. (1980). *Nature* **286**, 442.

Chow, Y. W. *et al.* (1971). *Bull. Am. Phys. Soc.* **16**, 641.

Debrunner, P. G., Schulz, C. E., Feher, G., and Okamura, M. Y. (1975). *Biophys. J.* **15**, 226a.

Dickson, D. P. E., and Cammack, R., (1974). *Biochem. J.* **143**, 763.

Dickson, D. P. E., Johnson, C. E., Cammack, R., Evans, M. C. W., Hall, D. O., and Rao, K. K. (1974). *Biochem. J.* **139**, 105.

Dickson, D. P. E., and Rottem, S. (1979). *Eur. J. Biochem.* **101**, 291.

Evans, E. H., Carr, N. G., Rush, J. D., and Johnson, C. E. (1977). *Biochem. J.* **166**, 547.

Evans, E. H., Rush, J. D., Johnson, C. E., and Evans, M. C. W. (1979). *Biochem. J.* **182**, 861.

Giberman, E. *et al.* (1974). *J. Phys.* **35**, C6-371.

Groves, J. L., Potasek, M. J., and De Pasquali, G. (1973). *Phys. Lett.* **42A**, 493.

Guest, L. (1976). *Ann. Occup. Hyg.* **19**, 49.

Guest, L. (1978). *Ann. Occup. Hyg.* **21**, 151.

Haffner, H., Andl, A., Appel, H., Büche, G., Holmes, K. C., and Morris, S. (1976). *J. Physique* **37**, C6-223.

Hewitt, P. J., and Hicks, R. (1973). *Ann. Occup. Hyg.* **16**, 213.

Huynh, B. H., Papaefthymiou, G. C., Yen, C. S., Groves, J. L., and Wu, C. S. (1974). *Biochem. Biophys. Res. Commun.* **60**, 1295.

Huynh, B. H., Münck, E., and Orme-Johnson, W. H. (1979). *Biochim. Biophys. Acta* **579**, 192.

Inoue, K., and Nath. A. (1977) *Bioinorg. Chem.* **7**, 159.

Johnson, C. E. (1971). *Phys. Today* **24**, 35.

Johnson, C. E., Knowles, P. F., and Bray, R. C. (1967). *Biochem. J.* **103**, 10C.

Johnstone, B. M., and Boyle, A. J. F. (1967). *Science* **158**, 389.

Johnstone, B. M., and Yates, G. K. (1974). *J. Acoust. Soc. Am.* **55**, 584.

Kellershohn, C., Chevalier, A., Rimbert, J. N., and Hubert, C. (1974). *Proc. Ampère Congr.,* 18*th, Nottingham* p. 289.

Kellershohn, C., Chevalier, A., Rimbert, J. N., and Hubert, C. (1975). *Proc. Int. Biophys. Congr. 5th, Copenhagen, IUPAB* P-629.

Kellershohn, C., Rimbert, J. N., Chevalier, A., and Hubert, C. (1976). *J. Physique* **37**, C6-185.

Kellershohn, C., Rimbert, J. N., Fortier, D., and Maziere, M. (1979). *J. Physique* **40**, C2-505.

Kündig, W., Bömmel, H., Constabaris, G., and Lindquist, R. H. (1966). *Phys. Rev.* **142**, 327.

Lang, G. (1970) *Rev. Biophys.* **3**, 1.

Lang, G., and Marshall, W. (1966). *Proc. Phys. Soc.* **87**, 3.

Lemberg R., and Legge, J. W. (1949). "Hematin Compounds." Wiley (Interscience), New York.

Lovenberg, W. (ed.) (1974). "Iron-Sulfur Proteins," Vols. I and II. Academic Press, New York.

Lovenberg, W. (ed.) (1977). "Iron-Sulfur Proteins," Vol. III. Academic Press, New York.

Marshall, J. H. (1968). *Phys. Med. Biol.* **13**, 15.

Moshkovskii, Y. S., Makarov, E. F., Zavarzing, A., Vedenna, N. Y., Mardanyan, S. S., and Goldanskii, V. I. (1966). *Biofizika* **11**, 357.

Moshkovskii, Y. S. (1968). *In* "Chemical Applications of Mössbauer Spectroscopy" (V. I. Goldanskii and P. H. Herber, eds), p. 524. Academic Press, New York.

Mullinger, R. N. *et al.* (1975). *Biochem. J.* **151**, 75.

Münck, E., Rhodes, H., Orme-Johnson, W. H., Davis, L. C., Brill, W. J., and Shah, V. K. (1975). *Biochim. Biophys. Acta* **400**, 32.

Nath, A., Harpold, M., Klein, M. P., and Kündig, W. (1968). *Chem. Phys. Lett.* **2**, 471.

Oberley, L. W., and Erhardt, J. C. (1975). *J. Chem. Phys.* **63**, 2329.

Oberley, L. W., Herskowitz, V., and Erhardt, J. C. (1974). *Phys. Lett.* **50A**, 77.

Ofer, S., Cohen, S. G., Bauminger, E. R., and Rachmilewitz, E. A. (1976). *J. Physique* **37**, C6-199.

Papaefthymiou, G. C., Huynh, B. H., Chow, Y. W., Groves, J. L., Yen, C. S., and Wu, C. S. (1973). *Bull. Am. Phys. Soc.* **18**, 671.

Rao, K. K., Cammack, R., Hall, D. O., and Johnson, C. E. (1971). *Biochem. J.* **122**, 257.

Rao, K. K. *et al.* (1972). *Biochem. J.* **129**, 1063.

Rhode, W. S. (1971). *J. Acoust. Soc. Am.* **49**, 1218.

Rhode, W. S., and Robles, L. (1974). *J. Acoust. Soc. Am.* **55**, 588.

Robles, L., Rhode, W. S., and Geisler, C. D. (1976). *J. Acoust. Soc. Am.* **59**, 926.

Rush, J. D., Dickson, D. P. E., Johnson, C. E., Hewitt, P. J., and Lam, H. F. (1975). *Phys. Med. Biol.* **20**, 128.

Smith, B. E., and Lang, G. (1974). *Biochem. J.* **137**, 169.

Spartalian, K., Oosterhuis, W. T., and Smarra, N. (1975). *Biochim. Biophys. Acta* **399**, 203.

Spartalian, K., Lang, G., and Yonetani, T. (1976). *Biochim. Biophys. Acta* **428**, 281.

Thompson, C. L. *et al.* (1974). *Biochem. J.* **139**, 97.

Tsang, C. P., Bogner, L., and Boyle, A. J. F. (1976). *J. Chem. Phys.* **65**, 4584.

Winterhalter, K. H. *et al.* (1972). *J. Mol. Biol.* **70**, 665.

Yen, C. S., Groves, J. L., Papaefthymiou, G., Huynh, B. H., Swerdlow, P. H., and Wu, C. S. (1973). *Bull. Am. Phys. Soc.* **18**, 670.

5

Oxygen Transport and Storage Materials

K. Spartalian[†] and G. Lang

Department of Physics
The Pennsylvania State University
University Park, Pennsylvania

I. Introduction

The basic energy source of aerobic life involves the transfer of electrons from highly reduced nutrient molecules to oxygen. By means of a complicated system of coupled reactions, the chemical energy thereby released is transformed into the standard cellular energy currency, the energy-rich phosphate bonds of adenosine triphosphate (ATP). Since this oxidative phosphorylation is an intracellular process, a limit on its rate is

[†]Present address: Department of Physics, The University of Vermont, Burlington, Vermont.

APPLICATIONS OF MÖSSBAUER
SPECTROSCOPY, VOL. II

imposed by the rate of oxygen delivery. In lower life forms, the low speed of oxygen diffusion in liquids imposes a limit of about 1 mm on the distance between living material and the surface that is exposed to air or oxygen-charged water. Evolution of larger animals has been made possible by the development of the circulatory system, which transports oxygen from organs developed for its acquisition to sites at which it is utilized, returning carbon dioxide for disposal. At the oxygen partial pressure of the atmosphere at sea level, only about 0.6 milliliter of the gas dissolves in 100 milliliter of water. Although the circulatory systems of some lower forms (e.g., crayfish) are able to make do with simple solution of oxygen, the more advanced and successful animals have evolved oxygen-binding pigments, which allow an order of magnitude increase in the oxygen content of the circulating fluid, make possible a more complete oxygen uptake and discharge, and facilitate the operation of various regulatory and control mechanisms.

The well-known iron-containing pigment hemoglobin[†] is by no means the only biological oxygen carrier. A copper protein, hemocyanin, serves this function in certain types of the marine worms—a group that appears to be nature's testing ground for oxygen-binding pigments. Hemerythrin, a protein in which oxygen is bound between two iron atoms, serves another class of marine worms. In spite of its name, it is not a hemoprotein. Chlorocruorin is yet another oxygen-binding pigment of the marine worms; in it the iron is incorporated in a prosthetic group rather similar to heme. Hemoglobin, present in the blood of all warm-blooded animals with rather minor variations, is also found in lower animals and even in plants. Its purpose is not solely the distribution of oxygen. The marine worm *Sabella* has chlorocruorin in its blood and hemoglobin in its muscle. In the root nodules of leguminous plants the hemoglobin, leghemoglobin, does not circulate. Since the nitrogen-fixing process involves proteins that are oxygen sensitive, it is possible that leghemoglobin serves merely to preserve an anaerobic atmosphere. The myoglobin of mammalian muscle is very similar to hemoglobin in its structure and oxygen-binding mechanism; its function is not known with certainty, being connected with oxygen storage in the cell or possibly acting to increase the effective diffusion rate into and within the cell. The cooper of hemocyanin is not amenable to study by Mössbauer spectroscopy, and to our knowledge chlorocruorin has not been examined using this method. Our discussion will be concentrated on hemoglobin, the most studied pigment, myoglobin, less studied and in many ways similar, and hemerythrin, which bears little relation to these two.

[†]Throughout this chapter the abbreviation Hb designates deoxyhemoglobin; complexes are designated as HbO_2, HbCO, etc. A similar scheme is used for deoxymyoglobin (Mb) and its complexes.

A.   Myoglobin

Because of its relative simplicity, it is most convenient to consider first the structure and function of myoglobin. The basic functional group of this protein, and with small variations all heme proteins, is the heme group, shown in Fig. 1. This nearly planar molecule is approximately 14×17 Å and has a molecular weight of 616. In myoglobin the heme is located in a pocket in a protein of molecular weight 17,000, a folded polypeptide chain made up of 153 amino acid residues. The heme group iron is bound to the nitrogen of the imidazole ring of a histidine residue. This so-called proximal histidine is the 93rd residue, counting along the chain from the amino end. The bond to the iron is the only strong bond between heme and the protein globin, the porphyrin plane of the heme being more or less propped against the side of the protein pocket and held by a large number of relatively weak bonds. The hydrophilic propionic acid chains at the bottom of the figure are positioned near the mouth of the pocket, while the other groups and most of the residues lining the pocket are hydrophobic. On the far side of the protein pocket, a second histidine (the distal histidine) approaches the iron from the other side of the heme, but does not bind to it. Oxygen can bind in the somewhat restricted space between the distal histidine and the iron, as can a number of other small molecules. In the physiologically active form of myoglobin, the iron remains ferrous and the molecule shuttles between a state in which the iron has no sixth ligand and a state in which the ligand is a molecule of oxygen. In the former the iron is high-spin ($S = 2$) ferrous, while in the latter the net spin is zero. The ferrous form of myoglobin also binds CO rather strongly, forming a complex with no unpaired spin. Oxidation of the heme iron to the ferric state makes possible the formation of a number of high- and

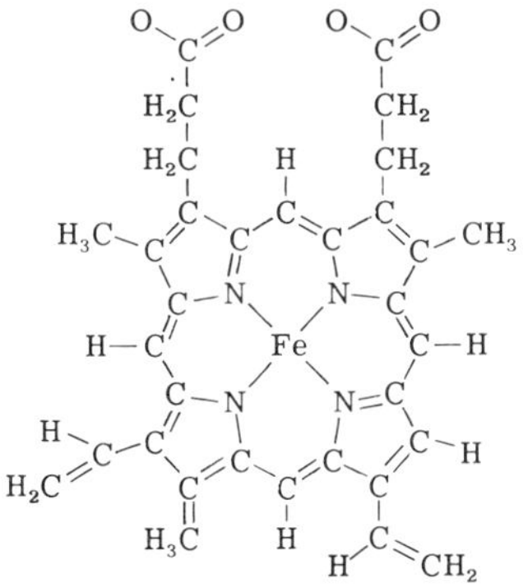

Fig. 1.   The heme molecule. The proximal histidine nitrogen binds to the fifth ligand position. The sixth position is available to small molecules, e.g., O_2, CO, etc.

low-spin complexes which have no physiological function but which may be useful models for a number of ferric heme protein enzymes.

The three-dimensional structure of myoglobin has been determined by x-ray diffraction to a resolution of 2 Å, and the location and orientation of the O_2 molecule have recently been determined (Perutz *et al.*, 1974). Because of the relative difficulty of extracting and purifying myoglobin, its use as a simplified model for the study of hemoglobin has not been as great as might be expected. Evidence of this is seen in the fact that while hundreds of variations of the amino acid sequence in hemoglobin are known, counting species variation and mutants, the corresponding number for myoglobins is only a few dozen. In Mössbauer spectroscopy a similar relative neglect of myoglobin exists, no doubt for the same reason.

B. Hemoglobin

The hemoglobin molecule, molecular weight 64,000, is a tetramer of four separate globular units, identical in pairs and each containing a single amino acid chain. In normal human hemoglobin (HbA) these units are called the α and β chains, and the tetramer is written as $\alpha_2\beta_2$. The beta chains are rather similar to myoglobin and have 146 amino acid residues. The alpha chains are slightly less similar and have 141 residues. The tetrameric form of the hemoglobin molecule is the key to a number of physiologically important interactions that can modify its oxygen affinity, while at the same time opening a Pandora's box of complications and sometimes uncontrolled variables to plague the experimenter.

The tetramer of hemoglobin is capable of assuming two quaternary structures designated as R and T. The compact relaxed (R) form is stable in the fully oxygenated material. Here the four subunits fit closely together in a somewhat tetrahedral arrangement, disposed about a twofold symmetry axis. The chains have a large number of close contacts, with more contacts between unlike chains than between like ones. With the oxygen removed, the chains distort slightly and move out from the symmetry axis and at the same time rotate. The resulting more open structure is called the tense (T) conformation and has an essentially empty hole along the twofold axis, which is large enough to admit solvent and a variety of small molecules. The activation of the conformational transition results from a distortion that is related to changes in the bond distances at and near the oxygen-binding iron site. The conformational transformation provides a means by which the oxygen-binding sites may communicate and gives rise to the well-known cooperative effect in hemoglobin oxygen uptake. It is observed that in hemoglobin the affinity for oxygen increases when some of the possible sites are occupied. As a result, the plot of oxygen bound as a function of the oxygen partial pressure is sigmoid in shape. The usual situation is that at the partial oxygen pressure that occurs in the lungs, the hemoglobin is almost fully loaded; the benefit of the cooperative effect is realized in the tissues where 30–50% unloading of the hemoglobin can be achieved without the necessity of attaining excessively low oxygen concentrations.

In addition to the homotropic interactions in which oxygen binding on one site can affect other oxygen-binding sites, the change in conformation, by altering the exposure of various protein groups to the solvent and distorting their local environments, provides the opportunity for heterotropic interactions—ones in which the binding of other molecules of various kinds affects the oxygen binding and vice versa. An important interaction of this type is the Bohr effect, in which an increase of pH increases oxygen affinity and, conversely, a loss of oxygen makes the hemoglobin bind more protons and lower the solution acidity. This is a physiologically useful phenomenon because the CO_2 produced by metabolism in the tissues tends to lower pH and force the unloading of oxygen where it is needed. A second important heterotropic interaction, demonstrated only in 1967, is the effect of organic phosphates, particularly 2, 3-diphosphoglycerate (DPG) in depressing the oxygen affinity of human and a number of other mammalian hemoglobins. This molecule binds to the ends of the β chains and stabilizes the T conformation. It is found that DPG concentration increases in the course of high altitude acclimitization of humans. This causes an increase in oxygen delivery efficiency by promoting a more effective unloading of hemoglobin in the tissues.

The oxygen-binding properties of the common heme proteins may be summarized (somewhat oversimplified) as follows. Monomeric forms, such as myoglobin and separated chains of hemoglobin, have high oxygen affinity and simple bimolecular hyperbolic oxygen uptake curves and are relatively little affected by pH and the presence of small molecules in solution. The tetrameric hemoglobin molecule is much more complicated. Oxygen affinity at a given site can be affected by binding of oxygen at other sites, resulting in the sigmoidal uptake curve, where the molecule has low affinity for the first one or two oxygen molecules but the last O_2 binds with affinity about that of myoglobin. Oxygen binding is also affected by interactions of small molecules at sites that are remote from the heme, H^+ and DPG being the most noteworthy. It is probable that most, but not all, modifications of oxygen binding occur via the T to R transformation in quaternary structure. Some cooperativity in oxygenation is seen in $\alpha\beta$ hemoglobin dimers, and there are even reports of cooperative binding in monomeric lamprey hemoglobin.

The advantages and limitations of Mössbauer spectroscopy as a biological probe have been discussed elsewhere (Oosterhuis and Spartalian, 1976).

In the present chapter we shall see how Mössbauer spectroscopy has been used by various researchers in order to elucidate the electronic structure of various complexes of myoglobin and hemoglobin. While Mössbauer spectroscopy can be a useful tool in the study of biological macromolecules, to the investigator entering the study of hemoglobin and myoglobin the profusion of experimental results is mind boggling. This is probably most true for physicists who are by training and inclination more accustomed to simpler and better defined problems with far fewer variables. The basic experimental material is highly complicated, having thousands of atoms in a functioning unit with some of them, such as the amino acid side chains on the exterior of the molecule, without well-defined position and orientation. The wide variety provided by species difference and hundreds of mutations is further complicated by the fact that these materials function in nature at high concentration and very complicated chemical mixtures, while most experiments are carried out in dilute solutions in conditions that are far different from the natural state and often from each other. Few of the results are compellingly clear-cut. Even the x-ray structure measurements, no doubt the most visual and directly comprehensible measurements in this field, when made on materials of such high complexity yield results that must be taken with a number of reservations in mind. Thus physicists, and most biological Mössbauer work has been done by physicists, should tread softly and speak conservatively about results they achieve and interpretations they make in this field. There is an obvious danger inherent in excessively rigid application of techniques and theories developed on simple materials to the flexible, variable, and varied materials of living systems. It behooves us who are used to crystals that break with a snap to remember that protein crystals break with a squish.

II. Oxygenated Complexes

In order to put the Mössbauer measurements on oxyhemoglobin into perspective, we must first give a historical account of the early magnetic measurements and the theoretical models that developed concurrently with these measurements. In the following discussion, although we shall refer exclusively to oxyhemoglobin, it is understood that the same arguments apply to oxymyoglobin as well. The first measurements of the magnetic properties of oxyhemoglobin were made by Pauling and Coryell in 1936, followed by Taylor and Coryell (1938) and Coryell et al. (1939). They established that oxyhemoglobin is diamagnetic at room temperature, and they set the framework within which models for the electronic structure were proposed.

The most widely accepted model for the oxygen binding to hemoglobin is that of Pauling. In its most recent form (Pauling, 1977), the dioxygen molecule lies above the heme plane with the Fe–O bond length of 1.72 Å, the O–O bond length at 1.27 Å, and the bond angle at 114°. Moreover, there is a resonance between the structures:

$$
\begin{array}{cc}
\text{O}^- & \text{O} \\
\diagup & \parallel \\
\text{O}^+ & \text{O} \\
\parallel & \vert \\
\text{Fe} & \text{Fe}
\end{array}
$$

The iron is in the low-spin ferrous state with all six 3d electrons paired in the three planar t_{2g} orbitals.

The Griffith model proposed in 1956 (Griffith, 1956) has the dioxygen molecule parallel to the heme plane and π bonded to the $|xz\rangle$ orbital of the iron:

$$
\begin{array}{c}
\text{O} \!=\!\!=\!\! \text{O} \\
\vert \\
\text{Fe}
\end{array}
$$

The oxidation state of the iron is again low-spin ferrous ($S = 0$).

In 1964 Weiss proposed an alternative model (Weiss, 1964) wherein the dioxygen has a net spin $S = \frac{1}{2}$, antiferromagnetically coupled to a spin $S = \frac{1}{2}$, residing on the iron atom, which is in the low-spin *ferric* state. The spins stay antiferromagnetically coupled up to room temperature in order to preserve the observed diamagnetism at the active site.

In Gray's two-electron oxidative addition model (Gray, 1971), a seven-coordinate complex is formed upon oxygenation, with the iron in the intermediate-spin Fe(IV) state:

$$
\begin{array}{ccc}
\underset{\text{N} \;\; \text{N}}{\underset{\diagdown \; \diagup}{\overset{}{\text{Fe}^{(\text{II})}}}} + \text{O}_2 \rightarrow &
\text{Fe}^{(\text{IV})} & 2^-
\end{array}
$$

Finally, the ozone model proposed by Goddard and Olafson (1975) and later expanded by "*ab initio* quality calculations" (Olafson and Goddard, 1977) has the iron in an intermediate spin $S = 1$ state coupled to the oxygen molecule spin $S = 1$ to yield a singlet $S = 0$ state. The coordination number in this case is six, with the dioxygen molecule in the "bent" configuration as in the Pauling model.

The above models are certainly in agreement with the magnetic susceptibility measurements because after all they were proposed subsequent to these measurements in order to explain the observed diamagnetism. All but one of the available states of the iron atom are put forth as possibilities, and the high-spin ferric state might well be in consideration, too, if it were not for the difficulties of canceling five unpaired spins. Mössbauer spectroscopy, by probing directly the environment of the iron nucleus, provides additional and valuable information about the active site of oxyhemoglobin, despite the diamagnetism of the site. The structure of the electronic orbitals, if given by a theoretical model, can be used to calculate the electric field gradient at the nucleus for comparison with the experimental values. Thus, the plethora of models for oxygen binding can be reduced by excluding those that are incompatible with the Mössbauer results.

The outstanding feature of the Mössbauer spectra of oxyhemoglobin and oxymyoglobin in zero field is their characteristic temperature dependence (Fig. 2). These spectra exhibit a relatively large quadrupole splitting

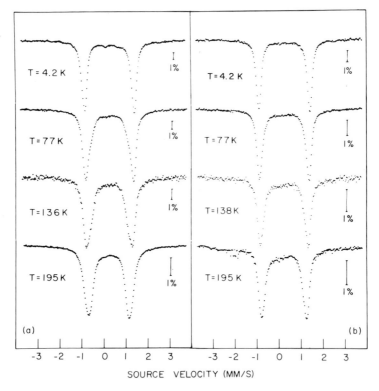

Fig. 2. Zero-field spectra of oxyhemoglobin (a) and oxymyoglobin (b) at temperatures as shown. With increasing temperature the separation between the two lines collapses while the asymmetric lines broaden.

(2.25 mm/s at 4.2 K) that decreases by about 20% as the temperature is raised to 200 K. The decrease is accompanied by an asymmetric broadening of each absorption peak individually, while the left-to-right symmetry of the spectrum about its centroid of energy is preserved (Spartalian and Lang, 1976). Similar behavior is exhibited by a model compound (Spartalian et al., 1975) capable of reversible oxygenation. Mössbauer spectra of this compound at selected temperatures are shown in Fig. 3. Spectra of oxyhemoglobin, oxymyoglobin, and of the same model compound in a high external magnetic field (Fig. 4) confirm the diamagnetism at low temperature and indicate by computer fits to the data (solid line) that the principal component of the efg (see, e.g., Cohen, 1976) is negative, while

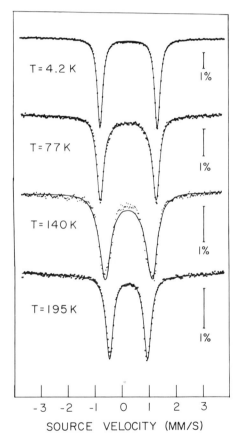

Fig. 3. Zero-field spectra of model compound FeO_2 (N–Me imid)P at temperatures as shown. Note the similarity of these spectra with the ones in Fig. 1. The solid lines are calculations based on the relaxation model discussed in Spartalian et al. (1975).

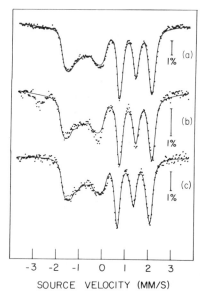

SOURCE VELOCITY (MM/S)

Fig. 4. Mössbauer spectra of oxyhemoglobin (a), oxymyoglobin (b), and model compound $FeO_2(N-Me\ imid)P$ (c) at 4.2 K in a transverse applied magnetic field of 6 T. The middle spectrum contains ferric impurities. Solid lines show least-squares fits that yielded the asymmetry parameter $\eta \approx 0.2$ in all cases.

the asymmetry parameter η in each case is small ($\eta \approx 0.2$). Figure 5 shows the variation of the quadrupole splitting with temperature for oxyhemoglobin, oxymyoglobin, and some oxygenated model compounds capable of reversible oxygenation. Table I contains the values of the quadrupole splitting at selected temperatures. We point out that the oxyheme complex $OxyCoP-(OMe)_2(C_5H_5N)$ was investigated by Marchant *et al.* (1972) with the use of Mössbauer *emission* spectroscopy. It was the first report of a model compound showing the characteristic temperature dependence of the zero-field Mössbauer spectra previously observed in oxyhemoglobin and indicating thereby that the globin does not strongly influence the electronic structure of the iron in the protein.

It has been shown (Lang, 1970; Kirchner and Loew, 1977) that the Pauling and Griffith models can be compatible with Mössbauer results. The models of Weiss and Gray have not been sufficiently developed to permit calculations of the efg parameters. The ozone model is incompatible with the Mössbauer results; this model predicts an axial efg whose largest component is positive and lies in the heme plane. The component normal to the heme, defined by the authors as the z direction, is negative, but this component is not the one of largest magnitude. At this point we must emphasize that, unless the axis of quantization is determined by other

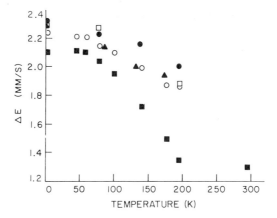

Fig. 5.   Quadrupole splitting as a function of temperature for the oxygenated materials referenced in Table I. Symbols are (O), oxyhemoglobin; (●), Oxymyoglobin; (□), $FeO_2(THT)P$; (■), $FeO_2(N-Me\ imid)P$; (▲), oxyheme-$(OMe)_2(C_5H_5N)$.

considerations, when Mössbauer spectroscopists define electric field gradients, the z direction is, by convention, one along which the local efg tensor has its largest component. As such, the z axis may or may not be along an axis of crystallographic or molecular symmetry or even along an axis defined by electron paramagnetic resonance measurements of g values. With almost all heme dioxygen complexes, difficulties of interpretation arise when the oxidation state of the iron is considered. Because in the Pauling model the iron is in the low-spin ferrous state with all six 3d electrons paired in the three planar t_{2g} orbitals, in principle there should be no 3d electron contribution to the quadrupole interaction. Moreover, if

TABLE I

Quadrupole Splittings[a] of Oxygenated Heme Proteins
and Model Compounds at Various Temperatures

	4.2 K	77 K	195 K	Reference
Oxyhemoglobin	2.25	2.13	1.89	Spartalian and Lang (1976)
Oxymyoglobin	2.31	2.23	2.00	Spartalian and Lang (1976)
Cytochrome P450	2.15	2.12	2.07	Sharrock et al. (1973)
		(82 K)	(200 K)	
$FeO_2(N-Me\ imid)P$	2.11	2.04	1.39	Spartalian et al. (1975)
$FeO_2(THT)P$	2.29	2.27	1.87	Spartalian and Lang (1976)
Oxyheme $(OMe)_2(C_5H_5N)$	2.28	2.14	1.94	Marchant et al. (1972)
		(83 K)	(172 K)	

[a] In Mössbauer units of mm/s.

there were such a contribution because of valence–electron delocalization by covalency effects, it ought to be temperature independent since there can be no orbitals available for promotion of electrons to higher energy states as shown by the diamagnetism at room temperature.

Attempts to explain quantitatively the temperature dependence of the zero-field spectra have so far met with limited success. Any such explanation requires the adoption of a model for oxygen binding and, at present, the Pauling model is widely accepted. Cianchi *et al.* (1976), following Lang's 1970 suggestion that the dioxygen molecule is able to rotate about the Fe-O axis, developed a simple quantitative description for the reduction of the observed quadrupole splitting with increasing temperature by assuming that the oxygen molecule rotates in an adiabatic potential with four harmonic oscillator-like potential wells centered along the direction of the pyrrole nitrogens. Although the agreement between calculation and experiment in the model of Cianchi *et al.* is impressive, this model can be criticized on the grounds that it has no provisions for the observed asymmetric broadening of the absorption lines, but instead it predicts symmetric lines of natural linewidth. The treatment of Kirchner and Loew (1977) starts by assuming rotational motion of the dioxygen, uses molecular orbital theory to account for the size and the temperature dependence of the quadrupole splitting correctly, but it also implies symmetric line shapes. It should be noted that the published values for the quadrupole splitting were based on fits of Lorentzian line shapes to lines that, at intermediate temperatures, are far from being Lorentzian. Lorentzian fits were done as first attempts to characterize spectra numerically; any effort to establish a detailed theory must not neglect the line shapes, especially since they appear to be a distinctive feature of the spectra.

Such a theory was presented by Spartalian *et al.* (1975) in the interpretation of the zero-field spectra of $FeO_2(N-Me \text{ imid})P$. These were analyzed in terms of a dynamic relaxation model based on the published crystal structure (Collman *et al.*, 1974). Briefly, this model is as follows: The dioxygen molecule may make transitions among the four possible orientations or conformational states observed in the crystallographic studies. These orientations are equivalent in opposing pairs that are separated in energy by an amount E_0; each pair produces a different efg at the nucleus. As the temperature is varied, the oxygen molecule relaxes between orientations with a characteristic relaxation rate that depends on temperature. Furthermore, the contribution of each orientation to the efg at a given temperature must be weighted by the appropriate Boltzmann factor. The above model was mathematically formulated by adapting the relaxation theory of Tjon and Blume (1968), and the experimental spectra were least-squares fitted by computer to obtain the pertinent parameters. The solid lines in Fig. 3 are the results of such fits. On the basis of the qualitative similarities of the zero-field spectra of the model compound and the oxyproteins, Spartalian and Lang (1976) concluded that conformational motion of oxygen is also the case in the latter. They did not provide, however, a rigorous mathematical treatment of the protein spectra.

Recent measurements of Cerdonio et al. (1977, 1978) have shown that contrary to all previous measurements, above 90 K there are paramagnetic contributions to the magnetic susceptibility of oxyhemoglobin. These contributions indicated to the authors the existence of an excited triplet state at about 150 K above the diamagnetic ground state. The result of Cerdonio et al. seems to militate against the conformational excitation hypothesis and, in fact, the temperature dependence of the quadrupole splitting is not incompatible with promotion of an electron from the $S = 0$ ground state to an $S = 1$ excited state. One wonders whether the singlet–triplet separation of 150 K reported by Cerdonio et al. bears any similarity to the separation $E_0 = 147$ K between the two conformational states in the relaxation model of Spartalian et al. (1975). Additional evidence against the conformational hypothesis is provided by the x-ray measurements on oxyhemoglobin by Heidner et al. (1976) and on oxymyoglobin by Phillips (1978). These measurements indicate that the dioxygen molecule is sterically hindered from rotating in the heme pocket. In conclusion, we may state that the question whether conformational or electronic excitation is responsible for the decrease of the quadrupole splitting in the oxyproteins and the model compounds cannot be answered by Mössbauer spectroscopy alone. Further magnetic studies such as additional susceptibility measurements on the model compounds or observation of NMR paramagnetic shifts ought to clarify the issue. The possibility of acquiring information on the oxygenated complexes through additional Mössbauer experiments (e.g., in high applied fields at elevated temperatures) is a point that will be discussed in Section VI.

III. Deoxygenated Complexes

A. *Deoxyhemoglobin, Deoxymyoglobin, and Synthetic Analogs*

In the deoxygenated form, hemoglobin and myoglobin have their iron in the ferrous $S = 2$ state. In this state the Fe^{2+} ion is too large to fit into the space between the porphyrin nitrogens, and so it moves out of the porphyrin plane in the direction of the proximal histidine. Recent measurements indicate that the displacement is smaller than originally was thought,

with the iron-mean porphyrin plane distance being 0.33 Å in oxymyoglobin and 0.55 Å in deoxymyoglobin (Phillips, 1978). In the Perutz model of
cooperativity (Perutz et al., 1974), the iron displacement produces a distortion of the protein structure which propagates in hemoglobin to the
subunit contact interfaces, and alters the R-to-T equilibrium as described
in Section I. The quaternary structure, on the other hand, affects the
interfacial interactions and, via structural distortions, alters the iron environment in both the oxy and deoxy sites. A complete understanding of
the subtleties of heme–heme interaction will require first an understanding
of the basic nature of the iron ion in the deoxy material, and this is the
topic on which we now concentrate.

Mössbauer measurements on deoxyhemoglobin and deoxymyoglobin
have been made by a number of investigators, and the results of zero-field
measurements are summarized in Tables II and III. Briefly, in each case an
isomer shift of about 0.9 mm/s is observed at 4.2 K. This is within but on
the low side of the range seen in high-spin ferrous materials (Greenwood
and Gibb, 1971); its low value would be characteristic of considerable
d-electron back donation to the ligands, relative to the more strongly ionic
typical high-spin ferrous compounds. The variation of the isomer shift with
temperature is about as one would expect on the basis of second-order
Doppler effect (Cohen, 1976) in a material with moderately low Debye
temperature. Values in the literature are in general agreement with the
exception of some (Huynh et al., 1974) that apparently are uncorrected for

TABLE II

Zero-Field Mössbauer Parameters of Human Deoxyhemoglobin
at Selected Temperatures[a]

T (K)	ΔE (mm/s)	δ (mm/s)	Reference
4.2	2.38 ± 0.02	0.83 ± 0.02	
81.4	2.32 ± 0.02	0.83 ± 0.02	
130	2.16 ± 0.02	0.85 ± 0.02	Huynh et al. (1974)
193	1.97 ± 0.02	0.80 ± 0.02	
4.2	2.392 ± 0.004	0.928 ± 0.010	
74.0	2.330 ± 0.005	0.920 —	Trautwein et al. (1976)
98.5	2.278 ± 0.004	0.919 —	
146.5	2.158 ± 0.004	0.903 —	
4.2	2.40	0.925	
35	2.39	0.923	
77	2.35	0.928	Kent (1979)
140	2.18	0.899	

[a] Isomer shifts with respect to metallic iron.

TABLE III

Zero-Field Mössbauer Parameter for Deoxymyoglobin
at Selected Temperatures[a]

T (K)	ΔE (mm/s)	δ (mm/s)	Species	Reference
4.2	2.22	0.920		
48.0	2.21	0.918		
77.0	2.17	0.910	Horse	Kent *et al.* (1977)
140.0	1.95	0.878		
195.0	1.78	0.842		
4.2	2.29	—		
45.0	2.25	—		
80.0	2.20	—	Sperm whale	Eicher *et al.* (1976)
145.0	2.00	—		
200.0	1.84	—		
77	2.286 ± 0.017	0.900 ± 0.017	Sperm whale	Gonser *et al.* (1974)
83	2.02	0.92	Sperm whale	Trautwein *et al.* (1970)

[a] Isomer shifts with respect to metallic iron.

adventitious variations in source temperature. In any case, detailed consideration of the temperature dependence of isomer shift has not proved to be fruitful thus far.

Reported values of the quadrupole splitting and its temperature dependence are not in perfect agreement, but the general features appear to be consistent. Variations are partly due to small species differences (see Section IV for more on this point). It is also likely that variations of pH, ionic strength, and freezing rate cause slight differences. There appears to be a tendency for the left line of the quadrupole pair of Hb to be broader than the right, possibly implying that the α and β subunits have slightly differing quadrupole splittings and chemical shifts.

A number of theoretical attempts have been made to explain in detail the quadrupole splittings of Mb and Hb and their temperature dependence. Initially these were carried out assuming that the heme normal must be a symmetry axis, a twofold or possibly fourfold rotation axis. Eicher and Trautwein (1969), for example, were able to account for their Mössbauer measurements of rat hemoglobin using a model in which the $^5B_2$ state (extra electron in $|xy\rangle$) lay 247 cm^{-1} below a ^3E state. This gave a positive efg with principal direction along the heme normal, or rather along the axis of assumed near-fourfold symmetry.

The first experiment bearing directly on the question of efg orientation was the measurement by Trautwein of the quadrupole-split Mössbauer spectrum of an enriched single crystal of Mb (Gonser *et al.*, 1974). In this experiment the intensity ratio I_h/I_l of the high and low energy absorption lines is determined as a function of gamma-beam direction. This, in turn, is related to the orientation of the efg tensor within the crystal. An unfortunate complication is the presence of two inequivalent iron sites in each unit cell, which are related by a twofold screw axis, the crystalline b axis. An initial interpretation was made under the assumption that the efg has a zero or near-zero asymmetry parameter. The data then implied that the quadrupole interaction is positive, i.e., $V_{zz} > 0$, where V_{zz} is the principal second potential derivative of largest magnitude. Two solutions were possible. The preferred one had the z direction in the heme plane but still implied that there is approximate C_{2v} symmetry in the structure, with the heme normal a twofold axis.

In examining the literature concerning Mössbauer studies of hemoglobin and myoglobin, especially the theoretical treatments, the reader should pay particular attention to coordinate system orientation. The convention of making the heme normal a z axis is a common one, and there is no reason that this should necessarily be compatible with the V_{zz} convention. Theoretical treatments require coordinate systems, and there is a strong temptation to identify the z axis of the calculation system with the heme normal. The reader may safely assume that hemoglobin and myoglobin do not know the alphabet and be on the lookout for unjustified inferences to the contrary.

Following the single crystal measurements, a number of theoretical treatments were carried out attempting to account for the quadrupole interaction and in some cases for the observed susceptibility as well. Huynh *et al.* (1974) were able to fit their human deoxyhemoglobin quadrupole-splitting data and the susceptibility measurements of Nakano *et al.* (1971) by using low-lying ^5E and $^5B_2$ states with a $^1A_1$ state at about 300 cm^{-1}. Their calculated efg is positive, with principal value lying in the xy plane. They claim that this is supported by the single crystal myoglobin results, but the association of their z direction with the heme normal appears arbitrary, especially in view of the very large rhombic distortion. A more reasonable interpretation might be that the single crystal results, assumed to apply to hemoglobin, imply that the x direction of the calculation of Huynh *et al.* (1974) lies in the heme plane. Agreement may be claimed with respect to the positive efg sign, though later developments (see below) suggest it is in fact negative. A similar energy level scheme was found by Eicher *et al.* (1976) who treated both hemoglobin and myoglobin, finding positive efg in each calculation, and $\eta = 0.5$ and 0.7, respectively. They claim agreement with the efg orientation as found from the single crystal experiment but again have no independent way to orient the axes of their calculation with the heme structure. In a more ambitious effort, Bade and Parak (1976) have attempted simultaneous least-squares fits to both susceptibility and quadrupole data on human hemoglobin. They do not report the sign of the efg, but the preferred solution is in the neighborhood of that of Huynh *et al.* (1974) and should have the same sign.

Zimmermann (1975) pointed out that when two sites are present and related as in the Mb crystal, an infinity of solutions for the efg orientation are possible. There exists a corresponding range of values of η, the asymmetry parameter, and both signs of the efg normally are possible and consistent with the single crystal data. On the basis of this information, Maeda *et al.* (1976) reinterpreted the single crystal data and produced a solution with V_{zz} positive and $\eta = 0.4$, in agreement with the existing electronic calculations. Both the heme normal and the Fe–N directions were near-principal axes, with z, the axis of largest second potential derivative, lying in the heme plane. Theoretical treatments were now in seeming agreement with Mössbauer and susceptibility results. A common feature of all the theories is that they appeared to demonstrate that a solution would not be possible within the ordinary ^5D configuration. It is possibly an oversimplification, but the difficulty can be characterized as follows: In order to produce the observed decrease of ΔE as temperature increases, a first excited orbital state should lie at about 300 cm^{-1}. With the symmetry imposed, and within ^5D, this implies that spin–orbit coupling will strongly mix the orbital states, resulting in a ground quintet of states in which the individual members will correspond to different spatial distributions of electronic charge. A characteristic feature of this is the appearance of a maximum in the ΔE versus T curve at a few tens of degrees (Ingalls, 1964). No such maximum is observed. In order to account for the overall shape of the ΔE versus T curve, it was concluded that a state of different spin multiplicity must lie rather low in energy. With the additional degrees of freedom thus provided, it was possible to account satisfactorily for the available data.

As the result of a series of studies of Mb in high magnetic field, Kent *et al.* (1977) were able to determine the sign and asymmetry parameter of Mb. In a diamagnetic complex the sign and asymmetry parameter of a quadrupole interaction can be determined in a straightforward manner, even in a powdered or frozen solution specimen. The spectrum is measured in high magnetic field and is fitted numerically, using a Hamiltonian in which the sign (V_{zz}) and the value of η are varied to achieve the best fit. The magnetic hyperfine interaction with the external applied field is known completely, and the calculation is straightforward. In the case of Mb, the interpretation is complicated by the paramagnetic contributions to the magnetic interaction. Lacking a theoretical model to provide this internal contribution, Kent *et al.* (1977) applied a phenomenological treatment in which it was assumed that at a given temperature there exists a factor ω_i, which specifies H_i^{int}/H_i^{app}, the ratio of the corresponding components of internal and applied field. Satisfactory fits were obtained with the ω and efg principal axis systems coincident; the ω_i were found to decrease with increasing T in the expected manner. The fits implied a negative V_{zz} with $\eta \approx 0.7$, in contradiction with the assumptions underlying all published theoretical treatments. Although this additional knowledge would ordinarily reduce the efg orientations to four possibilities (Zimmermann, 1975), the particular data values happen by chance to give a solution at the intersection of two families. Thus there remains only the twofold uncertainty always imposed by the presence of two inequivalent sites in the unit cell. The possible orientations are tabulated in Kent et al. (1977), and we show them graphically in Fig. 6, adapted from the thesis of Kent (1979). The orientation shown in the upper drawing has the interesting feature that the plane of the proximal histidine is approximately a principal plane of the efg.

In the thesis of Kent (1979) and in a related paper (Kent et al., 1979), the high field measurements have been extended to hemoglobin and to two synthetic analogs: These materials exhibit a great similarity in their Mössbauer spectra observed in high field at temperatures greater than 30 K. The anisotropy of the internal field factors ω_i is about what one would expect if the extra electron were in an orbital of form $|3z^2 - r^2\rangle$, with

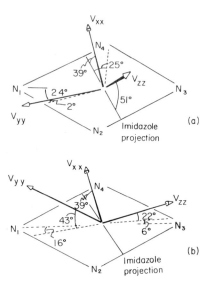

Fig. 6. (a) and (b): the two possible orientations of the electric field gradient tensor relative to the heme nitrogens in myoglobin. Some of the principal axes have been reversed in direction for clarity. The heme plane is viewed from the distal side.

little unquenched orbital angular momentum. Such an orbital would also be capable of producing the observed negative V_{zz}. If this description is accepted, the efg alignment of Fig. 6a is reasonable for it implies that the extra electronic charge, concentrated along the z axis, will neatly avoid the five nitrogen ligands. A less satisfying feature of Kent's interpretation is concerned with the strength of the internal field. It is possible to argue, in general terms, as follows: The susceptibility measurements on Hb and Mb (Nakano et al., 1971; Alpert and Banerjee, 1975) yield a value of n_{eff} at moderate temperatures (≈ 50 K or greater), which is close to the $S = 2$ free-spin value. This implies that little unquenched orbital angular momentum is present. Consequently, it is a good approximation to neglect orbital contributions to the hyperfine field, assume the direction-averaged dipolar contribution is zero, and attribute the direction-averaged internal field to the contact interaction. If this is done, the apparent hyperfine field per unpaired electron becomes about 5 T, in contrast to the more generally accepted value of about 11 T or 12 T. The smallest value for this quantity, which comes easily to mind, is the 8.5 T of $K_3Fe(CN)_6$ (Oosterhuis and Lang, 1969). For ferric hemoglobin fluoride it is 11 T. The discrepancy may be an independent indication that a state of different spin multiplicity plays an important role in the electronic description of the predominantly $S = 2$ iron complex in the deoxygenated proteins.

A final observation should be made about Kent's results. Although similar spectra were observed for proteins and synthetic analogs at temperatures over 30 K, the spectra varied widely at low temperature. One of the synthetic models exhibited a spectrum with distinct lines at 4.2 K in a 6 T field and was rather well simulated in a spin Hamiltonian treatment with $S = 2$, $D = -5$ cm^{-1}, and E/D small. The appearance of internal field at very low values of applied field provided an independent indication that the lowest state is degenerate or nearly degenerate. In contrast, the corresponding myoglobin spectrum is very diffuse in appearance, and the myoglobin susceptibility measurements imply a spin Hamiltonian parameter $D = +5$ cm^{-1}. Thus it seems possible in very similar geometries to have very similar angular momentum and spatial distribution of charge and spin in the ground quintet as a whole, while at the same time having quite different internal structure, i.e., different character when the substates are considered individually.

B. Anhydrous Hemoglobin and Myoglobin

Mössbauer spectra of anhydrous hemoglobin (AHb) were first recorded by Grant et al. (1967). They observed two-component spectra, each component occupying about 50% of the total area; one component was attributed to the usual spectrum of deoxyhemoglobin and the other to the low-spin $S = 0$ spectrum of hemochrome presumably formed by ligation of the distal histidine to the sixth coordination position. On the basis of their experiments, Grant *et al.* concluded that anhydration denatures one particular species of the hemoglobin chains, leaving the other species of chains unaffected. Trautwein *et al.* (1970) continued these experiments by recording Mössbauer spectra about 80 K of AHb, anhydrous myoglobin (AMb), and bispyridine hemin, the latter being an analog to the low-spin hemochrome component. They reproduced the results of Grant *et al.*, but they also observed that the spectra from AMb show no denaturation effects because no spin quenching upon anhydration was observed. Moreover, the bispyridine-hemin spectra exhibited quadrupole splittings and isomer shifts similar to the hemochrome component. Thus, they concluded, the chains whose tertiary structure remains unaffected in AHb must be the ones closest resembling Mb, viz., the α chains, while the β chains become six coordinated with the distal histidine occupying the sixth ligand position. In order to verify this conclusion, experiments on separated chains were performed by Chow and Mukerji (1975) and Papaefthymiou *et al.* (1975). These experiments showed that essentially the same 50–50 ratio of denatured to normal components found in whole AHb was also present in each case of the separated anhydrous α and β chains. This result seems to overrule the hypothesis of Trautwein *et al.* that one species of chains undergoes denaturation. Papaefthymiou *et al.* attributed the absence of denaturation effects in AMb to the hardiness of its tertiary structure so that anhydration does not appreciably disrupt hydrophobic bonding with subsequent formation of hemochrome.

IV. Special Topics in Hemoglobin

A. *Isolated and Selectively Enriched Deoxyhemoglobin Chains—Cooperativity*

The cooperativity of the hemoglobin tetramer leads to the question whether all four iron sites are identical or whether there are subtle differences among them. It is reasonable to assume that such differences would be reflected in the electronic structure at the sites. With recent advances in biochemical techniques, it has become possible to separate the chains into α and β subunits and investigate each dimer by itself. More importantly, it has also become possible to enrich the sites selectively with ^{57}Fe so that the sites can be separately investigated in the fully functional tetramer by the Mössbauer effect.

Mössbauer spectra of isolated subunits of deoxyhemoglobin (Huynh *et al.*, 1974) showed no difference in the values of the quadrupole splitting and isomer shift from those of the tetramer. On the basis of their measurements, the authors concluded that there is no difference in the electronic structure of the iron in the three cases. Susceptibility measurements, however (Alpert and Banerjee, 1975), showed that at low temperature only the susceptibility of the isolated α and β dimers is somewhat higher than that of the tetramer. A difference in the isolated dimers was also observed by subsequent Mössbauer measurements (Trautwein *et al.*, 1976) that showed consistently higher values for the quadrupole splitting of the isolated subunits than the tetramer. In the same work, Trautwein *et al.* also reported the Mössbauer parameters of the hemoglobin tetramer with selectively enriched α chains and found no difference between it and fully enriched deoxyhemoglobin. Thus the authors concluded that the α and β subunit association is accompanied by a measurable ligand modification of the heme iron.

An important feature of the mechanism of cooperativity proposed by Perutz is the motion of the heme iron away from the heme plane as the proximal histidine pulls on the iron upon deoxygenation (Perutz, 1972; Perutz *et al.*, 1974). In order to study the histidine–metal linkage with Mössbauer spectroscopy, Srivastava *et al.* (1977) recorded zero-field Mössbauer emission spectra of oxy- and deoxyhemoglobin in which the iron was substituted with ^{57}Co. The values of the Mössbauer parameters obtained by the authors were different from those obtained by conventional absorption spectroscopy. The emission spectra showed temperature-independent quadrupole splittings at about 2.15 mm/s with isomer shifts at 0.35 mm/s for oxy ^{57}CoHb; deoxy ^{57}CoHb also gave temperature-independent quadrupole splittings at about 1.84 mm/s with isomer shifts at about 0.6 mm/s. On the basis of these measurements, the authors argue that the ^{57}Fe daughter atom in ^{57}CoHb finds itself in an environment different from the one in the natural protein and stays "frozen" in the heme plane, the globin not permitting the iron to move to its normal position out of the plane. Thus Srivastava *et al.* conclude that their emission studies do not support Perutz's mechanism of cooperativity. In view of the very large linewidths reported and the numerous complications inherent in emission experiments (Wickman and Wertheim, 1968), we would regard the conclusion as lacking firm experimental support.

B. *Photodissociation Experiments*

The oxy and carbonyl complexes of hemoglobin and myoglobin have been known to photodissociate, and the photodissociated component can be stabilized by lowering the temperature below 20 K (Austin *et al.*, 1975).

Mössbauer spectra of photodissociated HbO_2, MbO_2, and HbCO have shown that the iron is in the high-spin ferrous state, as expected from these proteins with an empty sixth ligand position (Spartalian et al., 1976). The photodissociated components formed upon irradiation of HbO_2 and HbCO are closely similar. Both photoproducts have larger quadrupole splitting and smaller isomer shift than the ordinary deoxygenated protein. The values of the quadrupole splitting and isomer shift for deoxyhemoglobin, its isolated subunits, and its photodissociated forms are shown in Table IV and plotted in Fig. 7. Included for comparison are myoglobin and its photoproducts. The striking similarity between the Mössbauer parameters of the isolated β chain sample (β_{SH}) and photodissociated hemoglobin might be interpreted as follows: The isolated subunits are in the R (low oxygen affinity) state and so is the six-coordinated HbO_2 or HbCO prior to photodissociation. After photodissociation, the protein is expected to re-

TABLE IV

Mössbauer Parameters at 4.2 K of Deoxyhemoglobin, Its Isolated Subunits, Its Photodissociated Forms, and the Mass Hybrid $Hb(\alpha - 57, \beta - 56)$[a]

	ΔE (mm/s)[b]	δ (mm/s)[b]	Reference
Hb	2.392	0.928	Trautwein et al. (1976)
Hb	2.419	0.918	Spartalian et al. (1976)
$Hb(\alpha - 57, \beta - 56)$	2.381	0.938	Trautwein et al. (1976)
Hb	2.203 (2.198)	0.911 (0.911)	Spartalian et al. (1976)
α_{SH}	2.422	0.911	Trautwein et al. (1976)
$MB^*(O_2)$	2.382 (2.389)	0.905 (0.899)	Spartalian et al. (1976)
$Mb^*(CO)$	2.279 (2.276)	0.888 (0.889)	Spartalian et al. (1976)
β_{SH}	2.492	0.882	Trautwein et al. (1976)
$Hb^*(O_2)$	2.551 (2.552)	0.898 (0.883)	Spartalian et al. (1976)
$Hb^*(CO)$	2.255 (2.548)	0.882 (0.878)	Spartalian et al. (1976)

[a] Deoxymyoglobin and its photodissociated forms are included for comparison.
[b] Isomer shifts with respect to metallic iron. The myoglobin parameters are weighted averages. Numbers in parentheses indicate results from a Voigt profile fitting procedure.

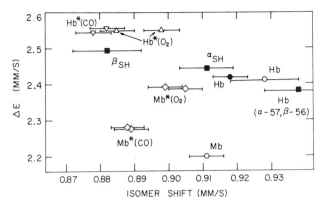

Fig. 7. Quadrupole splitting plotted against isomer shift for the materials listed in Table IV. Open symbols are taken from Spartalian *et al.* (1976), shaded symbols from Trautwein *et al.* (1976). Note the grouping of photodissociated hemoglobin with the isolated β subunits and of tetrameric hemoglobin with the isolated α subunits. Whenever applicable, both Lorentzian and Voigt profile results are shown.

main "frozen" in the R state at 4.2 K. If we now make the conjecture that the β subunits photodissociate before the α subunits, we ought to expect a similarity between the parameters of β_{SH} and Hb*(CO). A verification of this hypothesis can be obtained from Mössbauer spectra of a hemoglobin sample showing 100% photodissociation. In the experiments of Spartalian *et al.* (1976), at most 60% photodissociation was reported in HbCO. The Mössbauer parameters of photodissociated myoglobin are close to those of the isolated α_{SH} subunits. We should note that the amino acid sequence of myoglobin is similar to that of a hemoglobin α chain.

C. *Hemoglobin Variants. Species Survey. Sickle-Cell Hemoglobin*

In the above discussion of hemoglobin, we have purposely not identified the sources of the hemoglobin samples used by the various researchers in the field. In most cases human adult hemoglobin was used, however, one naturally wonders whether species variation is responsible for the differences in the Mössbauer parameters reported by the various authors. A species survey (Rhynard, unpublished) examined the Mössbauer spectra of oxygenated and reduced whole blood cells from cow, sheep, dog, chicken, rabbit, cat, rat, and trout. Zero-field spectra were recorded at 4.2, 77, and 195 K. None of the examined species showed any visible trends or stood out by itself in any way. No particular species showed consistently different values of its Mössbauer parameters from the others. The temperature dependence of the spectra from the oxygenated samples exhibited the same features as those of oxyhemoglobin mentioned in Section II; the spectra of the deoxygenated samples were similar to those of deoxyhemoglobin mentioned in Section III. The ranges of the Mössbauer parameters in this species survey are shown in Table V. It is indicated that the amino acid substitutions among the species make little, if any, difference to the electronic structure at the iron site.

The difference between sickle-cell hemoglobin (HB-S) and normal hemoglobin (Hb-A) is the substitution of the glutamic acid residues at the sixth position in the two β chains of Hb-A by valine. This causes the deoxy Hb-S molecules to form fibrous aggregates resulting in the "sickling" of the erythrocyte. A comparison study of the Mössbauer spectra of deoxygenated Hb-S and deoxygenated Hb-A was done over the temperature range from 80 to 250 K by Yen *et al.* (1976). The spectra of deoxy Hb-S were reported no different from those of normal deoxyhemoglobin. The only difference was an additional 10% component of hemochrome present in the sample of deoxy Hb-S; it was attributed to partial anhydration of the Hb-S molecule during sickling. Since the valine substitution occurs some distance away from the heme, it is not surprising that the Mössbauer spectra remain unaffected.

TABLE V

Lowest and Highest Quadrupole Splittings in the Species Survey[a]

	T (K)	ΔE (mm/s)	δ (mm/s)
	4.2	2.243 (rat) 2.255 (sheep)	0.269 0.270
Oxygenated cells	77	2.108 (dog) 2.165 (rat)	0.272 0.269
	195	1.841 (chicken) 1.953 (trout)	0.230 0.234
	4.2	2.368 (rat) 2.413 (dog)	0.933 0.929
Deoxygenated cells	77	2.279 (rat) 2.327 (dog)	0.921 0.916
	195	1.858 (trout) 1.970 (rabbit)	0.864 0.878

[a] Isomer shifts with respect to metallic iron.

V.   Less Common Heme and Nonheme Oxygen Carriers

A.   *Hemerythrin*

Hemerythrin is the oxygen-binding pigment of the sipunculid marine worms. It is an octamer of total molecular weight 108,000; each subunit contains two irons and is capable of binding one molecule of oxygen (Klotz and Klotz, 1955; Klotz and Keresztes-Nagy, 1963). In spite of the name, it is not a heme protein. The nature of the iron site in the oxy and deoxy forms, as well as in some ferric compounds, has been investigated by Mössbauer spectroscopy (Okamura *et al.*, 1969; Garbett *et al.*, 1971). With the help of ESR observations and some rough susceptibility measurements, a satisfactory understanding of the general nature of the system has been achieved. Observation of a number of ferric hemerythrin compounds yielded isomer shifts of about 0.5 mm/s at 4.2 K and quadrupole splittings near 2.0 mm/s with little or no temperature dependence. These were interpreted as high-spin ferric in a highly distorted environment because of the size of the quadrupole splitting, which is unusually large for this state. Lack of magnetic hyperfine interaction was interpreted as indicating antiferromagnetic coupling between iron sites, and is consistent with susceptibility observations. The coupling was interpreted as indicating that the iron sites are very close, probably bridged by the bound small ion (NCS^-, N_3^-, Cl^-, or F^-). Susceptibility results were in agreement in that they showed a decrease of susceptibility upon formation of these compounds from the deoxygenated protein. Oxyhemerythrin was found to exhibit two quadrupole doublets with quadrupole splittings 1.9 and 1.0 mm/s, both with isomer shift 0.5 mm/s. The greater intensity observed initially in the outer pair was identified as resulting from (ferric) methemerythrin contamination, and it was demonstrated that in oxyhemerythrin the two components were of equal intensity to within the experimental uncertainty of about 10%. In view of the similarity of isomer shifts and the lack of indication of unpaired spin, the oxygenated complex was interpreted as consisting of a pair of high-spin ferric iron atoms, bridged in an asymmetric way by μ-oxo and μ-peroxo ions, the former deriving from the water. Charge balance was attributed to the agency of hydroxyl ions or possibly ionized groups on the adjacent protein. In the deoxy form, a single quadrupole pair was seen at 4.2 K, with isomer shift 1.17 mm/s and quadrupole splitting 2.89 mm/s. The quadrupole splitting was essentially undiminished at 77 K. This is characteristic of high-spin ferrous iron with fairly large crystal field splitting, and was so interpreted. Some unpaired electron spin was detected in susceptibility measurements at higher temperatures, but at 4.2 K in strong applied fields no evidence of unpaired spin was seen in Mössbauer measurements. This was interpreted as indicating little or no iron–iron spin coupling, but a spin–orbit induced zero-field splitting parameter D which is large and negative, resulting in a ground doublet polarizable in a single direction only. It is not obvious to us why a large positive D is not equally likely. Hemerythrin demonstrates a case in which a modest program of measurement and qualitative theoretical treatment can make a significant contribution to the overall understanding of a biomolecule since little was known to begin with. Unfortunately, such situations are not very common because Mössbauer sample requirements are usually not particularly easy to satisfy relative to those of other physical methods.

B. Cytochrome P-450$_{cam}$

The enzyme cytochrome P-450$_{cam}$ of *Pseudomonas putida* serves to hydroxylate camphor and transfers electrons to molecular oxygen in the process. Although its function is neither oxygen transport nor storage, it does bind oxygen to a heme-iron site and has states analogous to oxy- and deoxyhemoglobin. It was additionally the first high-spin ferrous heme protein subjected to thorough study by high field Mössbauer spectroscopy (Sharrock *et al.*, 1973; Champion *et al.*, 1975). In the presence of bound camphor and oxygen, the 4.2 K quadrupole splitting of 2.15 mm/s was observed, with isomer shift 0.31 (Sharrock *et al.*, 1973). These values are fairly close to those of oxyhemoglobin, and P-450 differs mainly in having very little variation of its quadrupole splitting with temperature. If the interpretation of the ΔE variation in oxyhemoglobin in terms of conformal excitation is, in fact, correct, these results would suggest that the oxygen is more firmly positioned in P-450, as would be fitting in a protein which is chemically modifying rather than merely transporting the molecule.

Reduced P-450 has been found to have quadrupole splitting of 2.42 mm/s at 4.2 K and isomer shift 0.82 mm/s (Champion *et al.*, 1975). This isomer shift is 0.1 mm/s lower than that of deoxyhemoglobin and deoxymyoglobin, but the most striking feature is a very small temperature effect on ΔE. A number of measurements were made in applied magnetic fields and over a range of temperature, and these were all fitted reasonably well using a spin Hamiltonian. In contrast to myoglobin, the quadrupole interaction was found to be positive. Although a fairly large number of spin Hamiltonian parameters were used, a single set was found to cover all conditions of temperature and field. In particular, the fits were rather good at large H/T, where relaxation rate is irrelevant, and large T, where relaxation is almost certainly very fast. These cases are illustrated in Figs. 5 and 6 of Champion *et al.* (1975). It is interesting to compare some features of cytochrome P-450 with those of deoxymyoglobin and deoxy model compounds (Kent *et al.*, 1979). The temperature-independent ΔE of P-450 is, as stated earlier, indicative of a large crystal field splitting E_c between orbital levels. The parameter D, which describes zero-field splittings within the ground quintet, is expected to be proportional to λ^2/E_c, and so it should be smaller for P-450. In contrast, the apparent value of D in myoglobin is 5 cm^{-1} (Nakano *et al.*, 1971) and that of the 2-MeImFeTPP is -5 cm^{-1} (Kent *et al.*, 1979), while $D = 20$ K or 14 cm^{-1} for P-450. This apparent paradox in fact reinforces the observation of Kent (1979) that the observed zero-field splitting and the apparent absence of orbital angular momentum in his compounds are difficult to reconcile with the crystal field splitting implied by the observed behavior of the quadrupole splitting. In addition, the inclination of the zero-field splitting axes to the efg axes in cytochrome P-450 is in keeping with our recent unpublished spin-Hamiltonian simulation of the low-temperature high field Mössbauer spectrum of deoxymyoglobin using the D value found from susceptibility (Nakano *et al.*, 1971), but assuming a zero-field splitting axis transverse to the efg V_{zz} axis. Thus, although cytochrome P-450 is not identical to the other oxygen-binding heme compounds we have discussed, it seems to confirm that, in all cases, we are dealing with complicated low symmetry situations, which probably involve low-lying states of different spin multiplicity.

VI. Summary and Conclusions

Even though oxy- and deoxyhemoglobin and myoglobin have been exhaustively studied by Mössbauer spectroscopy, and a number of other methods as well, it is clear from the above discussion that the electronic structure of the iron in these proteins still eludes us. In oxyhemoglobin, susceptibility measurements, while capable of detecting the existence of a paramagnetic excited state, yield magnetic information that is averaged over all directions in a solution specimen. As such, the usefulness of this information is limited. If the spin–triplet excited state in oxyhemoglobin becomes unequivocally established by additional independent measurements, then the Mössbauer results from oxyhemoglobin would certainly provide crucial clues regarding the orbital nature of this spin–triplet. It is conceivable that one can find a number of spin–triplet orbital states in d^6 that would reproduce the temperature dependence of the quadrupole splitting. In order to find the actual orbital state, one would have to look for its signature in the magnetic hyperfine field at the nucleus. This is usually accomplished by examining the temperature dependence of the magnetic hyperfine field in high applied fields (~ 6 T). In the particular case of oxyhemoglobin, however, the assumed spin–triplet state is at 150 K above the ground state; when the temperature is high enough to populate this excited state appreciably, its contribution to the internal hyperfine field becomes vanishingly small. The determination of the electronic structure of oxyhemoglobin is, therefore, a laborious task. A full theoretical treatment of the problem ought to involve a mathematical formulation of the observed large quadrupole splitting in the ground state, formally a t_{2g}^6 configuration; it would involve the correct choice of an excited spin–triplet state explaining the temperature dependence of the quadrupole splitting and susceptibility, and, in order to account for the characteristic line shape of the zero-field spectra, it would also involve the consideration of some relaxation process between the ground and excited states. It seems to us that the information needed to fulfill this task is incomplete; with what is available now one can at best eliminate as unacceptable candidates some of the spin–triplet orbital states in d^6.

We can safely state that the electronic structure of deoxymyoglobin is close to being determined. We have seen how the possible orientations of the efg principal axes have been established with respect to the heme axes. We have also seen the development of the various theoretical treatments, their shortcomings, and the limitations imposed by these treatments on the electronic structure. We are fairly confident that a complete determination of the electronic structure of deoxymyoglobin is feasible with the existing information. Once the electronic model of deoxymyoglobin is established, the electronic charge and spin distributions in relation to the heme plane will be known; this is an important step toward correlating electronic structure with biological function. Our understanding of deoxyhemoglobin is not as complete. Single crystal spectra have not been recorded, and, therefore, the orientation of the efg principal axes with respect to the heme axes is unknown. The problem is further complicated by the existence of four iron sites per protein molecule. Even though all four sites appear to be equivalent in the tetramer, additional measurements in high applied fields are needed not only on the fully enriched tetramer, but also on selectively enriched α and β chains (mass hybrids) in tetrameric deoxyhemoglobin.

Collectively considered, the studies of the various low-spin (oxygenated) and high-spin (deoxygenated) ferrous complexes by Mössbauer spectroscopy have shown significant similarities among members in each group. It is clear that given the coordination of the iron ion, the electronic structure is nearly invariant from sample to sample. Amino acid substitutions far removed from the iron site, or even the absence of the globin as in the case of model compounds, do not seem to alter the spectra appreciably. It is obvious that Mössbauer spectroscopy is the method to use for the investigation of the electronic structure of the iron in a variety of environments. Its importance lies in its ability to provide a wealth of magnetic information, especially where other magnetic measurements fall short, viz., in integral-spin systems. Without Mössbauer spectroscopy our progress in understanding the electronic structure of ferrous hemoglobin, myoglobin, and their synthetic analogs would be severely limited. Above and beyond that, Mössbauer spectroscopy can examine problems relating to protein tertiary and quaternary structure changes when the effects of such changes on the iron site can be direct and immediate, i.e., when the conformation, coordination, or the oxidation and spin states of the iron are affected.

In closing, we feel compelled to make the regretful observation that as our knowledge in heme proteins expands, the easy experiments have already been done. The days of [57]Fe-enriched rats, spun-down blood cells, and back-of-the-envelope calculations are behind us. We are now faced with the challenge of using involved computations to simulate the Mössbauer spectra and predict the electronic structure of samples painstakingly obtained through complex biochemical preparation.

Acknowledgments

We wish to thank Dr. D. Rhynard for communicating to us the results of the hemoglobin species survey. The authors' studies reported here were supported by NIH grant HL-16860.

References

Alpert, Y., and Banerjee, R. (1975). *Biochim. Biophys. Acta* **405**, 144–154.

Austin, R. H., Beeson, K. W., Eisenstein, L., Frauenfelder, H., and Gunsalus, I. C. (1975). *Biochemistry* **14**, 5355–5373.

Bade, D., and Parak, F. (1976). *Biophys. Struct. Mech.* **2**, 219–231.

Cerdonio, M., Congiu-Castellano, A., Mogno, F., Pispisa, B., Romani, G. L., and Vitale, S. (1977). *Proc. Nat. Acad. Sci. U. S.* **74**, 398–400.

Cerdonio, M., Congiu-Castellano, M., Calabrese, L., Morante, S., Pispisa, B., and Vitale, S. (1978). *Proc. Nat. Acad. Sci. U. S.* **75**, 4916–4919.

Champion, P. M., Lipscomb, J. D., Münck, E., Debrunner, P., and Gunsalus, I. C. (1975). *Biochemistry* **14**, 4151–4158.

Chow, Y. W., and Mukerji, A. (1975). *Biochem. Biophys. Res. Commun.* **62**, 986–989.

Cianchi, L., Mancini, M., and Spina, G. (1976). *Lett. Nuovo Cimento* **16**, 505–508.

Cohen, R. L. (1976). *In* "Applications of Mössbauer Spectroscopy" (R. L. Cohen, ed.), Vol. I, p. 1. Academic Press, New York.

Collman, J. P., Gagne, R. R., Reed, C. A., Robinson, W. T., and Rodley, G. A. (1974). *Proc. Nat. Acad. Sci. U. S.* **71**, 1326–1329.

Coryell, C. D., Pauling, L., and Dodson, R. W. (1939). *J. Phys. Chem.* **43**, 825–839.

Eicher, H., and Trautwein, A. (1969). *J. Chem. Phys.* **50**, 2540–2551.

Eicher, H., Bade, D., and Parak, F. (1976). *J. Chem. Phys.* **64**, 1446–1455.

Garbett, K., Johnson, C. E., Klotz, I. M., Okamura, M. Y., and Williams, R. J. P. (1971). *Arch. Biochem. Biophys.* **142**, 574–583.

Goddard, W. A., and Olafson, B. D. (1975). *Proc. Nat. Acad. Sci. U. S.* **72**, 2335–2339.

Gonser, U., Maeda, Ya., Trautwein, A., Parak, F., and Formanek, F. (1974). *Z. Naturforsch.* **29b**, 241–244.

Grant, R. W., Cape, J. A., Gonser, U., Topol, L. E., and Saltman, P. (1967). *Biophys. J.* **7**, 651–658.

Gray, H. (1971). *Adv. Chem. Ser.* **100**, 365–389.

Greenwood, N. N., and Gibb, T. C. (1971). "Mössbauer Spectroscopy," p. 117. Chapman and Hall, London.

Griffith, J. S. (1956). *Proc. R. Soc. London Ser. A* **235**, 23–36.

Heidner, E. J., Ladner, R. C., and Perutz, M. F. (1976). *J. Mol. Biol.* **104**, 707–722.

Huynh, B. H., Papaefthymiou, G. C., Yen, C. S., Groves, J. L., and Wu, C. S. (1974). *J. Chem. Phys.* **61**, 3750–3758.

Ingalls, R. (1964). *Phys. Rev.* **133**, 3A, A787-A795.

Kent, T. A. (1979). Doctoral Dissertation, The Pennsylvania State Univ.

Kent, T. A., Spartalian, K., Lang, G., and Yonetani, T. (1977). *Biochim. Biophys. Acta* **490**, 331–340.

Kent, T. A., Spartalian, K., Lang, G., Yonetani, T., Reed, C. A., and Collman, J. P. (1979). *Biochim. Biophys. Acta* **580**, 245–258.

Kirchner, R. F., and Loew, G. H. (1977). *J. Am. Chem. Soc.* **99**, 4639–4647.

Klotz, I. M., and Keresztes-Nagy, S. (1963). *Biochemistry* **2**, 445–452.

Klotz, I. M., and Klotz, T. A. (1955). *Science* **121**, 477–480.

Lang, G. (1970). *Q. Rev. Biophys.* **3**, 1 − 60.

Maeda, Y., Harami, T., Trautwein, A., and Gonser, U. (1976). *Z. Naturforsch.* **31b**, 487–490.

Marchant, L., Sharrock, M., Hoffman, B. M., and Münck, E. (1972). *Proc. Nat. Acad. Sci. U. S.* **69**, 2396–2399.

Nakano, N., Otsuka, J., and Tasaki, A. (1971). *Biochem. Biophys. Acta* **236**, 222–233.

Okamura, M. Y., Klotz, I. M., Johnson, C. E., Winter, M. R., and Williams, R. J. P. (1969). *Biochemistry* **8**, 1951–1958.

Olafson, B. D., and Goddard, W. A. (1977). *Proc. Nat. Acad. Sci. U. S.* **74**, 1315–1319.

Oosterhuis, W. T., and Lang, G. (1969). *Phys. Rev.* **178**, 439–456.

Oosterhuis, W. T., and Spartalian, K. (1976). *In* "Applications of Mössbauer Spectroscopy" (R. L. Cohen, ed.), Vol. I, p. 141. Academic Press, New York.

Papaefthymiou, G. C., Huynh, B. H., Yen, C. S., Groves, J. L., and Wu, C. S. (1975). *J. Chem. Phys.* **62**, 2995–3001.

Pauling, L. (1977). *Proc. Nat. Acad. Sci. U. S.* **74**, 2612–2613.

Pauling, L., and Coryell, C. D. (1936). *Proc. Nat. Acad. Sci. U. S.* **22**, 210–216.

Perutz, M. F. (1972). *Nature (London)* **237**, 495 − 499.

Perutz, M. F., Ladner, J. E., Simor, S. R., and Ho, C. (1974). *Biochemistry* **13**, 2163–2173.

Phillips, S. E. V. (1978). *Nature (London)* **273**, 247–248.

Sharrock, M., Münck, E., Debrunner, P. G., Marshall, V., Lipscomb, J. D., and Gunsalus, I. C. (1973). *Biochemistry* **13**, 258–265.

Spartalian, K., and Lang, G. (1976). *J. Phys.* **37**, Coll. C6, C6195-C6197.

Spartalian, K., and Lang, G., Collman, J. P., Gagne, R. R., and Reed, C. A. (1975). *J. Chem. Phys.* **63**, 5375–5382.

Spartalian, K., Lang, G., and Yonetani, T. (1976). *Biochim. Biophys. Acta* **428**, 281–290.

Srivastava, T. S., Tyagi, S., and Nath, A. (1977). *Proc. Nat. Acad. Sci. U. S.* **74**, 4996–5000.

Taylor, D. S. and Coryell, C. D. (1938). *J. Am. Chem. Soc.* **61**, 1263–1268.

Tjon, J. A., and Blume, M. (1968). *Phys. Rev.* **165**, 456–461.

Trautwein, A., Eicher, H., and Mayer, A. (1970). *J. Chem. Phys.* **52**, 2473–2477.

Trautwein, A., Alpert, Y., Maeda, Y., and Marcolin, H. E. (1976). *J. Phys.* **37**, Coll. C6, C6191–C6193.

Weiss, J. J. (1964). *Nature (London)* **202**, 83–84.

Wickman, H. H., and Wertheim, G. K. (1968). *In* "Chemical Application of Mössbauer Spectroscopy" (V. I. Goldanskii and R. H. Herber, eds.), p. 548. Academic Press, New York.

Yen, C. S. *et al.* (1976). *Biochim. Biophys. Acta* **453**, 233–239.

Zimmermann, R. (1975). *Nucl. Instrum. Methods* **128**, 537–543.

METALLURGY

6

Analysis of Phases and States in Metallic Systems via Mössbauer Spectroscopy

Ulrich Gonser

Fachbereich Angewandte Physik
Universität des Saarlandes
6600 Saarbrücken, West Germany

Moshe Ron

Department of Materials Engineering
Technion
Haifa, Israel

APPLICATIONS OF MÖSSBAUER
SPECTROSCOPY, VOL. II

I. Introduction

In physical metallurgy the qualitative and quantitative identification of phases and states is of great significance; in addition, knowledge is required regarding defects of atomic dimension such as vacancies, interstitials, impurities, dislocations, various types of boundaries, etc., and of larger dimension such as precipitates, grain size, texture, etc. Here the term *phases* will not be restricted to crystallographic phases; in general, other phases such as order–disorder phases and magnetic phases will be considered. The term *states* will be used to indicate certain positions or configurations of atoms, on surfaces, for instance, as well as associations of atoms, defects, etc. The quantitative aspects of the phases and states combined with an understanding of the defects determine, to a great extent, the usefulness and applicability of metals and alloys in physical metallurgy and technology. In the first half of this century *macroscopic methods* were used for determining the mechanical, electrical, magnetic, optical, and thermodynamical properties of metals and alloys. By means of such methods as electrical resistivity, susceptibility, and specific heat measurements, average or integral values representing the whole sample are obtained, but it is often difficult to deduce information regarding specific phases, states, and defects.

In the past two decades a wealth of new methods useful in physical metallurgy have been developed. Some of these methods are *microscopic* and consist of a selected "point probe", which permits investigation of the immediate surroundings of an atom, its state, and its phase. Important advances have been made by one such method that was discovered twenty years ago—Mössbauer spectroscopy. The achievements of this method in application to physical metallurgy are documented in the "Mössbauer Effect Data Index", which is published every year (Stevens and Stevens, 1966–1976) and have been reviewed in a number of articles (Gonser, 1966, 1971; Zemčik, 1971, 1975; Janot, 1972, 1977; Ron, 1973; Flinn, 1973; Fujita, 1975; Schwartz, 1976).

II. Mössbauer Spectroscopy in Physical Metallurgy

In Mössbauer spectroscopy the point of observation is the resonating nucleus. More than 80 nuclei with more than 100 excited states have been utilized for Mössbauer spectroscopy. However, one isotope, Fe^{57}, has taken the lion's share of the experimental applications. More than 50% of the work in the field is performed with Fe^{57}. This isotope has the most favorable parameters for Mössbauer spectroscopy, although the natural abundance of Fe^{57} is only 2.19%. Physical metallurgists might say that it is kind of nature to see to it that the best Mössbauer isotope is an isotope of the most important element in their field. Naturally, this handy method has been utilized for many scientific and practical problems in physical metallurgy, while on the other hand, physical metallurgy has contributed significantly to the advancement of the method. This can be demonstrated, for instance, by inspecting the evolutionary sequence of refinements in the preparation of the most common sources. Usually, metallic matrices are chosen in which the parent isotopes are diffused, implanted, Coulomb excited, or produced by nuclear reaction. In the early days of Mössbauer spectroscopy, fcc stainless steel was used for Co^{57} sources. However, broadening of the spectrum occurs because of the distribution of small isomer shifts and quadrupole splittings in the manifold of environments in this alloy, and so stainless steel was replaced by a Cu matrix. Here, however, the thermodynamical properties are unfavorable. That is, the solubility of Co or Fe in Cu is unmeasurably small at room temperature. As a result, Co–Fe precipitation has frequently occurred, sometimes being observable after one or two years as a broadening of the single line or as a shoulder on it, or even as small side lines arising from magnetic hyperfine interactions. In the next period this difficulty was circumvented by the use of Pd or, less frequently, Pt and Cr matrices. However, problems were encountered with Pd and Pt due to the K and L x rays, respectively, and with Cr in the fabrication of appropriately thin layers, due to the brittleness of the metal. Eventually, sources in a Rh matrix became popular after the metallurgical knowhow was developed.

Mössbauer spectroscopy certainly cannot compete with ordinary methods of chemical analysis. However, with its high resolution of the electric and magnetic hyperfine interactions (isomer shift, magnetic hyperfine splitting, and electric quadrupole splitting), it provides valuable information regarding phases, states, and defects.

In the phase analysis of a multiphase alloy by Mössbauer spectroscopy, it is necessary that the phases containing resonance atoms be distinguishable on the basis of differences in the values of at least one of the hyperfine parameters (electric monopole, magnetic dipole or electric quadrupole). In the analysis, one correlates the fraction of a spectral component with the fraction of a particular phase present. Thus the distribution or the solubility of the resonance atoms is determined. (See Chapters 1 and 2 in Volume I of this treatise.)

The recoil-free resonance technique has the advantage that it is nondestructive. In the thin absorber approximation, the resonance signal of a homogeneous phase containing resonating isotopes is proportional to the amount present. Quantitative determinations require thickness corrections

(Williams and Brooks, 1975), and this means that the Debye–Waller factor f and, in some cases, the effects of preferred orientations of crystallites (texture) (Pfannes and Gonser, 1973; Gonser and Pfannes, 1974) and of polarization (Housley et al., 1969) have to be taken into account. Reference spectra of the various phases in least-squares fit Lorentzian representation, which can then be subtracted in appropriate steps from the measured spectrum, are useful. This stripping technique (Muir, 1968) is continued until every part of the measured spectrum is assigned to a specific phase.

The guidelines in physical metallurgy are the constitution diagrams indicating the phases existing in thermal equilibrium at various temperatures. The appropriate phase diagrams are given at the beginning of the following sections.

Three systems have been chosen as examples and as typical representatives: Cu–Fe, Al–Fe, and Ti–Fe. Many studies have been carried out in these systems, and extensive data have been published.

Also, these systems rank high in intrinsic interest and in applicability of the Mössbauer effect to the field of physical metallurgy. In addition, these systems allow one to demonstrate the advancement and sophistication that have been achieved in Mössbauer spectroscopy in recent years.

III. The Cu–Fe System

The Cu–Fe phase diagram (Hansen and Anderko, 1958) is shown in Fig. 1. This system is rather simple. No intermetallic phase exists, and relatively small mutual solubilities for both components are observed at elevated temperatures. We focus our attention on the copper-rich side to be seen in Fig. 1b. According to the phase diagram, the solubility of Fe in Cu is about 4 wt% or at.% at the melting point and drops sharply with temperature. Below 400°C the solubility becomes extremely or unmeasurably small. Extrapolation from high temperature data yields a rough estimate of ~ 0.027 at.% at 400°C. By fast quenching techniques, supersaturated solid solutions can be obtained. Subsequent annealing at elevated temperatures causes association of Fe atoms and precipitation.

Copper with a small addition of iron can be regarded as a kind of model matrix for various phases and states of Fe and, consequently, a large number of investigations have been carried out on this system (Gonser et al., 1963, 1965, 1966; Ridout et al., 1964; Bennett and Swartzendruber 1970; Window 1970, 1971, 1972; Hornstein and Ron, 1974; Campbell and Clark, 1974; Williamson et al., 1974; Campbell et al., 1976; Longworth and

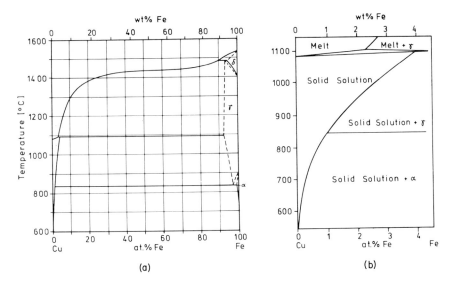

Fig. 1.   (a) The Cu–Fe phase diagram. (b) The copper-rich side is enlarged.

Jain, 1978; Clark, 1979). The following phases and states will be considered:

(A)   Fe in supersaturated solid solution;
(B)   Fe associations (dimers, trimers, etc., and clusters);
(C)   Fe-impurity associations;
(D)   Fe in coherent fcc precipitates;
(E)   Fe in surface states of fcc precipitates;
(F)   Fe in the antiferromagnetic state as fcc γ-Fe;
(G)   Fe in the ferromagnetic state as bcc α-Fe ($\gamma \to \alpha$ transformation after plastic deformation);
(H)   Fe in metastable phases (films);
(I)   Fe in the oxidized state.

A.   Fe in Supersaturated Solid Solution

At high temperatures, samples with low concentrations of Fe are solid solutions. When quenched rapidly into water, such samples produce spectra exhibiting mainly a single Mössbauer line corresponding to Fe in solution with an isomer shift (IS) of 0.225 mm/s relative to α-Fe at room temperature. If the measurement is made with a Co^{57}–Cu source at the same temperature as the quenched absorber, the single resonance line should be exactly symmetric and located at zero velocity. This is because the resonance conditions and atomic environments of source and absorber are equal and no shift can occur. Any additional resonance indicates that an association of Fe atoms has taken place and that the alloy deviates from randomness.

Figure 2 is the spectrum of a Cu–0.2 at.% Fe absorber after solution annealing at 970°C and quenching into water. As shown, the spectrum can be decomposed into two components: a central line corresponding to that of Fe in very dilute solid solution (that is, isolated Fe atoms with 12 nearest-neighbor Cu atoms) and a quadrupole resonance doublet indicating associations of Fe atoms. The intensity of the single resonance line (isolated Fe atoms) relative to that of the doublet (associations of Fe atoms) is critically dependent on the quenching conditions. Slower quenching rates reduce the intensity of the single line because more associations are created during the quenching process. Note that in order to produce a solid solution in this system, not only should the quenching be fast, but also the temperature from which the quenching starts should be well below 1000°C. This reduces the number of vacancies in thermal equilibrium at the start of the quench and thereby reduces the number of vacancies that can be stabilized during the quenching process. Because vacancies still have a certain mobility, even at room temperature, they can produce Fe associations as they go on their way to their final sinks.

The copper-rich side of the Cu–Co phase diagram is very similar to the copper-rich side of the Cu–Fe phase diagram. Therefore, very similar considerations are of great importance for the fabrication of Co^{57}–Cu sources having single resonance emission lines without any additional

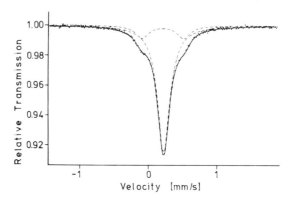

Fig. 2.   Mössbauer spectrum of a Cu–0.2 at.% Fe absorber after solution annealing at 970°C and quenching into water. Both absorber and Co^{57}–Rh source were at room temperature. The velocity scale has been adjusted to α-Fe (Krischel *et al.*, 1980).

components, as has already been mentioned. The states of fcc Co precipitates in Cu–Co alloys were investigated by Nasu et al. (1968).

At low temperatures, the resonance spectrum of Fe^{57} in a Cu matrix can be used as a probe of the local impurity magnetization which is proportional to the measured hyperfine field (Frankel et al., 1967; Steiner et al. 1974a, b). Thus information regarding the Kondo state is obtainable. In a dilute alloy, below some critical temperature that is characteristic of the alloy, anomalies occur because of strong coupling between the local electrons and the conduction electrons. This produces spin-compensated states or bound states, which can be destroyed by applying a high magnetic field. The nature of the Kondo state has been deduced mainly from measurements made by macroscopic techniques in which anomalies, such as a minimum in the electrical resistivity or a maximum in the specific heat, are observed. The microscopic observation by the Fe^{57} resonance has also contributed to the understanding of the Kondo state.

At high temperatures (in the vicinity of 1000°C), diffusion broadening of the spectrum of an iron–copper alloy has been observed (Knauer and Mullen, 1968). Two years after the discovery of Mössbauer spectroscopy, a broadening of the resonance line proportional to the diffusivity was predicted (Singwi and Sjölander, 1960). This broadening represents an apparent decrease in the time scale of the measurements. The inherent time scale is the lifetime of the excited state, but can be less if the atom is moving rapidly from one side to another. Thus the mean jump frequency of a resonance atom can be studied by measuring the diffusive line broadening (see Chapter 8).

B. Fe Associations (Dimers, Trimers, etc., and Clusters)

Quenching rate, temperature, Fe concentration, and subsequent annealing conditions will influence the occurrence of associations of Fe atoms. The temperature and quenching rate for the sample that produced the spectrum of Fig. 2 was chosen in such a way that besides Fe in solution, mainly dimers—nearest-neighbor Fe pairs—would be formed. Only a few associations of higher order such as trimers, etc., or of clusters might have occurred. Thus, the additional quadrupole split lines can be attributed to the presence of dimers. Isomer shifts (IS) and quadrupole splittings (ΔE_q) of Fe dimers at room temperature are listed in Table I. It has been deduced (Window, 1972) that the electric field gradient EFG is negative for Fe nearest-neighbor pairs in Cu and Au. This is consistent with the model that the electrons are localized along the Fe–Fe axis, making the Fe–Fe attraction stronger than the Fe–Cu bond. The relative areas of the two spectral components in Fig. 2—a single central line of about 88% and a quadrupole doublet of about 12%—are roughly proportional to the corresponding fraction of Fe atoms in monomers and dimers (trimers, etc.), respectively. This analysis is based on the thin absorber approximation and on the assumption that the effective Debye–Waller factor is isotropic and the same for both monomers and dimers. The latter assumption of the invariance of the Debye–Waller factor of the Fe atoms involved in various phases or states is usually made in this type of analysis.

A Cu alloy containing some Fe in solution and cooled from a high temperature, with considerably slower rates than the sample of Fig. 2, contains in its spectrum large resonance contributions outside the lines corresponding to Fe monomers and dimers. From this one deduces strong deviations from randomness. A large fraction of the spectrum now consists of many weak components having different quadrupole splittings and different isomer shifts corresponding to different arrangements of Fe atoms in clusters. In the analysis of small clusters of Fe atoms, the assumption of additivity is sometimes made; that is, the substitution of an iron atom for a copper atom as a nearest neighbor to an iron atom adds a certain incremental isomer shift, and an iron atom with n iron atoms as neighbors has n times as much incremental isomer shift as an atom with one-iron nearest neighbor. In this case—Fe clusters in a Cu matrix—the increment is about -0.03 mm/s for each additional Fe atom.

Attempts have been made to evaluate the EFG tensor, which gives rise to the quadrupole interaction, by making a point charge calculation of the contributions from individual neighbors. However, in metals the screening effects are difficult to take into account, and, in addition, complications arise due to the manifold of configurations of nearest-neighbor Fe atoms formed in the preprecipitation clustering process. Therefore, accurate fittings of spectra are not easy to accomplish or, one might say, that the resolution of the method is insufficient for this type of analysis. It has been assumed (Window, 1971) that Guinier–Preston zones are formed as platelets of Fe atoms on {100} planes; the platelets being a planar arrangement of Fe atoms, each having four nearest-neighbor Fe atoms with four

TABLE I

Isomer Shifts Related to α-Fe at Room Temperature

State of Fe in Cu	IS (mm/s)	ΔE_Q (mm/s)
Fe in solution	0.222	0
Fe pairs (dimer)	0.189	0.639 $(-)$
γ-Fe	-0.088	0

nearest-neighbor Cu atoms above the plane, and another four Cu atoms below it. According to the argument given earlier concerning the electron distribution in the Fe–Fe and Fe–Cu bonds, the electrons of the Fe atoms in such a cluster will tend to be localized in the {100} plane, that is, by symmetry considerations, equatorial or perpendicular to the principal axis of the EFG. This would suggest a sign of the EFG opposite to the case of the dimer, that is, a positive EFG.

The presence of an electric field gradient at the sites of the Fe nuclei having Fe nearest neighbors indicates that the local cubic symmetry is removed even though the symmetry of the lattice as a whole remains cubic. Thus the microscopic method of Mössbauer spectroscopy and the macroscopic method of x-ray diffraction give different indications about the symmetry because of differences in the space and area they are observing and probing.

In a sample containing small clusters of Fe atoms, the relative intensity of the line corresponding to Fe in solution can be increased by cold rolling. The plastic deformation causes a breakup of the Fe associations, producing a larger fraction of isolated Fe atoms (Window, 1972; Hornstein and Ron, 1974).

C. Fe-Impurity Associations

The formation of associations of Fe atoms in a copper matrix can be influenced strongly by additional impurity atoms as has been shown with Al, Ti, Mn, Ni, Zn, Rh, Pd, Pt, and Au (Window, 1971). In these cases the behavior deviates further from that expected for a random distribution of Fe atoms. The third metal also causes additional variation in the environment of the resonance atom by distorting the screening charges, for instance, or by forming different virtual bound states. These variations produce additional changes in isomer shift and quadrupole splittings. Some particularly interesting examples of Fe-impurity associations have been found in aluminum (see Section IV).

D. Fe in Coherent fcc Precipitates

After quenching, aging below the solubility temperature of iron-containing copper causes precipitation according to the phase diagram shown in Fig. 1. But even below the temperature of the $\gamma \rightarrow \alpha$ transition, the iron precipitates not in the thermodynamical stable bcc-α phase but rather coherently with the fcc copper matrix as fcc γ-Fe. Annealing time, Fe concentration, and temperature determine the size of the precipitates.

Electron microscopy indicates that the Fe precipitates have a spherical or cubic shape. The fraction of Fe atoms on the surfaces decreases rapidly during the growing process of the γ-Fe precipitates. With long annealing times, precipitates more than 1000 Å in diameter have been obtained. It turns out that this is a rather convenient way to stabilize relatively pure γ-Fe for investigations at low temperatures.

Altogether there are three possible ways to obtain γ-iron for low temperature studies:

(1)   coherent precipitation of fcc iron,

(2)   extention of the γ-phase region by alloying as, for instance, in austenitic steel, and

(3)   epitaxial evaporation of Fe on appropriate surfaces (see below).

A sample containing 3 at.% Fe was prepared by solution annealing at 1000°C and subsequently aged at 660°C for about one week. The Mössbauer spectrum of this sample shown in Fig. 3a consists essentially of a single resonance line corresponding to γ-Fe having an isomer shift of −0.088 mm/s relative to α-Fe at room temperature.

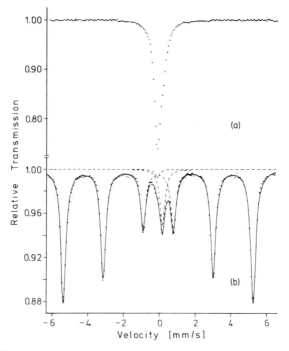

Fig. 3. Mössbauer spectra before (a) and after (b) cold rolling of a Cu sample containing γ-Fe precipitates, obtained with absorber and source (Rh) at room temperature. The γ → α transformation is evident by the appearance of the hyperfine pattern of the α-phase. The velocity scale has been adjusted to α-Fe.

In general, the Mössbauer data can be analyzed and interpreted in terms of the processes leading from supersaturated Fe solution to γ-Fe precipitates in the following way: Fe in solution and γ-Fe precipitates of large dimensions exhibit single resonance lines with isomer shifts of 0.222 and -0.088 mm/s, respectively, relative to α-Fe at room temperature. Of the 12 nearest-neighbor atoms that surround an Fe resonance atom, all are Cu atoms in the first case and all are Fe atoms in the second. The small second-nearest-neighbor interactions can be neglected. The two cases represent the initial and final states of a process where an Fe atom is originally in solution and eventually finds itself in the interior of a γ-Fe precipitate. Any resonances with isomer shifts other than the two limiting single-line values indicate the presence of clusters and surface states of Fe atoms having one or more Cu nearest neighbors. The isomer shift ranges from the positive value to the negative value, and the change is approximately proportional to the number of nearest-neighbor Fe atoms. In addition, the configuration of the neighboring Fe atoms will determine the magnitude and the sign of the EFG of the various possible associations and surface states formed in the precipitation process. The resulting spectra are superpositions of all the resonances originating from the manifold of configurations.

E. Fe in Surface States of fcc Precipitates

If the fcc γ-precipitates are relatively small—less than about 40 Å— plastic deformation by cold rolling will not transform the metastable γ-phase to the thermodynamically stable α-phase, in contrast to large γ-Fe precipitates. Nevertheless, the spectra of samples containing γ-Fe precipitates of about 25-Å size (≈ 1500 atoms) change significantly after cold rolling, as shown in Fig. 4 (Williamson et al., 1976). Before cold rolling, about one-third of the atoms in precipitates of that size are surface atoms. Considering the fact that large γ-Fe precipitates appear to be cubic with $\{100\}$ habit (Easterling and Weatherley, 1969) and that preprecipitates (Guinier–Preston zones) also occur on $\{100\}$ planes (Window, 1971), we may assume that 25 Å precipitates have cubic shapes. The Fe atoms on perfect $\{100\}$ interfaces are surrounded by eight Fe and four Cu nearest-neighbor atoms, schematically indicated in Fig. 4a. The Fe atoms on surface sites will produce quadrupole splittings due to the presence of EFG's with axial symmetry. The principal line in Fig. 4a corresponds mainly to Fe atoms in the interior of the fcc Fe precipitates—that is, to Fe atoms with 12 Fe nearest neighbors and, therefore, an isomer shift of -0.088 mm/s, relative to α-Fe at room temperature. In addition, in small precipitates a quadrupole doublet can be identified, one of whose lines

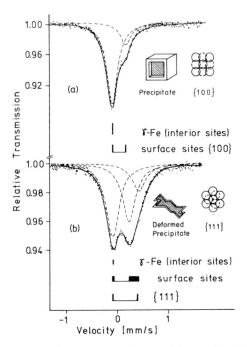

Fig. 4. Mössbauer spectra of a copper sample containing small γ-Fe precipitates at room temperatures before (a) and after (b) cold rolling. The atomic arrangements of the surface states corresponding to the line positions and the quadrupole splittings are indicated schematically on the right (Williamson *et al.*, 1976).

shows up as a shoulder, while the other line is hidden under the main resonance line. This quadrupole component will diminish with increasing size of the precipitates because the fraction of surface atoms relative to interior atoms decreases. Thus the spectral quadrupole component of large precipitates will effectively vanish (see Fig. 3a). In the small precipitates there will be some odd surface sites such as edges, corners, and deviations from the cube shape. Those will give rise to characteristic isomer shifts and splittings, but the method is incapable of resolving these specific sites.

After cold rolling, characteristic changes are observed as shown in Fig. 4b. In general, the resonances are isomer shifted toward the positive velocity side, indicating that on the average the Fe atoms have obtained an increased number of Cu nearest-neighbor atoms. During cold rolling plastic deformation occurs, and dislocations will traverse coherent precipitates on the {111} slip planes of fcc Cu. Thus we might expect substantial alteration of the shapes of the precipitates such that many are no longer cubic but rather elongated and flattened. In addition, the deformation process will produce the cold-rolling texture of copper: (011) [21$\bar{1}$]

(Wassermann and Greven, 1962). Significant in this case is the fact that by the movement of the dislocations through the precipitates, Fe {111} planes will be exposed and become surface planes. The Fe atoms on {111} interfaces are surrounded by nine Fe and three Cu nearest-neighbor atoms, schematically indicated in Fig. 4b. These surface atoms will produce quadrupole-split resonance lines with axially symmetric EFG. For these atoms the localization of electrons should be similar to that of Fe–Fe dimers. However, the sign of the EFG should be opposite to that for the dimers because the electrons are localized toward the six Fe–Fe nearest neighbors, and thus there will be an increase in electron concentration in the surface plane perpendicular or equatorial to the principal axis of the EFG. In addition, the localization should be more pronounced than in the case of {100} interface atoms with only four planar nearest neighbors, and therefore a larger EFG of the {111} interface is expected. Considering the small size of the precipitates and the shearing processes by the dislocations, a significant fraction of the Fe atoms will be located on odd (edges or corners) or imperfect surface sites.

The additional resonances on the positive velocity side—represented by two dashed resonance lines in Fig. 4b—have been indentified as consisting mainly of quadrupole-split components due to atoms in these new surface states. The second line is hidden under the line close to zero velocity. The relative line intensities change when the sample is rotated, giving proof that a quadrupole interaction is involved and that, in addition, a preferred orientation of the principal axes of the EFGs—texture—is present. A more refined analysis of such a spectrum has to take into account the distribution and orientation of EFGs and the distribution of isomer shifts corresponding to the variety of Fe surface configurations formed by the plastic deformation process.

F. Fe in the Antiferromagnetic State as fcc γ-Fe

Coherent γ-Fe precipitates in copper, austenitic steel, and epitaxial Fe on Cu {100} surfaces exhibit line broadening at temperatures below 77 K (Gonser et al., 1963). As an example, the spectra of epitaxially evaporated γ-Fe thin films of 18-Å thickness sandwiched between {100} Cu layers are shown in Fig. 5 at room temperature and liquid helium temperature (Keune et al., 1977). The spectrum at room temperature (a) is typical for γ-Fe. The weak satellite line most likely corresponds to odd sites or defects at the surfaces. The broadening in the spectrum (b) at 4.2 K is caused by the nuclear Zeeman effect due to antiferromagnetic ordering of fcc γ-Fe. Although the internal magnetic field is rather small and the characteristic six-line magnetic hyperfine pattern is not resolved, it was possible to estimate the strength of the magnetic field as about 18 kOe. The locations of the lines and the assumed relative line intensities in the fitting are indicated by the stick diagram; thus the spectrum can be understood as the superposition of six magnetic hyperfine lines and, in addition, the weak line. The dependence of the linewidth on temperature is shown in Fig. 6. The onset of line broadening indicates antiferromagnetic ordering. A similar curve is obtained in a simple fashion by measuring transmission only at the resonance peak using a constant velocity. Such a thermal scan is rather convenient for determining Néel and Curie temperatures. In this particular case a smearing of the transition is evident. This distribution of Néel temperatures might be explained by the size distribution of the island structure and defects of the 18-Å Fe film.

Magnetic susceptibility measurements have established the existence of antiferromagnetic ordering in stainless steel (Kondorskii and Sedov, 1959). Neutron diffraction results indicated the presence of spin structure in coherent γ-precipitates in copper (Abrahams et al., 1962) and in stainless steel (Ishikawa et al., 1970). In alternating ferromagnetic sheets in {100} planes, the spin vectors are oriented along ⟨100⟩ directions. The Néel temperatures, as determined by Mössbauer spectroscopy, are about 40 K for stainless steel, depending somewhat on composition, and 67 K for

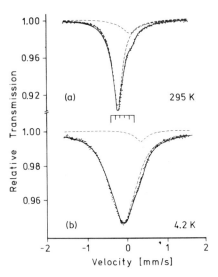

Fig. 5. Mössbauer spectra of an epitaxially evaporated γ-Fe thin film of 18-Å thickness (a) at room temperature and (b) at liquid helium temperature obtained with a Co^{57}–Rh source. The stick diagram indicates the line positions of the magnetic hyperfine pattern in the antiferromagnetic state (Keune et al., 1977).

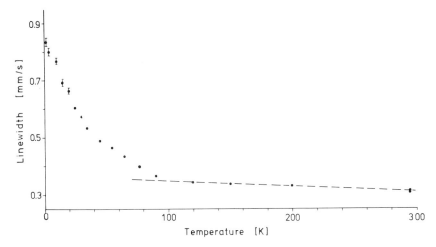

Fig. 6. Resonance linewidth versus temperature for the γ-Fe thin film of Fig. 4 (Keune *et al.*, 1977).

γ-iron precipitates (Gonser *et al.*, 1963). The Néel temperature, as a function of the size D of the precipitates, was also established (Williamson *et al.*, 1974; Williamson and Keune, 1975) and can be seen in Fig. 7. It has been pointed out (Ettwig and Pepperhoff, 1975) that the Néel temperature of 67 K for large γ-iron precipitates has to be considered a lower limit because copper impurities may remain in the coherent strained lattice. Indeed, from other measurements an upper limit of 90 K has been suggested for "pure" γ-Fe.

Some interesting and seemingly contradictory observations have been made concerning the magnetic ordering of fcc γ-Fe films: ferromagnetism was found using macroscopic methods (Wright, 1971; Gradmann *et al.*, 1976), while antiferromagnetism was established using the microscopic

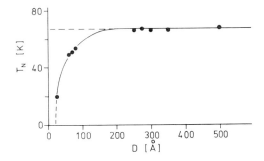

Fig. 7. Néel temperature T_N versus size D of γ-Fe precipitates in copper (Williamson and Keune, 1975).

Mössbauer effect (Keune *et al.*, 1977). At first glance it seems that these results are incompatible. However, there is an explanation for this conflict. In the first two cases the films were oriented parallel to {110} and {111} planes, while in the Mössbauer experiments the films were oriented on {100} planes. Thus one is tempted to conclude that a thin film of fcc γ-Fe can be either ferro- or antiferromagnetically ordered and the factor determining the ordering is the orientation of the film. Concurrently with the film orientation, adjustments in the lattice parameter at the coherent Cu–Fe interface may occur, and consequently magnetostriction might influence the ordering. According to the Bethe–Slater curve, the magnetic ordering is extremely sensitive to the lattice spacings: from Debye–Scherrer spots of "pure" γ-Fe precipitates with effectively no constraints a lattice parameter of 3.588 Å at room temperature has been obtained (Newkirk, 1957). At low temperature it becomes antiferromagnetically ordered. γ-Fe, fully coherent with the copper lattice—3.615 Å at room temperature— seems to occur on {110} and {111} surfaces and to be ordered ferromagnetically; thus the increase of the lattice parameter by only about 1% seems to be sufficient to change the sign of the exchange integral and stabilizing the ferromagnetic ordering.

G. Fe in the Ferromagnetic State as bcc α-Fe (γ → α Transformation after Plastic Deformation)

As we have seen, very fine coherent γ-Fe precipitates do not transform to the α-phase by cold rolling. A certain critical size is required ($\gtrsim 40$ Å) to produce the thermodynamical α-phase on plastic deformation (Gonser *et al.*, 1966). To demonstrate this, the 3 at.% Fe sample containing large γ-Fe precipitate, as shown in Fig. 3a, was used again. After cold rolling to about 50% of the original thickness, the changes in the spectrum are significant, as can be seen in Fig. 3b. The γ → α transformation can be followed quantitatively by phase analysis of the spectrum: the relative intensity of the six-line hyperfine pattern corresponds to the α-phase (91%), while the centerline corresponds to the remaining paramagnetic γ-phase, Fe in solution, clusters, and, possibly, very small α-phase particles exhibiting superparamagnetism (see Chapter 1). Further cold rolling increases the relative intensity of the hyperfine pattern only slightly. The contribution of the centerline remains about 9%.

The experiment described above seems suitable for use in a laboratory course to demonstrate to students the ease of phase analysis by this method. Three spectra should be taken corresponding to the three stages: Fe in solution (quenched after solution annealing), γ-Fe precipitates (after aging), and γ → α transformation (after plastic deformation).

H.   *Fe in Metastable Phases (Films)*

By fast quenching, metastable phases of supersaturated solutions can be produced in concentration ranges far outside the lines that bound the equilibrium phases in the Cu–Fe diagram of Fig. 1. However, supersaturation can be achieved more effectively by simultaneous vapor deposition of Fe and Cu on cold substrates to produce films (Keune *et al.*, 1974). These films are rather stable up to temperatures of about 300°C. This fact might be explained by the similarity of the atomic radii of Cu and Fe.

Using Mössbauer spectroscopy and electron microscopy, single phases were determined *in situ* and identified in the concentration range up to about 40 at.% of one of the constituents. These phases were bcc alloys in the range from 0 to 40 at.% Cu and fcc alloys from about 60 to 100 at.% Cu. Only in the concentration range of about 40 to 60% Cu has a mixture of the two phases been observed. The bcc alloys exhibit a magnetic hyperfine pattern, as shown in Fig. 8a. Surprisingly, the lines are relatively sharp, and the spectrum is well fitted by assuming two superimposed hyperfine patterns with slightly different internal fields. The relative line

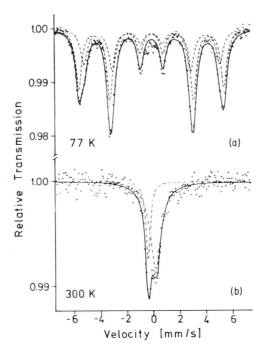

Fig. 8.   (a) Mössbauer spectrum of a Fe–26.4 at.% Cu thin film obtained at 77 K using a source (Cu) at room temperature. (b) Mössbauer spectrum of a Cu–15.6 at.% Fe thin film obtained with both absorber and source (Pd) at room temperature (Keune *et al.*, 1974).

intensity ratios of nearly $3:4:1:1:4:3$ indicate that the magnetization lies in the plane of the film. The fcc films are paramagnetic at room temperature as seen in Fig. 8b. It is interesting to note that the resonance line at -0.09 mm/s has the same isomer shift as γ-Fe precipitates at room temperature. At lower temperatures, complicated and unresolved hyperfine patterns are observed.

I. Fe in the Oxidized State

The oxide phases produced by internal and external oxidation can be followed conveniently by the Mössbauer effect (Gonser et al., 1966). Internal oxidation occurs in Cu samples containing some Fe during aging at elevated temperatures ($\sim 800°C$) *in vacuo* $> 10^{-4}$ mm Hg. The processes involve the formation of FeO (wustite) first and Fe_3O_4 (magnetite) next. The terminal oxidation product has been identified as $CuFeO_2$ (delafossite), which also occurs in slags from copper smelting furnaces (Muir and Wiedersich, 1967).

The occurrence of oxide phases in external oxidation processes (corrosion) is strongly dependent on parameters such as oxygen partial pressure, temperature, and humidity. Conversion electron Mössbauer spectroscopy (CEMS) has been proven to be very useful in the investigation of this type of problem. The sensitivity of CEMS is so high that resonance lines are observable for monolayers. A typical CEMS spectrum of about eight atomic layers of Fe^{57} deposited on a Cu substrate is shown in Fig. 9 (Petrera et al., 1976). The isomer shift is typical of Fe^{3+}. However, it is not possible to identify the corresponding bulk phase from this spectrum if it exists at all. The sensitivity of CEMS was demonstrated by these measure-

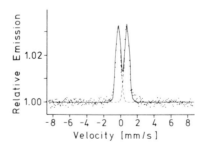

Fig. 9. CEMS spectrum of an oxidized Fe^{57} film of approximately eight atomic layers deposited on a Cu substrate. Absorber and source (Rh) at room temperature (Petrera et al., 1976).

ments. Specifically, it was shown that a monolayer of Fe^{57}, corresponding to $\sim 1.8 \times 10^{15}$ atoms/cm^2, would give an effect of about 1.8% if the spectrum consisted of a single resonance line with natural linewidth.

IV. The Al–Fe System

In contrast to the Cu–Fe system (Fig. 1), the Al–Fe phase diagram shown in Fig. 10 (Hansen and Anderko, 1958; Warlimont et al., 1969) exhibits a number of phases: $FeAl_3$ (Fe_4Al_{13}), Fe_2Al_5, $FeAl_2$, $FeAl$, and Fe_3Al and ordered structures. Because of the interesting properties of the Al–Fe phases, as well as the small electronic absorption of the 14.4-keV γ-rays in Al, this system has been extensively studied by Mössbauer spectroscopists.

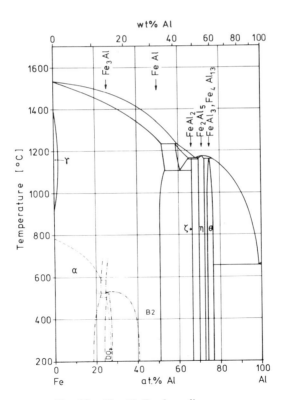

Fig. 10. The Al–Fe phase diagram.

A. Fe in fcc Al

The solubility of Fe in Al is extremely small, being only 0.005 wt% Fe at 450°C (Nishio et al., 1970). It was possible by Mössbauer spectroscopy to measure quantitatively the solubility from 450°C up to the melting point and to derive the intrinsic entropy change and the partial enthalpy change. In dilute Al alloys, Fe atoms are located on substitutional sites, and from the absence of any quadrupole interaction one can deduce that the distortion by an Fe impurity—smaller in size than an Al atom—is cubicly symmetric. Fe in Al has been used as a probe to investigate associated point-, line- and two-dimensional lattice defects. Particularly, Al interstitials introduced by radiation damage at low temperatures (~ 4 K) show interesting properties and have been studied in detail (Vogl et al., 1976). When the Al interstitials become mobile at about 40 K, they can be trapped by Co^{57} or Fe^{57} impurities and then released again at higher temperatures. By trapping, mixed Fe–Al dumbbells are formed and exhibit a characteristic line in the spectrum. In addition, the Fe atoms in the mixed dumbbells show an unusual dynamic behavior, as indicated by the temperature dependence of the Debye–Waller factor. A cage model has been suggested where the impurity changes the dumbbell partner of the adjacent face-center atoms in a sixfold configuration with a low activation energy of only 0.017 eV, so that even at low temperature the jump frequency of the process is very high. As a result, the electric field gradient becomes effectively zero when averaged over residence times, short compared to the lifetime of the excited state. At very low temperatures one might expect quadrupole interaction as a result of the freezing-in of particular dumbbell pairs.

Associations of impurity (Fe^{57}) atoms with vacancies, dislocations, and grain boundaries (Janot and Gibert, 1977; Preston et al. 1979; Nasu et al. 1980) have been observed.

B. Al–Fe Phases

Intermetallic compounds and the various states in the processes leading to phases in the form of precipitates have been investigated in Al-rich Fe alloys (Nemoshkalenko et al., 1968; Preston and Gerlach, 1971; Preston, 1971; Nasu et al., 1974; Janot et al., 1974).

The spectrum of a splat-quenched Al-1 at.% Fe sample is shown in Fig. 11a. This spectrum is similar in appearance and in interpretation to Fig. 2 of a quenched Cu–Fe sample. That is, one can distinguish a single line, corresponding to monomers, and a quadrupole line, mainly due to dimers.

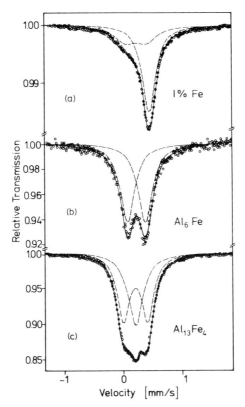

Fig. 11.   Mössbauer spectra at room temperature obtained with a Co^{57}–Cu source. The velocity scale has been adjusted to α-Fe. (a) Splat-quenched Al–1 at.% Fe specimen. The single line represents Fe in solution, and the doublet represents the dimers. (b) Al_6Fe precipitates. (c) $Al_{13}Fe_4$ precipitates (Nasu and Gonser, 1973).

The frequency of occurrence of dimers is rather large, which indicates a deviation from randomness in the solid solution. Iron associations, mostly dimers, have also been reported in an investigation of Fe^{57} implantation into Al (Sawicka and Sawicki, 1977). These measurements were performed by means of CEMS.

The agglomeration of iron and the development of phases can be enhanced and advanced by aging, thermal annealing, radiation, cold working, etc. Chill casting of an Al–0.5 at.% Fe alloy has been shown to form the phase Al_6Fe, which appears to exist only as a metastable precipitate in an Al matrix (Fig. 11b). The structure is orthorhombic and is isomorphous with Al_6Mn.

Usually the annealing of Al-rich Fe alloys leads to $Al_{13}Fe_4$ or Al_3Fe precipitates. This intermetallic compound has a monoclinic structure, space group c2/m (Black, 1955). In physical metallurgy, the observed phases generally have broad ranges of existence. Because of this, the stoichiometry of the phases is not well defined and discrepancies in nomenclature occur frequently in the literature. In this case the two phases $Al_{13}Fe_4$ and Al_3Fe are basically identical, but it seems that the chemistry-oriented scientist prefers Al_3Fe, while the physicist prefers $Al_{13}Fe_4$. The spectrum of a sample containing $Al_{13}Fe_4$ precipitates is shown in Fig. 11c. Iron in various lattice sites in this phase produces a single line in the center superimposed on quadrupole components.

C. Order–Disorder in FeAl and Fe₃Al

From Fig. 12 we might derive the order–disorder states of these phases. In the state of complete disorder at high temperatures, sublattices cannot be distinguished, and all sites are randomly populated by the constituents. The structure can be represented by an α-Fe elementary cell with a lattice parameter of $\frac{1}{2}$ and a volume of $\frac{1}{8}$ of the structure shown in Fig. 12.

The ordered state of the stoichiometric compound FeAl is represented by the CsCl or B2 structure. It consists of two interpenetrating simple cubic lattices: sublattice A occupied entirely by Fe atoms and sublattice D occupied entirely by Al atoms (disregard the use of two different symbols for atoms in the D sublattice in Fig. 12). The size of the elementary cell is again $\frac{1}{8}$ of the volume shown in Fig. 12.

The structure of the ordered state of Fe_3Al is also shown in Fig. 12. In this case sublattice A contains only Fe atoms, each having four Fe and

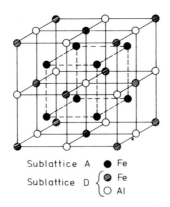

Sublattice A ● Fe

Sublattice D $\begin{cases} \oslash\ Fe \\ \bigcirc\ Al \end{cases}$

Fig. 12.   Structure of Fe₃Al (DO₃).

four Al nearest neighbors, while sublattice D contains alternately Al and Fe atoms, each having eight Fe nearest neighbors (nn). The crystallographic ordering indicated in Fig. 12 is called DO_3. An intermediate pseudoordered state might be considered, as, for instance, where the D sublattice only (or preferentially) becomes disordered.

In these alloys disorder can be achieved through high temperature, plastic deformation, and radiation damage. The state of order is also strongly influenced by the stoichiometry.

In ordered FeAl no magnetic moment is associated with the Fe atoms, and therefore a single-line spectrum is observed. After severe cold working by crushing, the spectrum exhibits coexisting paramagnetic and magnetic hyperfine lines (Huffman and Fisher, 1967; Wertheim and Wernick, 1967). These results were explained by considering the near-neighbor configuration of the partially disordered alloys. The dislocations, having passed through the crystal during the deformation process, have destroyed the order mainly by producing antiphase domain boundaries that form configurations of magnetic moment-bearing Fe atoms. By analyzing the fractions (relative intensities) of the superimposed paramagnetic and hyperfine spectral components, it was concluded that Fe atoms having three or more Fe nearest neighbors exhibit hyperfine interactions and are capable of ferromagnetic behavior.

In the ordered Fe_3Al intermetallic compound the Fe atoms possess magnetic moments. From neutron-scattering experiments, the magnetic moments for the A and D sites are known to be 1.46 and 2.14 μ_B, respectively (Nathans et al., 1958). The Mössbauer spectrum of ordered Fe_3Al at room temperature is shown in Fig. 13a (Czjzek and Berger, 1970). It exhibits two superimposed six-line patterns with internal magnetic fields of 210 kOe (A site with four nn Fe atoms) and 294 kOe (D site with eight nn Fe atoms). Thus the magnetic moments and the internal fields are roughly proportional. However, it should be noted that the field at the D site with eight nn Fe atoms is lower than the 330 kOe value for α-Fe, which also has eight nn Fe atoms. The smaller field in the alloy reflects the influence of the next-nearest Al neighbors. The intensity ratio of the two superimposed magnetic hyperfine patterns is approximately 2 : 1, as one expects from the relative sublattice occupancy of the two components.

Radiation damage effects and disorder after neutron-capture reactions $Fe^{56}(n, \gamma)$ Fe^{57} in Fe_3Al are evident from the significant differences between the spectra of Fig. 13a and b. In the latter case, only one magnetic hyperfine pattern with broad lines and an average internal magnetic field of 243 kOe can be distinguished effectively. The recoil energy imparted to the Fe^{57} nuclei by the nuclear reaction is up to 549 eV, well above the threshold displacement energy, which is in the order of 25 eV in metals. A

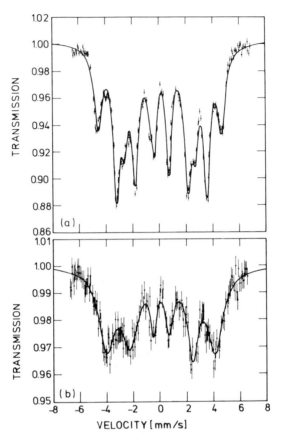

Fig. 13. Mössbauer spectra of Fe_3Al obtained at room temperature. (a) Absorption spectrum with Cu:Co[57] source, fitted with 12 Lorentzian lines. (b) (n, γ)-target spectrum with $Na_4Fe(CN)_6 \cdot 10\ H_2O$ absorber, fitted with six Lorentzian lines. The velocity scale has been adjusted to α-Fe (Czjzek and Berger, 1970).

collision model was advanced in the analysis, and an attempt was made to obtain information regarding the final positions and possible configurations and distributions of the resonating Fe[57] atoms in terms of interstitials and vacancies at various distances. The contributions of the point defects and also of the disordering to the hyperfine patterns were derived by the fitting of the experimental curve.

The Fe-rich side of the phase diagram in Fig. 10 has been extended by drawing in dotted lines to indicate suggested ordered phases (Warlimont *et al.*, 1969). This part of the diagram has attracted considerable interest, and it was at first thought that Mössbauer spectroscopy would be an ideal tool to advance our knowledge of order–disorder states and phases (Cser *et al.*, 1967, 1970; Huffman and Fisher, 1967; Wertheim and Wernick, 1967;

Lesoille and Gielen, 1970; Hergt *et al.*, 1970; Huffmann, 1971; Losiyevskaya and Kuzmin, 1972; Shiga and Nakamura, 1976). However, in the course of the work done in this area, the complexity of the problems and the limitations of the method became evident (see Schwartz, 1976, Vol. 1).

For the analysis of phases, the interpretation of the results in the α-region (up to about 18 at.% Al) is rather straightforward: the alloys can be represented by simple dilution of Al impurities in the bcc lattice. These give Mössbauer spectra which resemble the spectrum of α-Fe except that several slightly different hyperfine patterns are seen (Stearns, 1966). The observed variation in the hyperfine field reflects differences in the local configurations of impurity (Al) atoms relative to Fe atoms and originates in the spatial variation of the spin polarization of the 4s conduction electrons.

Beyond 20 at.% Al, and particularly between 30 and 45 at.% Al, these alloys yield complex and unresolved spectra that strongly depend on the history and treatment of the samples. Except for the ordered phases FeAl and Fe_3Al, these spectra are characterized by various spectral components and by a wide distribution of hyperfine fields, which are described as ferromagnetic, antiferromagnetic, and paramagnetic. One might be tempted to attribute each of the superimposed subspectra—representing certain (magnetic) states—to a distinct phase. In doing so, one would neglect the fact that Mössbauer spectroscopy probes only the immediate surrounding of certain atoms, while a phase is spatially extended in symmetry and order. For instance, in FeAl after plastic deformation, a magnetic state is observed by means of the magnetic hyperfine broadening of the Mössbauer resonances. This state consists of rather local configurations of Fe magnetic moment at antiphase domain boundaries, but it does not represent what we commonly call a phase.

From a phase analysis point of view, the result so far are inconclusive, and it remains a challenge to check systematically and complete the phase diagram and to get further information regarding order–disorder states and phases and the nature of their magnetic interactions.

V. The Ti–Fe System

A. *Titanium Base Alloys*

Titanium alloys play a role of increasing technological importance due to their high strength-to-weight ratio and excellent corrosion resistance at low and intermediate temperatures. Their service at high temperature is limited by metallurgical instabilities such as appearance of intermetallic or metastable phases after prolonged exposure.

The element titanium exists in two allotropic forms: an hcp α-phase at temperatures below 855°C and a bcc β-phase above that temperature. Alloying of Ti with Al, Sn, or the interstitial elements O, N, and H stabilizes the α-phase, which has valuable properties such as good toughness, ductility, and formability. Alloying elements such as Mo, V, Cr, Mn, Fe, etc., stabilize the β-phase and cause it to be retained at room temperature. This effect is the basis of the commercial $\alpha + \beta$ alloys, which combine high strength and acceptable levels of toughness and ductility. Various kinds of $\alpha + \beta$ alloys have been extensively developed in order to achieve the best combinations of strength, toughness, ductility, and metallurgical stability. The metallurgy of $\alpha + \beta$ Ti alloys is quite complicated, and sophisticated methods must be used for the control and study of phase composition, concentration of alloying elements and microstructure and their effect on formability and mechanical properties. In titanium alloys, besides the equilibrium phases, a number of metastable phases are encountered.

A metastable ω-phase, which forms at intermediate temperatures, is known to have a detrimental effect on the mechanical properties of $\alpha + \beta$ Ti alloys. Temperatures and times of thermal treatments, as well as cooling rates, should be adjusted to minimize the amount of ω-phase. The temperature and concentration ranges of the ω-phase existence, typical of an $\alpha + \beta$ Ti alloy, are shown in Fig. 14.

Developments in the metallurgy of Ti alloys depend on improved knowledge of the complex and sometimes unique reactions and phase transformations in these alloys. Detailed analysis of phase transformations in Ti alloys is difficult due to the shortcomings of the various detection techniques currently used. Often the introduction of a more advanced

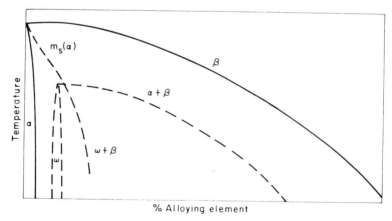

Fig. 14.   Metastable phase diagram for a typical β-isomorphous system.

method permits better determination of phases and thus a deeper understanding of phase transformations. The early studies concerning the precipitation of the athermal ω-phase (ω-phase formed during quenching) are good examples of this (Hickman, 1969a; Williams, 1973; Williams *et al.*, 1973).

Routine x-ray diffraction techniques are effective in detecting phases encountered in Ti alloys, but their detectability threshold is a few volume percent. Moreover, the submicroscopic size of the particles that commonly segregate upon the decomposition of metastable phases may make the analysis or even the detection by x-rays difficult. For the same reason, optical metallography is of limited use in such studies.

Thin-foil electron microscopy has been successfully applied to the identification of phases and the study of phase transitions in Ti alloys (Williams, 1973; Williams *et al.*, 1973). Particles as small as 20 Å have been detected, and the morphology as well as the crystallographic relationships with the parent phases have been revealed by electron microscopy. Selected-area diffraction has been used for the identification of the ω-phase. However, these results must be interpreted with caution since a phenomenon known as diffuse streaking caused by the overlapping of diffraction spots makes the interpretation uncertain. Another shortcoming of electron microscopy is the ambiguity of identification due to thin-foil artifacts.

The Mössbauer method has been shown to be sensitive in detecting and identifying phases in the Ti–Fe alloys (Stupel *et al.*, 1974, 1976a). Amounts below the sensitivity of routine x-ray diffraction analysis have been detected in these alloys.

The characteristics of the Ti–Fe system are of great importance. The Ti–Fe system, although not used as the basis of a commercial alloy, can serve as a model system for $\alpha + \beta$ Ti alloys since iron is a strong β-stabilizer.

B. *Phases in the Ti–Fe Binary System*

The binary Ti–Fe equilibrium phase diagram (Fig. 15), which is fairly well established, is characterized by two congruently melting intermetallic compounds TiFe and TiFe$_2$ and limited solubilities at the Ti and Fe ends (Hansen and Anderko, 1958). The β-phase, a random solid solution, dissolves as much as 25 wt% Fe at 1100°C. It decomposes by a eutectoid reaction $\beta \rightarrow \alpha +$ TiFe, for which temperatures ranging between 585 and 615°C have been reported (Molchanova, 1965). For iron concentrations higher than 6 wt%, the β-phase is retained as a metastable phase at room temperature, even upon slow cooling. Upon aging below the eutectoid temperature, the β-phase decomposes through a number of steps including

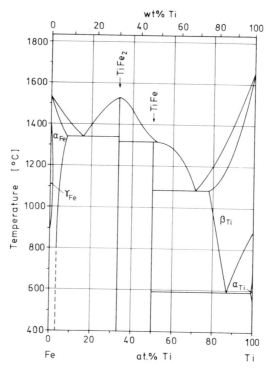

Fig. 15.   The Ti–Fe phase diagram.

the formation of a metastable ω-phase, which will be discussed later. Mössbauer spectra for α, β-phases and TiFe are shown in Fig. 16.

The structure and composition of the equilibrium phases TiFe and TiFe$_2$ are well established, while substantial discrepancies exist in regard to metastable phases. The extent of the solubility of iron in α-Ti(Fe) at and below the eutectoid temperature is still subject to controversy (Hansen and Anderko, 1958; Molchanova, 1965). The existence of a metastable Θ-phase, in the process of low temperature aging of supersaturated α-Ti(Fe), has been recently revealed by a Mössbauer study (Stupel et al., 1974, 1976a).

C.   Identification and Analysis of Ti–Fe Phases by Means of Mössbauer Spectroscopy

1.   The α-Fe(Ti) Solid Solution

The Mössbauer spectra of dilute polycrystalline iron alloys have been represented as a superposition of a number of six-line spectra, each with hyperfine magnetic field and isomer shift related to nearest neighbor (nn)

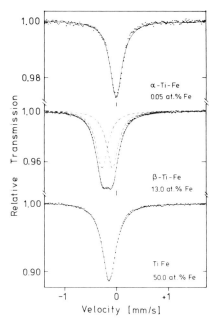

Fig. 16. Mössbauer spectra for α, β, and TiFe phases. The velocity scale has been adjusted to α-Fe.

and next nearest neighbor (nnn) solute concentrations. According to a work by Wertheim *et al.* (1964), the intensities of the component spectra are proportional to $W(ij)$—the fraction of iron atoms with i-nn and j-nnn solute atoms. The subject has been summarized by Schwartz (1976) who discussed the analysis of substitutional iron-based alloys and has pointed out that the effects of anisotropy and of effective thickness should not be neglected in quantitative phase analysis.

The maximum solubility of Ti in α-Fe at 1350°C is 6.3 wt%, according to Hansen and Anderko (1958). The hyperfine magnetic splittings of the α-Fe(Ti) spectra have been shown to consist of main peaks related to Fe atoms having only Fe (nn) and (nnn) neighbors and satellite peaks for those having Ti (nn) and (nnn) neighbors (Cranshaw, 1972; Huffman and Podgurski, 1975). The following hyperfine parameters have been determined by Huffman and Podgurski (1975) for a satellite spectrum of an Fe (0.68 at.% Ti) alloy: a magnetic hyperfine field of -307.3 kOe and an isomer shift (IS) of -0.03 mm/s relative to α-Fe. The satellite peak intensities (and spectral areas) were used to calculate the fractions of Fe atoms with various (nn) and (nnn) configurations. From these the concentration of Ti in the alloy was calculated, and the result agreed fairly well with the known concentration.

2. Intermetallic Compounds $TiFe_2$ and $TiFe$

The hexagonal Laves-phase compound $TiFe_2$ ($MgZn_2$ type) has been subject to several Mössbauer studies, mainly because of its interesting magnetic properties (Wertheim et al., 1969, 1970a, b; Bruckner et al., 1970). The bulk magnetic properties of this phase have been shown by Nakamichi (1968) to be very sensitive to deviations from stoichiometry. He suggested that in the iron-rich range of $TiFe_2$ (30.8–32.8 at.% Ti) the material is ferromagnetic, while in the titanium-rich range (33.3–35.2 at.% Ti) it is antiferromagnetic. He also suggested that in the ferromagnetic region the saturation moment is proportional to the deviation from stoichiometry. It has been shown (Wertheim et al., 1969) that the iron atoms occupy two nonequivalent sites. There are six h sites with mm symmetry, and iron atoms on these sites order antiferromagnetically. There are two a sites with 3 m symmetry, and iron atoms on these sites show only quadrupole splitting. However, the a-site atoms develop a hyperfine field when they are in a magnetic cluster. Apparently this happens at 20 K (see Fig. 17). On the titanium-rich side of $TiFe_2$, above the magnetic ordering temperature 275 K, the spectra for the a- and h-type of Fe atoms are identical and have the hyperfine parameters shown in Table II.

The intermetallic compound TiFe has been studied by Mössbauer spectroscopy (Rupp, 1970; Stupel et al., 1976a; Mielczarek and Winfree, 1975). TiFe of stoichiometric composition is known to have a CsCl structure, but the question whether the Fe and Ti atoms are randomly distributed or ordered is still open since no evidence for a superstructure has been published. In the Mössbauer spectrum TiFe shows a single line, the parameters for which are shown in Table II, as determined by Stupel et

TABLE II

Room Temperature Mössbauer Parameters
for the Stable Phases in the Ti–Fe
Binary System

Mössbauer parameters	α^a	$TiFe^a$	$TiFe_2$ [b] (2a and 6h sites)
IS^c	0.02	− 0.15	− 0.286
Γ	0.28	0.28	
QS			0.404

[a] From Stupel et al. (1976a).
[b] From Wertheim et al. (1970).
[c] All isomer shifts in mm/s with respect to α-Fe. Γ and QS units in mm/s.

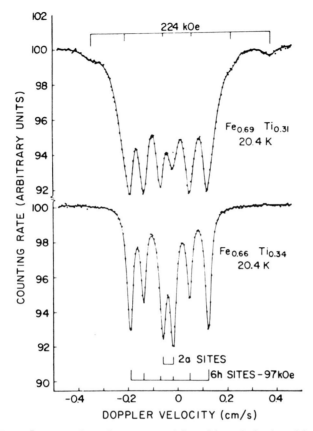

Fig. 17. Mössbauer absorption spectra of iron-rich and titanium-rich $TiFe_2$. The samples were immersed in liquid hydrogen (Wertheim *et al.*, 1970b).

al. (1976a). The linewidth is seen to be relatively narrow, $\Gamma = 0.28$ mm/s. A linewidth of $\Gamma = 0.27$ mm/s has also been reported by Swartzendruber and Bennett (1970). The relatively narrow linewidth and the absence of a quadrupole splitting support the assumption that the CsCl-type structure of TiFe is ordered. Mössbauer measurements in high magnetic fields have shown that no magnetic moment is associated with the Fe atoms in TiFe (Rupp, 1970).

3. α_m- and α-Ti(Fe) Solid Solutions

Contrary to experience with the intermetallics TiFe and $TiFe_2$, the evaluation of the Mössbauer characteristics of the various phases in the Ti-rich end of the Ti–Fe system has been subject to discrepancies and controversy that have been discussed by Stupel *et al.* (1976a). Conceivably, these discrepancies should be attributed to the coexistence of stable and metastable phases due to the sluggishness of the reactions in the Ti–Fe system. In addition, a source of confusion may be the improper use, sometimes made, of the term "phase." In Mössbauer spectroscopy it is common to use the term "phase" to describe a species having distinct Mössbauer parameters, rather than in the metallurgical sense.

For the purpose of phase identification and analysis, a series of Ti–Fe alloys has been prepared and heat treated with the objective of obtaining precisely controlled phase quantities (Stupel *et al.*, 1977a). The phases have been examined by Mössbauer spectroscopy and x-ray diffraction in parallel. In these alloys the Mössbauer method has been shown to be able to detect amounts that are below the sensitivity of routine x-ray diffraction analysis. For instance, in titanium alloys with a concentration of $C_{Fe} < 0.5$ wt% at temperatures below 600°C, the amount of precipitated TiFe cannot exceed 0.01 by weight. Such amounts are hardly discernible by x-ray diffraction. On the other hand, the relative Mössbauer spectral area for a given phase i is proportional to $X_i C_i$, the product of the fractional weight X_i of the phase, and concentration of iron, C_i, within the phase. An evaluation based on the Ti–Fe equilibrium phase diagram shows that within the range $0.05 \leqslant C_{Fe} \leqslant 0.5$ wt%, the relative Mössbauer spectral

TABLE III

Room Temperature Mössbauer Parameters for Metastable Phases in the Ti–Fe Binary System

Treatment	wt% Fe in alloy	$IS^b_{\alpha_m}$ (mm/s)	IS^b_α (mm/s)	IS^b_θ (mm/s)	Γ_θ (mm/s)	IS^b_ω (mm/s)	Γ_ω (mm/s)	IS^b_β (mm/s)	QS_β (mm/
Quench from 1000°C	0.25	−0.01							
Aged at 335°C for 6 h	0.25		0.02	−0.28	0.38				
Quench and plastic deformation (71%)	0.25	0.00				−0.64	0.42		
Quench and plastic deformation and aging at 335°C for 40 h	0.25		0.02			−0.58	0.44		
Homogenized	8							−0.19	0.20
Homogenized	20							−0.17	0.22

*a*Accuracy: ±0.005 mm/s.
*b*All isomer shifts in mm/s with respect to α-Fe.

areas for Ti(Fe) and TiFe phases are $0.1 \leqslant X_\alpha C_\alpha / X_{\mathrm{TiFe}} C_{\mathrm{TiFe}} \leqslant 0.5$. Thus both phases can be detected and analyzed, provided that the alloying iron is enriched in the Fe^{57} isotope.

Room-temperature Mössbauer spectra for a series of Ti–Fe alloys of varying iron concentration, quenched from 1000°C, are shown in Fig. 18. The approximate positions of the lines attributed to α- , α_m , β- , and ω-phases are marked by bar diagrams on the top of the spectra. Room-temperature Mössbauer parameters determined for the equilibrium phases are given in Table II and for the metastable ones in Table III.

The metastable β-Ti alloys convert into α by a martensitic type of phase transformation. The structure of the martensitic product α_m has been the subject of conflicting evaluations and nomenclature in the literature (Williams, 1973). In titanium and in many titanium alloys, the martensite forms as a result of cubic-to-hexagonal transformation and is characterized by a habit plane near $(334)_\beta$ (Blackburn, 1970). Structures other than hcp have also been reported. In particular, a fcc martensite has been reported

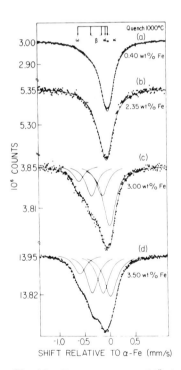
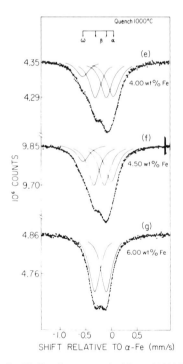

Fig. 18. Room-temperature Mössbauer spectra for Ti–Fe alloys, quenched from 1000°C with the following iron content: (a) 0.4 wt% Fe, (b) 2.5 wt% Fe, (c) 3.0 wt% Fe, (d) 3.5 wt% Fe, (e) 4.0 wt% Fe, (f) 4.5 wt% Fe, (g) 6.0 wt% Fe.

for a Ti–3wt% Fe alloy. A Mössbauer analysis by Stupel *et al.* (1974, 1976a) indicated two α-phases. One is α_m, an athermal martensite obtained by quenching Ti–Fe alloys of $C_{Fe} < 2.7$ wt% from 1000°C. The other is an equilibrium α-phase, which forms in Ti–Fe alloys after aging. Interestingly, no α_m martensite forms in alloys of $C_{Fe} > 2.7$ wt%. These two phases differ in their isomer shifts (see Tables II and III) but show identical x-ray diffraction patterns characteristic of hcp structures.

The difference between the isomer shift of the α- and α_m phases is positive

$$\Delta IS = IS_\alpha - IS_{\alpha_m} > 0,$$

implying a decrease in $|\Psi_S(0)|^2$, the electron density at the Fe^{57} nucleus.

As is known from high-pressure Mössbauer experiments, in a dilute Ti–Fe alloy (Edge *et al.*, 1965) the sign of this change, $\Delta IS > 0$, corresponds to a relief in pressure. This may reflect the situation in which compressive stresses imposed upon the Fe atoms in the quenched Ti matrix are relieved upon annealing as a result of lattice relaxation. Due to the nonuniformity of stresses, in the quenched Ti matrix, a distribution of small deviations from a mean value of the isomeric shift may be expected. Such a distribution results in a line broadening which has, in fact, been observed in several alloys in the as-quenched state (Stupel *et al.*, 1977a; Hornstein and Ron, 1974). The smaller linewidth Γ_α of the equilibrium

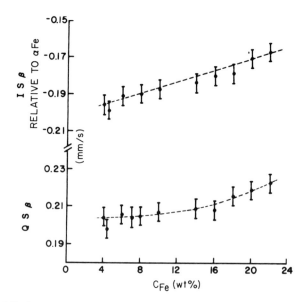

Fig. 19. Mössbauer parameters of retained β-phase versus Fe concentration of the alloy (C_{Fe}). $(IS)_\beta$ and $(QS)_\beta$ versus C_{Fe} (Stupel *et al.*, 1977b).

α-phase, as compared with Γ_{α_m}, the linewidth of the α_m-phase, supports the above interpretation.

The solubility of iron in α-Ti(Fe) is discussed below.

4. The β-Ti(Fe) Phase

Iron is known to be a strong β-stabilizer in Ti–Fe alloys. High iron concentration (> 2.7 wt% Fe) depresses the martensitic start temperature T_s to below room temperature (Williams, 1973). The metastable β-phase is then retained when quenched to room temperature. In the range of compositions ($2.7 < C_{Fe} \leqslant 5$ wt%) the α-, β-, and athermal ω-phases were observed to exist together in the quenched alloys, whereas in alloys with $C_{Fe} > 5.5$ wt% only the β-phase was observed (Stupel et al., 1976a).

Both the Mössbauer analysis and the Debye–Scherrer x-ray diffraction technique were able to detect the β-phase (see Table III). The Mössbauer spectra of the β-phase, for alloys containing more than 2.7 wt% Fe, display a doublet. The nonvanishing quadrupole interaction is an indication that the cubic symmetry of the lattice is distorted on a local scale.

From Fig. 19 and Table III, it can be seen that IS_β, as well as QS_β, depend on C_β, the iron concentration of the β-phase. The β-phase obtained by quenching from the $\alpha + \beta$ field (700°C) has the same Mössbauer parameters as the β-phase obtained by quenching from 1000°C. A β'-phase of different Mössbauer characteristics, obtained as a result of quenching from the $\alpha + \beta$ field, has been reported by Rupp (1970).

The β-Ti(Fe) is a superconducting phase with a high transition temperature T_c, which increases with the concentration of iron within the phase. It has been found by a Mössbauer study that the iron atoms have no localized magnetic moments (Rupp, 1970).

5. The Metastable Θ- and ω-Phases

A previously unknown metastable phase, appearing during the aging of supersaturated α_m-Ti(Fe) at temperatures above 280°C, has been revealed by a Mössbauer study by Stupel et al. (1974), (1976a) who have named it the Θ-phase. The nature of this phase is still unknown; suggestions such as GP zones, coherent precipitates, clusters of Fe atoms, and spinodal decomposition have been examined by Stupel et al. (1977a) and Ron et al. (1977). Figure 20 shows Mössbauer spectra for a 0.4 wt% Fe alloy aged at 320°C for various times. Next to the α-phase component is the Θ-phase component whose intensity can be seen to increase with aging time. The existence of the Θ-phase has not been indicated by either x-ray diffraction or electron microscopy, to the best of our knowledge.

A metastable phase, known as the athermal ω-phase, has been shown to form upon quenching metastable β-Ti alloys (Silcock et al., 1955; Hickman, 1969a; Baganataskii and Nosova, 1962). Athermal ω-phase formation

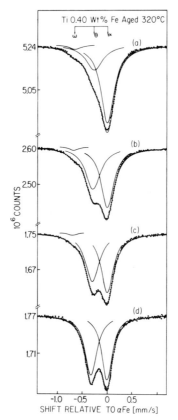

Fig. 20. Mössbauer spectra for Ti (0.4 wt% Fe) alloy, aged at 320°C for (a) 1 hr, (b) 3 hr, (c) 10 hr, and (d) for 180 hr.

occurs over a limited range of compositions depending on the particular alloy system considered (Williams, 1973). For morphological reasons, ω-phase formation cannot be assigned to the martensitic class of transformations but to a new class called displacement-controlled transformations (Williams et al., 1973; De Fontaine and Kikuchi, 1974). Both hexagonal and trigonal structures have been claimed for the ω-phase. The range of concentrations, over which the athermal ω-phase forms (upon rapid quenching) in the Ti–Fe system, was found to be $2.7 \leqslant C_{Fe} \leqslant 5$ wt% by a Mössbauer study (Stupel et al., 1976a).

The athermal ω-phase exhibits a single line in the Mössbauer spectra. The isomer shift IS_{ω} changes from -0.56 to -0.50 mm/s, depending on the iron concentration C_{Fe} of the alloy.

D. The Formation of Metastable Phases and Ranges of Their Existence

1. The Θ-phase

As mentioned earlier, the formation of the Θ-phase upon aging α-Ti(Fe) at 320°C has been revealed by a Mössbauer study (Stupel *et al.*, 1974). The nature of this phase has not so far been explored. The relative spectral area A_Θ related to the Θ-phase increases with the iron concentration of the alloy C_{Fe} for a given aging temperature and time. The Θ-phase, which has been shown to form upon aging of α-Ti(Fe) alloys at 280–320°C, was not discernible by x-ray diffraction or electron microscopy (Stupel *et al.*, 1974, 1976a, 1977a). The magnitude of the difference between the isomer shifts of the α- and Θ-phases ($\Delta IS = -0.26 \pm 0.025$ mm/s) indicates a considerable change in the s-electron density at the Fe^{57} nucleus. A change of such magnitude in the isomer shift can be caused either by a change in the crystal structure or by compositional changes. Since no changes in the crystal structure were found, compositional changes must have occurred which, however, can only be very local due to the limited migration of atoms at 280–320°C. The decomposition of a solid solution into two compositional components without change in the crystal structure can be explained by assuming a spinodal decomposition mode, viz., periodic fluctuations in composition forms creating iron-rich and iron-lean regions. The Θ-phase must be related to the iron-rich regions since the iron concentration of the Θ-phase is always higher than that of the α-phase. For Θ-phase formed during relatively slow quenching (athermal-Θ), it has been found that $C_\Theta \simeq 0.55$ wt% and $C_\alpha \simeq 0.22$ wt%; upon aging, C_Θ has been found to increase.

2. Athermal and Thermal α → Θ → ω Transitions

The athermal ω-phase has been shown to exist along with α- and β-phases in quenched alloys having concentrations in the range $2.7 \leqslant C_{Fe} \leqslant 5$ wt%. The isomer shift of the athermal ω-phase IS_ω increases linearly with increasing the iron concentration C_ω corresponding to the equation (Stupel *et al.*, 1977a)

$$IS_\omega = 0.037C_\omega - 0.672 \text{ mm/s.}$$

The formation of thermal ω-phase in the process of decomposition of a supersaturated α-Ti (0.22 wt% Fe) alloy has been revealed by a Mössbauer study (Stupel *et al.*, 1977a). The observed reaction can be described as

$$\alpha_m \rightarrow \alpha + \Theta + \omega \rightarrow \alpha + \Theta.$$

As seen from Fig. 21, a small amount of ω-phase appears after aging at

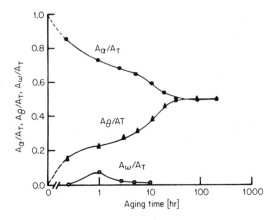

Fig. 21. Relative spectral areas (for Ti 0.22 wt% Fe) of the α, θ, and ω phases as functions of aging time at 320°C (Stupel *et al.*, 1977a).

320°C for 30 min, reaches a maximum at 60 min, and disappears after about 10 h.

3. $\beta \rightarrow \omega$ Decomposition

In retained metastable β-Ti(Fe) alloys, in the range of 5 wt% $\leqslant C_\beta \leqslant 20$ wt%, the ω-phase forms upon aging at temperatures up to 500°C (Hickman, 1969b; Erost *et al.*, 1954). The formation of the ω-phase during aging of β-Ti (7 wt% Fe) has been studied by means of Mössbauer spectroscopy (Stupel *et al.*, 1977b). The linear dependence of IS_ω on the Fe

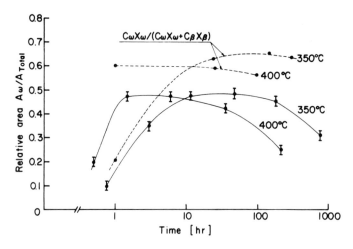

Fig. 22. A_ω/A_T versus isothermal aging time, continuous line——; $C_\omega X_\omega$ versus isothermal aging time, intermittant line--- (Stupel *et al.*, 1977b).

content of the quenched alloys was used to derive C_ω, the Fe concentration of the ω-phase formed in alloys aged at 350°C.

The formation of the ω-phase upon aging at 350°C for various times is clearly depicted by the change in the relative area A_ω in Fig. 22. The f factors for the ω- and β-phases have been found to be nearly equal at room temperature (Stupel, 1976), so that the relative Mössbauer spectral area can be compared with $C_\omega X_\omega/(C_\omega X_\omega + C_\beta X_\beta)$, determined from x-ray diffraction data by Hickman (1969b).

4. The $\Theta \to \omega$ Transition upon Plastic Deformation

In a recent Mössbauer study (Ron *et al.*, 1977), the ω-phase has been found to form as a result of the plastic deformation of an α-Ti (0.25 wt% Fe) in which the Θ-phase is present. Room-temperature Mössbauer spectra of an α-Ti (0.25 wt% Fe) alloy in the as-quenched state and cold worked by rolling to reductions of 47 and 71% are shown in Fig. 23. The positions of

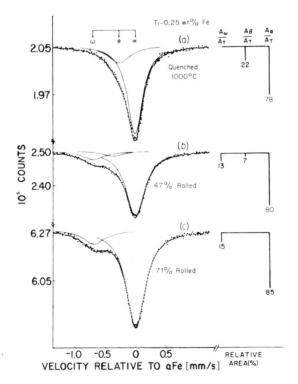

Fig. 23. Room-temperature Mössbauer spectra for a Ti (0.25 wt% Fe) alloy and bar diagrams in which the length of the bar is proportional to A_j/A_T with $j = \alpha, \theta, \omega$ and A_T the total spectral area. (a) As quenched from 1000°C, (b) cold worked by rolling to 47% reduction, (c) cold worked by rolling to 71% reduction (Ron *et al.*, 1977).

the characteristic lines of each phase are marked by a bar diagram at the top of the figure. Beside each spectrum in Fig. 23, a bar diagram is shown in which the length of the bar is proportional to A_j, where A_j is the relative spectral area of the jth phase, with $j = \alpha$, Θ, and ω. The relative area A_Θ decreases with the degree of cold work. At a reduction of 71%, the Θ-phase is no longer discernible, while the relative area A_ω has increased from almost zero to 0.15 and A_α has increased by 0.07.

The $\Theta \rightarrow \omega$ and $\Theta \rightarrow \alpha$ transitions have been observed upon cold work, whereas the reverse $\alpha \rightarrow \Theta$ and $\omega \rightarrow \Theta$ transitions have been observed upon aging. The combined effect of a decrease in specific volume during the $\alpha \rightarrow \Theta$ transition and a small displacement of atoms during plastic deformation has been suggested as the mechanism for the $\alpha \rightarrow \Theta \rightarrow \omega$ transitions. The formation of the ω-phase explains the known fact that both the toughness and the ductility of α-Ti(Fe) decrease upon cold working. The partial transformation $\Theta \rightarrow \alpha$ indicates the dissolution of a fraction of the metastable Θ-phase upon cold work. Similar effects have been observed for very small metastable precipitates in Cu(Fe) (Ron, 1973b; Hornstein and Ron, 1974) and in Cu(Cr) (Sargent and Purdy, 1974).

5. The Effect of Cold-Work on the α-Phase

It has been observed that the position of the Mössbauer line related to α-Ti(Fe) shifts in the negative direction upon cold-working. For a reduction in thickness of 80%, the ultimate change reaches a magnitude of $\Delta IS_\alpha = -0.015 \pm 0.005$ mm/s along with a line broadening of $\Delta \Gamma \approx 0.05$ mm/s.

The sign of this change, $\Delta IS_\alpha < 0$, implies an increase in $|\Psi_s(0)|^2$, the s-electron density at the Fe^{57} nucleus. This corresponds to an increase in pressure or a decrease in volume per Fe atom, if chemical changes are eliminated from consideration. These changes in IS_α and Γ_α have been interpreted as due to the accommodation of Fe atoms to lower energy positions in the compressive stress fields of dislocations.

On the other hand, it is known from high-pressure Mössbauer experiments (Ingalls et al., 1967) that for Ti(Fe), $dIS_\alpha/dp = (-7.1 \pm 1.1) \ 10^{-4}$ mm/s /kbar. Accordingly, the observed value of $\Delta IS_\alpha = -0.015$ mm/s corresponds to an increase of pressure of ~ 20 kbar or a decrease in volume per Fe atom of $\Delta V/V \sim 1.45\%$.

E. The Solubility of Iron in α-Ti(Fe)

Different estimates are given in the literature for the solubility of iron in α-Ti(Fe). At the eutectic temperature 585°C, the solubility of iron is 0.2 at.% according to Hansen and Anderko (1958), 0.4 at.% according to Molchanova (1965), and 0.04 at.% according to Raub et al. (1967).

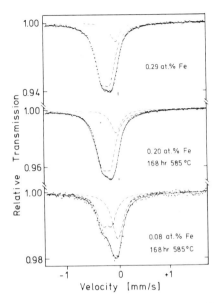

Fig. 24.   Room-temperature Mössbauer spectra of Ti–Fe alloys annealed at 585°C. Zero velocity corresponds to the isomer shift of α-Fe (Bläsius and Gonser, 1976).

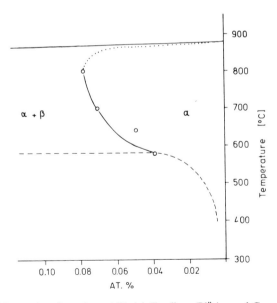

Fig. 25.   α-phase boundary of Ti-rich Fe alloys (Bläsius and Gonser, 1976).

The determination of the concentration of iron C_α in the α-phase, below the eutectic temperature, is difficult due to the appearance of metastable Θ- and ω-phases, as shown by Mössbauer studies.

However, for temperatures at and above 585°C, C_α has been determined with very high precision by means of a Mössbauer analysis (Bläsius and Gonser, 1976). Alloys annealed for long times in the $\alpha + \beta$ field have been quenched and measured at room temperature. The spectra shown in Fig. 24 have been analyzed as a superposition of an α-line and a β-doublet. The relative intensity I_β / I_α depends on C_{Fe}, the iron concentration of the alloy. The Fe concentration in the α-phase has been derived by means of the analysis in which one correlates the fraction of a spectral component with the fraction of a particular phase by applying the lever relationship. The results for four temperatures $T \geqslant 585°C$ are shown in Fig. 25.

F. Diffusion and Sintering in β-Ti(Fe)

Singwi and Sjölander (1960) predicted that a Mössbauer line would broaden by an amount proportional to the jump frequency of a diffusing Mössbauer atom. According to Mullen and Knauer (1969), the line broadening $\Delta\Gamma$ is related to the diffusivity by

$$\Delta\Gamma = \left(12\hbar/r_0^2\right)D,$$

where $\hbar$ is Planck's constant divided by 2π, D is the diffusion coefficient, and r_0 is the diffusion distance.

Mössbauer measurements of diffusivity corresponding to the above equation are generally in agreement with tracer diffusion results for alloys where correlation effects are negligible. The diffusivity of Fe, determined from Mössbauer line broadening in a Fe (3 wt% Si) alloy, has been found by Lewis and Flinn (1969) to agree well with tracer diffusion results.

Lewis and Flinn (1972) have also studied the diffusion of iron in β-Ti(Fe) alloys, by measuring the Mössbauer line broadening. The experimental results show good quantitative agreement with a model in which the diffusion mechanism involves complexes of extrinsic vacancies tightly bound to oxygen interstitials whose concentration is roughly 300 ppm by weight.

Mössbauer spectroscopy has been used along with x-ray diffraction and scanning microscopy for tracking the process of sintering a β-Ti (9 wt% Fe) alloy compact (Stupel et al., 1976b). The baking was carried out at two temperatures 800 and 1000°C. The phase changes occurring during baking at 800°C are depicted by the spectra in Fig. 26. After 5 min, satellites in the outer lines of the hyperfine splitting of α-Fe are seen along with a broad line in the central part of the spectra. The above changes indicate that titanium has diffused into iron simultaneously with the diffusion of

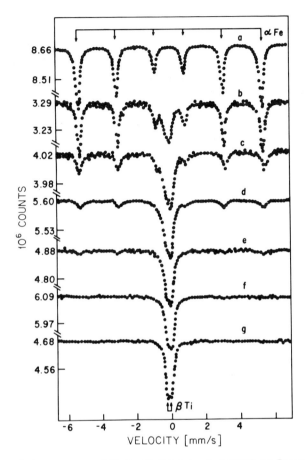

Fig. 26. Mössbauer spectra of Ti (9 wt% Fe) sintered at 800°C. (a) Pressed mixture of Ti and Fe powder, (b) after 5 min, (c) after 15 min, (d) after 45 min, (e) after 150 min, (f) after 6 hr, (g) after 384 hr (Stupel *et al.*, 1976b).

iron into titanium. Further, the α-Fe is seen to have disappeared from the spectra after less than 2.5 h. The last two sintering steps (Fig. 26f and g) were found to be the formation of TiFe and its transformation into β-Ti(Fe).

VI. Conclusions

Mössbauer analysis has a unique ability of yielding quantitative information on composition of phases, concentration of solute, metastable states, atomic configurations, etc. The results for the three binary systems Cu–Fe, Al–Fe, and Ti–Fe reveal the power of the method.

In the Cu–Fe system it has been possible to analyze the formation of Fe associations and precipitation from supersaturated solutions, surface states, oxidized states and observe transitions induced by plastic deformation, etc. Information relevant to physical phenomena such as spin structure, magnetic moments, Kondo effect, etc., has been obtained, as well.

In the Al–Fe system the method is helpful in the analysis of intermetallic phases and in the differentiation between associations of Fe impurities, vacancies, and other defects. Order–disorder reactions and configurations currently encountered in this system have been studied, and valuable information has been derived. Such information is of great help in the understanding of macroscopic properties of aluminium alloys and their processing.

Titanium alloys are of great technological importance; their physical metallurgy is quite complex, and modern methods are currently used for their analysis. The Ti-rich Ti–Fe alloys are quite typical $\alpha + \beta$ Ti alloys. The Mössbauer analysis yields quantitative data on phase composition, concentration of iron in various phases, and information on metastable phases and states. It is with the help of Mössbauer spectroscopy that a new metastable Θ-phase has been revealed and the ranges of formation and existence of this and other metastable phases have been derived. The maximum solubility of iron in the α-Ti phase (at $T > 585°C$) has been determined with high precision.

Much of the metallurgical information derived by Mössbauer spectroscopy is very difficult, if not impossible, to obtain by other methods. It is hoped that the adaptation of the Mössbauer method for routine analysis in physical metallurgy will become widespread.

Acknowledgments

Valuable discussions with Professor R. Preston and Dr. S. Nasu are gratefully acknowledged.

References

Abrahams, S. C., Guttman, L., and Kasper, I. S. (1962). *Phys. Rev.* **127**, 2052.
Baganataskii, I. A., and Nosova, G. I. (1962). *Phys. Metall. Metalloved* **13**, 92.
Bennett, L. H., and Swartzendruber, L. J. (1970). *Acta Metall.* **18**, 485.
Black, P. J. (1955). *Acta Crystallogr.* **8**, 43.
Blackburn, M. J. (1970). "The Science Technology and Application of Titanium" (R. I. Jaffe and N. Promisel, eds.). Pergamon, Oxford.
Bläsius, A., and Gonser, U. (1976). *Proc. Int. Conf. Appl. Mössbauer Effect, Corfu J. Phys.* **37**, C6-397.
Brückner, W., Kleinstück, K., and Schultze, G. E. R. (1970). *Phys. Status Solidi (a)* **1**, K1.

Campbell, S. J., and Clark, P. E. (1974). *J. Phys. F* **4**, 1073.

Campbell, S. J., Clark, P. E., and Hicks, T. I. (1976). *J. Phys. F* **6**, 249.

Clark, P. E., Cadogan, Y. M., Yazxhi, M. J., and Campbell, S. J. (1979). *J. Phys. F.* **9**, 379.

Cranshaw, T. E. (1972). *Proc. Int. Conf. Appl. Mössbauer Effect, Israel* p. 13.

Cser, L., Otanevich, J. M., and Pal, L. (1967). *Phys. Status Solidi* **20**, 581, 591.

Cser, L., Otanevich, J. M., and Pal, L. (1970). *Phys. Status Solidi* **42**, K147.

Czjzek, G., and Berger, W. G. (1970). *Phys. Rev. B* **1**, 957.

De Fontaine, D., and Kikuchi, R. (1974). *Acta Metall.* **22**, 1139.

Easterling, K. E., and Weatherley, G. C. (1969). *Acta Metall.* **17**, 845.

Edge, C. K., Ingalls, R., Debrunner, P., Drickamer, H. G., and Frauenfelder, H. (1965). *Phys. Rev.* **138A**, 729.

Erost, P. D., Parris, W. M., Hirsch, L. L., Doig, J. R., and Schwartz, C. M. (1954). *Trans. Am. Soc. Met.* **46**, 231.

Ettwig, H. H., and Pepperhoff, W. (1975). *Arch. Eisenhüttenwes.* **46**, 667.

Flinn, P. A. (1973). *Proc. Int. Conf. Mössbauer Spectrosc.*, *5th, Bratislava Czechoslovak AEC, Prague* Vol. 2, p. 275.

Frankel, R. B., Blum, N. A., Schwartz, B. B., and Kim, D. J. (1967). *Phys. Rev. Lett.* **18**, 1050.

Fujita, F. E. (1975). "Topics in Applied Physics" (U. Gonser ed.), Vol. 5, Mössbauer Spectroscopy, p. 201. Springer-Verlag, Berlin and New York.

Gonser, U. (1966). *Z. Metallkd.* **57**, 85.

Gonser, U., Grant, R. W., Muir, A. H., and Wiedersich, H. (1966). *Acta Metall.* **14**, 259.

Gonser, U. (1971). *In* "An Introduction to Mössbauer Spectroscopy" (L. May, ed.), p. 155. Plenum Press, New York.

Gonser, U., and Pfannes, H. D. (1974). *Proc. Int. Conf. Appl. Mössbauer Effect, Bendor. J. Phys.* **35**, C6–113.

Gonser, U., Meechan, C. J., Muir, A. H., and Wiedersich, H. (1963). *J. Appl. Phys.* **34**, 2373.

Gonser, U., Grant, R. W., Meechan, C. J., Muir, A. H., and Wiedersich, H. (1965). *J. Appl. Phys.* **36**, 2124.

Gradmann, U., Kümmerle, W., and Tillmanns, P. (1976). *Thin Solid Films* **34**, 249.

Hansen, M., and Anderko, K. (1958). "Constitution of Binary Alloys." McGraw Hill, New York.

Hergt, R., Wieser, E., Gengnagel, H., and Gladun A. (1970). *Phys. Status Solidi* **41**, 255.

Hickman, B. S. (1969a). *J. Mater. Sci.* **4**, 554.

Hickman, B. S. (1969b). *Trans. TMS-AIME* **243**, 1329.

Hornstein, F., and Ron, M. (1974). *Acta Metall.* **22**, 1537.

Housley, R. M., Grant, R. W., and Gonser, U. (1969). *Phys. Rev.* **178**, 514.

Huffman, G. P. (1971). *J. Appl. Phys.* **42**, 1606.

Huffman, G. P., and Fisher, R. M. (1967). *J. Appl. Phys.* **38**, 735.

Huffman, G. P., and Podgurski, H. H. (1975) *Acta Metall.* **23**, 1367.

Ingalls, R., Drickamer, H. G., and De Pasquali, G. (1967). *Phys. Rev.* **155**, 165.

Ishikawa, Y., Endoh, Y., and Takimoto, T. (1970). *J. Phys. Chem. Solids* **31**, 1225.

Janot, C. (1972). "L'Effet Mössbauer et ses Applications à la Physique du Solid et à la Metallurgie Physique." Masson, Paris.

Janot, C. (1977). *Proc. Int. Conf. Mössbauer Spectrosc., Bucharest, Romania* Vol. 2.

Janot, C., and Gibert, H. G. (1977). *J. Phys. F* **7**, 231.

Janot, C., Gibert, H. G., and Maugin, P. (1974). *J. Phys.* **35**, C1 49.

Keune, W., Lauer, J., and Williamson, D. L. (1974). *Proc. Int. Conf. Appl. Mössbauer Effect, Bendor. J. Phys.* **35**, C6-473.

Keune, W., Halbauer, R., Gonser, U., Lauer, J., and Williamson, D. L. (1977). *J. Appl. Phys.* **48**, 2976.

Kondorskii, E. I., and Sedov, V. L. (1959). *Sov. Phys. JETP* **8**, 1104.

Knauer, R. C., and Mullen, J. G. (1968). *Phys. Rev.* **174**, 711.
Krischel, K., Nasu, S., Gonser, U., Keune, W., Lauer, J., and Williamson, D. L. (1980). *Proc. Int. Conf. Appl. Mössbauer Effect, Portoroz. J. Phys.* **41** C1–417.
Lesoille, M. R., and Gielen, P. M. (1970). *Phys. Status Solidi* **37**, 127.
Lewis, S. J., and Flinn, P. A. (1969). *Appl. Phys. Lett.* **15**, 331.
Lewis, S. J., and Flinn, P. A. (1972). *Phil. Mag.* **26**, 977.
Longworth, G., and Jain, R. (1978). *J. Phys. F.* **8**, 351.
Losiyevskaja, S. A., and Kuzmin, R. N. (1972). *Russ. Metall.* **3**, 141.
Mielczarek, E. V., and Winfree, W. P. (1975). *Phys. Rev.* B **11**, 1026.
Molchanova, A. K. (1965). "Phase Diagrams of Ti Alloys," p. 82 Israel Program for Scientific Translations.
Muir, A. H. (1968). "Mössbauer Effect Methodology" (I. J. Gruverman, ed.), Vol. 4, p. 75. Plenum Press, New York.
Muir, A. H., and Wiedersich, H. (1967). *J. Phys. Chem. Solids* **28**, 65.
Mullen, J. G., and Knauer, R. C. (1969). "Mössbauer Effect Methodology" (I. J. Gruverman ed.), Vol. 5, p. 197. Plenum Press, New York.
Nakamichi, T. (1968). *J. Phys. Soc. Jpn.* **25**, 1189.
Nasu, S., and Gonser, U. (1973). *Proc. Int. Conf. Mössbauer Spectrosc., 5th, Bratislava Czechoslovak AEC, Prague* p. 311.
Nasu, S., Murakami, Y., Nakamura, Y., and Shinjo, T. (1968). *Scripta Metall.* **2**, 648.
Nasu, S., Gonser, U., Shingu, P. H., and Murakami, Y. (1974). *J. Phys. F* **4**, L 24.
Nasu, S., Gonser, U., and Preston, R. S. (1980). *Proc. Int. Conf. Appl. Mössbauer Effect, Portoroz. J. Phys.* **41**, C1–385.
Nathans, R., Pigott, M. T., and Shull, C. G. (1958). *J. Phys. Chem. Solids* **6**, 38.
Nemoshkalenko, V. V., Rasumov, O. N., and Gorskii, V. V. (1968). *Phys. Status Solidi* **29**, 45.
Newkirk, J. B. (1957). *Trans. AIME* **209**, 1214.
Nishio, M., Nasu, S., and Murakami, Y. (1970). *Nippon-Kinzoku-Gakkai-shi* **34**, 1173 (in Japanese).
Petrera, M., Gonser, U., Hasmann, U., Keune, W., and Lauer, J. (1976). *Proc. Int. Conf. Appl. Mössbauer Effect, Corfu. J. Phys.* **37**, C6-295.
Pfannes, H. D., and Gonser, U. (1973). *Appl. Phys.* **1**, 93.
Preston, R. S. (1971). *Metall. Trans.* **3**, 1831.
Preston, R. S., and Gerlach, R. (1971). *Phys. Rev.* B **3**, 1519.
Preston, R. S., Nasu, S., and Gonser, U. (1979). *Proc. Int. Conf. Appl. Mössbauer Effect, Kyoto. J. Phys.* **40**, C2–564.
Raub, E., Raub, Ch. J., Röschel, E., Compton, V. B., Geballe, T. H., and Matthias, B. T. (1967). *J. Less-Common Met.* **12**, 36.
Ridout, M. S., Cranshaw, T. E., and Johnson, C. E. (1964). *Proc. Int. Conf. Magn., Nottingham, England* p. 214.
Ron, M. (1973a). "Nuclear Techniques in the Basic Metal Industries," p. 493. IAEA, Vienna.
Ron, M. (1973b). "Nuclear Techniques in the Basic Metal Industries," p. 531. IAEA, Vienna.
Ron, M., Stupel, M. M., and Weiss, B. Z. (1977). *Acta Metall.* **25**, 1355.
Rupp, G. (1970). *Z. Phys.* **230**, 265.
Sargent, C. M., and Purdy, G. R. (1974). *Scripta Metall.* **8**, 569.
Sawicka, B. D., and Sawicki, J. A. (1977). *Proc. Int. Conf. Mössbauer Spectrosc., Bucharest, Romania* Vol. 2
Schwartz, L. H. (1976). "Application of Mössbauer Spectroscopy" (R. L. Cohen, ed.), Vol. 1, p. 37. Academic Press, New York.
Silcock, M. J., Davies, M. M., and Hardy, H. K. (1955). *Nature (London)* **175**, 731.
Shiga, M., and Nakamura, Y. (1976). *J. Phys. Soc. Jpn.* **40**, 1295.
Singwi, K. S., and Sjölander, A. (1960). *Phys. Rev.* **120**, 1093.

Stearns, M. B. (1966). *Phys. Rev.* **147**, 439.

Steiner, P., Hüfner, S., and Zdrojewski, W. V. (1974a). *Phys. Rev. B* **10**, 4704.

Steiner, P., Gumprecht, D., Zdrojewski, W. V., and Hüfner, S. (1974b). *Proc. Int. Conf. Appl. Mössbauer Effect, Bendor. J. Phys.* **35**, C6-523.

Stevens, J. G., and Stevens, V. E. (1966–1975); Muir, A. H., Ando, K. J., and Coogan, K. M. (1958–1965). "Mössbauer Effect Data Index." Hilger, London, Plenum Press, New York, and Wiley Interscience, New York.

Stupel, M. M. (1976). Ph. D. Thesis (in Hebrew).

Stupel, M. M., Ron, M., and Weiss, B. Z. (1974). *Proc. Int. Conf. Appl. Mössbauer Effect, Bendor. J. Phys.* **35**, C6-483.

Stupel, M. M., Ron, M., and Weiss, B. Z. (1976a). *J. Appl. Phys.* **47**, 6.

Stupel, M. M., Hornstein, F., Weiss, B. Z., and Ron, M. (1976b). *Metall. Trans.* **7a**, 689.

Stupel, M. M., Weiss, B. Z., and Ron, M. (1977a). *Acta Metall.* **25**, 667.

Stupel, M. M., Ron, M., and Weiss, B. Z. (1977b). *Metall. Trans.* (in press).

Swartzendruber, L. J., and Bennett, L. H. (1970). *J. Res. Nat. Bur. Std. Phys. Chem.* **74A**, 691.

Vogl, G., Mansel, W., and Dederichs, P. H., (1976). *Phys. Rev. Lett.* **36**, 1497.

Warlimont, H., Mühe, H., and Gengnagel, H. (1969). *Z. Angew. Phys.* **26**, 847.

Wassermann, G., and Greven, I. (1962) "Texturen Metallischer Werkstoffe," p. 183. Springer-Verlag, Berlin and New York.

Wertheim, G. K. Jaccarino, V., Wernick, J. H., and Buchanan, D. N. E. (1964). *Phys. Rev. Lett.* **12**, 24.

Wertheim, G. K., and Wernick, J. H. (1967). *Acta Metall.* **15**, 297.

Wertheim, G. K., Wernick, J. H., and Sherwood, R. C. (1969). *Solid State Commun.* **7**, 1399.

Wertheim, G. K., Buchanan, D. N. E., and Wernick, J. H. (1970a). *Solid State Commun.* **8**, 2173.

Wertheim, G. K., Wernick, J. H., and Sherwood, R. C. (1970b). *J. Appl. Phys.* **41**, 1325.

Williams, J. M., and Brooks, J. S. (1975). *Nucl. Instrum. Methods* **128**, 363.

Williams, J. C., (1973). *In* "Titanium Science and Technology" (R. I. Jaffe and H. M. Burke ed.), Vol. 3, p. 1433. Plenum Press, New York.

Williams, J. C., De Fontaine, D., and Paton, N. E. (1973). *Metall. Trans.* **4**, 2701.

Williamson, D. L., Keune, W., and Gonser, U. (1974). *Proc. Int. Conf. Magn.*, Vol. 1 (2), p. 246. "Nauka," Moscow.

Williamson, D. L., and Keune, W. (1975). *Proc. Int. Conf. Mössbauer Spectrosc., Cracow, Poland*, Vol. 1, p. 133.

Williamson, D. L., Nasu, S., and Gonser, U. (1976). *Acta Metall.* **24**, 1003.

Window, B. (1970). *J. Phys. C.* **3**, 323.

Window, B. (1971). *J. Phys. F* **1**, 533.

Window, B. (1972). *Phil. Mag.* **26**, 681.

Wright, J. G. (1971). *Phil. Mag.* **24**, 217.

Zemčik, T. (1971). *Proc. Conf. Mössbauer Spectrosc., Dresden, DDR* p. 103.

Zemčik, T. (1975). *Proc. Int. Conf. Mössbauer Spectrosc., Bratislava, 1973, Czechoslovak AEC, Prague* **2**, 275.

7

Iron-Carbon and Iron-Nitrogen Systems

Moshe Ron

Department of Materials Engineering
Technion
Haifa, Israel

I. Introduction

Iron base alloys have been in the service of mankind for about three and a half millennia and are without a doubt the foundation of our present technology.

APPLICATIONS OF MÖSSBAUER
SPECTROSCOPY, VOL. II

Due to the remarkable range of properties of iron and its alloys, the steel industry in all its aspects has grown to be the world's largest materials activity. A realistic evaluation of the fundamental phenomena that control the properties of steel will show that the interstitial nature of the major alloying elements is extremely important. The first investigations of the Fe–C system originated in 1868. Since then virtually every new investigation tool has been applied to study these alloys.

Experimental Mössbauer effect studies of interstitial iron alloys and steels started in 1966, a few years after the effect was discovered (Mössbauer, 1958). The isotope Fe^{57}, found at an abundance of 2.2% in natural iron, is the easiest one to experiment with among the resonant nuclei. A high recoil-free fraction at room temperature and the availability of a relatively long lifetime source (Co^{57}; 270 days) are among the favorable characteristics for using Fe^{57} for Mössbauer spectroscopy. The principles of the recoil-free γ-ray scattering—the Mössbauer effect—have been detailed in a number of monographs (Wertheim, 1964; Goldanskii and Herber, 1968; Greenwood and Gibb, 1971; Gonser, 1975). The application of the Mössbauer spectroscopy to the investigation of problems in metallurgy has been extensively reviewed (Gonser, 1968, 1971; Ron, 1973; Johns, 1973; Fujita, 1975). The sensitivity of the Mössbauer effect to atomic scale environment has given rise to its use for the investigation of the structure and the symmetry of interstitial atom sites in Fe–C and Fe–N martensite. Processes of interstitial C and N atoms agglomeration, clustering, ordering, and precipitation kinetics have been widely studied during the period since 1966. In recent years improved techniques have stimulated the investigation of the martensitic structure (DeCristofaro and Kaplow, 1977a; Gridnyev et al., 1977; Ino and Ito, 1978; Williamson et al., 1978; Cadeville et al., 1977).

Unlike other modern analytical techniques, such as electron microscopy, Auger spectroscopy, etc., no surface preparation or thinning is required for Mössbauer backscattering measurement, which is inherently nondestructive. Sachdev et al. (1979) and Sachdev (1977) have observed substantial differences in the phase composition, microstructure, substructure, and transformation temperatures in thin martensite films as compared with bulk martensite. For this reason the Mössbauer backscattering method is especially appropriate for martensite.

This chapter is confined to the Fe^{57} isotope since only iron alloys are considered. The first paragraphs are introductory to the analysis of spectra and to the Fe–C binary system. In subsequent paragraphs the research on Fe–C martensite is reviewed in an almost chronological order. Subsequently, aging, segregation, ordering, and precipitation of carbides are described. The investigations of the Fe–N system are described in the same order, although the volume of research work related to it is much smaller than that for the Fe–C system. A number of representative investigations of technical problems using Mössbauer spectroscopy conclude the chapter.

II. Fe–C Alloys

A. *Phases and Phase Transformations in the Fe–C System*

Iron exists in two allotropic forms. It crystallizes as the fcc lattice in the temperature range 906°C $< T_\gamma <$ 1401°C and as the bcc lattice in the range $T_{bcc} >$ 1401°C, $T_\alpha <$ 906°C. The reoccurrence of the bcc structure above 1401°C has been explained by allowing for magnetic and nonmagnetic terms in the expression for the change of free energy with temperature. In pure iron the $\gamma \rightarrow \alpha$ transformation occurs rapidly. However, the transition rate can be substantially slowed down by alloying. At temperatures slightly below the transformation temperature (small undercooling), the $\gamma \rightarrow \alpha$ transition can be bypassed, and at temperatures well below (large undercooling) the martensitic reaction occurs.

The Fe–C binary system is normally described in terms of Fe and the metastable carbide Fe_3C–cementite, as well as in terms of Fe and stable graphite C. The Fe–C binary phase diagram is shown in Fig. 1. The γ-phase (austenite) is a solid solution in which the interstitial carbon atoms are separated from each other by a repulsive interaction. The maximum solubility of C– in Fe–C austenite (in metastable equilibrium with Fe_3C) is 2 wt%, while in α-Fe it is only 0.02 wt%. At 723°C and carbon concentration of 0.8 wt%, the γ-phase undergoes an eutectoid decomposition $\gamma \rightarrow \alpha + Fe_3C$. The lamellar microstructure that originates from the eutectoid reaction is known as pearlite. The most stable iron–carbide, cementite–Fe_3C, has an orthorhombic unit cell of P_{nma} symmetry. The composition of cementite may vary slightly at high temperature. Fe_3C has a melting temperature of about 1600°C, is very hard ($\sim$ 800 HB), and has a ferromagnetic transition at $\sim$ 210°C. Graphite may form under certain conditions when the metastable cementite decomposes. This process is of practical importance for controlling the microstructure and properties of cast irons. The cementite structure accepts substitutionally nonmetals (such as N, O), which replace carbon, and metal atoms, which replace iron. Controlled precipitation of such carbides in steels alloyed by both interstitial and substitutional elements strongly affects the mechanical properties.

A number of solid-state reactions occur in the iron-rich portion of the Fe–C phase diagram (Fig. 1). These reactions provide the basis for

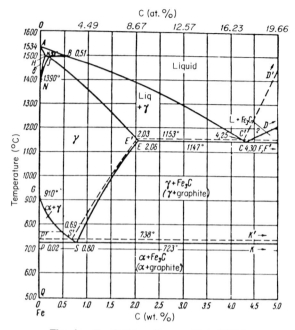

Fig. 1. Fe–C phase diagram (Vol, 1967a).

controlling the phase composition and microstructure of steels by heat treatment. If austenite is cooled fast enough so that formation of pearlite cannot start, a degree of undercooling can be attained at which the austenite transforms martensitically. Below a certain temperature M_s, the $\gamma \rightarrow \alpha_m$ (α-martensite) transformation occurs, and a fraction of the austenite transforms into martensite. In Fe–C alloys the M_s temperature decreases as the carbon concentration increases.

The martensitic transformation involves a macroscopic shear that produces a change in shape. This is accompanied by large strains, which lead to the platelike shape of the martensitic crystal. Internally, microplastic deformation results and a substructure forms with a high density of dislocations, stacking faults, twins, etc. The particular morphology and substructure that form depend on the composition and M_s temperature of the alloy. The microstructure of Fe–C martensite containing up to 0.6 wt% contains lathlike shear units that form on the $(111)_\gamma$ planes. The substructure of the lath martensite consists predominantly of tangled dislocations. For Fe–C alloys containing more than 1 wt% C, the M_s temperature is below 450 K and the morphology of the martensite is platelike. The martensite form on irrational habit planes ranging from $\{225\}_\gamma$ to $\{259\}_\gamma$, changing with increasing carbon content. The martensite plates are internally twinned and may have a midrib.

The octahedral symmetry of the interstitial carbon atoms in austenite is maintained in the martensite. However, in the bct lattice the carbon atom expands the octahedral void by pushing apart two iron atoms in a direction that becomes the c axis of the bct martensitic cell. As indicated by x-ray diffraction, the c axes align in parallel, leading to a bct structure. The ratio c/a increases with the carbon content of the alloy due to the increase in c parameter and decrease in a parameter. It has been suggested that at low temperatures a fraction of carbon atoms occupies tetrahedral interstices. This concept is currently the subject of debate among investigators.

Martensite is a metastable solid solution supersaturated by carbon. There is a high activation energy for nucleation of pure graphite because of large surface energy terms. Consequently, ordered structures, metastable and transition carbides with lower activation energies, form instead. The segregation of carbon is known to take place even at room temperature and has been experimentally observed by electrical resistivity measurements, electron microscopy, and Mössbauer spectroscopy. In addition to the migration of carbon to defects, the following ordered regions, transition, and metastable iron carbides are known to form during the aging and tempering processes of martensite.

Carbon-rich ordered domains with a local composition from Fe_4C, and a structure isomorphous with γ'-Fe_4N, form between ~ -10 and $80°C$, with a $\{012\}_\alpha$ habit plane.

The hexagonal ϵ-$Fe_{2.4}$ C-carbide forms between 80 and $270°C$, with various morphologies and habits. The formation of ϵ-carbide at temperatures lower than $80°C$ has also been reported.

Orthorhombic η-Fe_2C forms with a habit $\{100\}_\alpha$ in the temperature range $120–200°C$.

The monoclinic χ-Fe_5C_2 carbide forms with a $\{112\}_\alpha$ habit in the temperature range $250–400°C$.

The most stable iron carbide, Fe_3C, is orthorhombic and forms with $\{112\}_\alpha$ or $\{110\}_\alpha$ habits at temperatures above $300°C$.

B. Analysis of Mössbauer Spectra of Fe–C and Fe–N Alloys

The common procedure for the evaluation of experimental Mössbauer spectra has been using a least-squares-minimum computer fit to a superposition of Lorentzian shape lines. However, for accurate quantitative results, the effects of finite absorber thickness and differences in the f factors of the various components (phases, nonequivalent sites, etc.) must be considered (Goldanskii and Herber, 1968; Frauenfelder et al., 1962). Corrections for absorber (or scatterer) finite thickness and procedures for the evaluation of spectra by "blackness removal" were discussed by Schwartz in Chapter 2 in Volume I of this treatise. Complex Mössbauer patterns are sometimes encountered in Fe–C and Fe–N alloys. For instance, a sample consisting of martensite and precipitated carbides (or nitrides) may display a spectrum consisting of six or more sextuplet components. Frequently, lines are broadened due to metastable states and lattice relaxation caused by strains, which are inherently present in martensite.

A method of "difference spectra" was used to facilitate the differentiation between a Mössbauer pattern obtained before and after a certain treatment of martensite (Choo and Kaplow, 1973; DeCristofaro and Kaplow, 1977a; Albritton and Lewis, 1971). Changes generated by the appearance or disappearance of components hidden in the spectra, and otherwise undetectable, were revealed by the difference spectra.

The idealized Mössbauer pattern is given by the following convolution (Hanna and Preston, 1965):

$$P_{th}(E_s) = \int_{-\infty}^{\infty} I(E, E_s)\left\{1 - \exp\left(-k\sum(E)\right)\right\} dE, \qquad (1)$$

where $P_{th}(E)$ is the theoretical resonant absorption; $I(E, E_s)$ the Lorentzian energy distribution of the gamma radiation source; k is proportional to cross section for resonant absorption, recoilless fraction, number of nuclei/cm^2, and thickness of the absorber; and

$$\sum(E) = \sum_{i=1}^{N} \frac{g_i \Gamma_a^2}{4(E - E_i)^2 + \Gamma_a^2}. \qquad (2)$$

The last term is the sum of a series of Lorentzian curves, each representing one distinct absorption energy in the sample. Γ_a is the width of a Mössbauer line ($\Gamma_a = 0.095$ mm/s for ^{57}Fe); E_i and g_i are the positions and fractional weights of the ith line, respectively.

A method elaborated particularly for the analysis of the specific features of the Fe–C and Fe–N martensite and austenite spectra was used for computer analysis by DeCristofaro and Kaplow (1977a). They compensated empirically for instrumental broadening by convoluting the theoretical resonant absorption with a normalized Gaussian curve G, i.e.,

$$P(E_s) = \int_{-\infty}^{\infty} P_{th}(E)G(E - E_s)\,dE. \tag{3}$$

The appropriate width for the broadening was determined by comparison of $P(E_s)$ with the experimental Mössbauer spectrum of α iron. This procedure differs from other attempts to correct for instrumental broadening that have employed the use of artificially large linewidths in Eq. (2). The Gaussian form is more appropriate than a Lorentzian for instrumental effects and appears to account adequately for the experimental shapes.

While Eq. (3) was used as the basic theoretical model, certain spectra, particularly virgin martensite, are more complex. Not only are the readily distinguishable peaks in virgin martensite broader, but within each ferromagnetic spectrum the peak width varies as a function of the magnitude of magnetic splitting of the nuclear state, i.e., peaks farthest from the center of the spectrum are broader than the inner peaks. Close fits between model and experimental spectra for both iron–carbon and iron–nitrogen martensites were obtained by the following procedure: for each ferromagnetic spectrum, each individual absorption peak [each term in Eq. (2)] was broadened by convoluting it with a Gaussian curve whose width was proportional to the distance of the peak from the center of the spectrum. This is equivalent to assuming that a small spread of the effective field H associated with each of the separate spectra is included. Thus Eq. (2) would become

$$\sum(E) = \sum_{i=1}^{N}\left(\int \frac{g_i\Gamma_a^2}{4(E - E_i)^2 + \Gamma_a^2} H(E, E_i)\,dE\right), \tag{4}$$

where $H(E, E_i)$ represents a Gaussian curve whose width is a function of peak position E_i. Computer-generated patterns, based on the described method, closely fit the experimental spectra of Fe–C and Fe–N martensite and austenite, as will be seen below.

III. Fe–C Austenite

A. The Austenite Solid Solution

X-ray analyses made more than half a century ago have shown that Fe–C austenite is an interstitial solid solution of carbon in γ-iron (Westgren and Phragmen, 1922; Wever, 1924), γ-iron being the only polymorph that is capable of dissolving a large percentage (2 wt%) of interstitial carbon due to the large octahedral voids in the fcc structure. The volume of the average unit cell of austenite expands regularly with the carbon content (see Fig. 2), the lattice parameter being given by the equation

$$a(\text{in Å}) = 3.548 + 0.044\, C(\text{wt\%}). \tag{5}$$

The stability of austenite increases with carbon content. As a result, austenite may be retained on quenching to room temperature if the carbon content of the alloy is sufficiently high. The overall interaction energy between carbon atoms in austenite is one of repulsion, implying the tendency of carbon atoms in austenite to be separate from each other rather than to be statistically distributed. Carbon has been observed to behave in Fe–C austenite as a positively charged particle (Okabe and Guy, 1970, 1973). Also, elastic strain around an octahedral interstice may contribute to the blocking of 12 neighboring octahedral sites, which prevents them from admitting an interstitial carbon atom (Moon, 1963).

For a perfectly random solid solution, the probability that a given site will be empty is $(1 - x)$, whereas x (at.%) is the probability of a given site being occupied. In austenite each iron atom has six neighboring octahedral

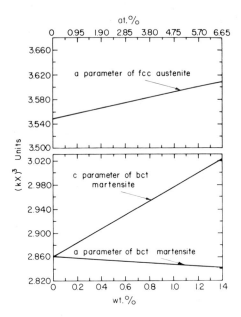

Fig. 2. Lattice parameters versus the concentration of carbon in Fe–C austenite and martensite (Roberts, 1953).

TABLE I

Fraction of Iron Atoms in Austenite Having No Interstitial Atom Neighbors[a]

	Fe–C	Fe–N
Composition (x)	0.0875	0.0956
Random model	0.577	0.547
Separated model	0.472	0.426
Measured value	0.495 ± 0.008	0.55 ± 0.008

[a] Taken from DeCristofaro and Kaplow (1977a).

interstitial sites. Thus in a perfectly random distribution, the probability that an iron atom will have no interstitial neighbors is $P_o = (1 - x)^6$; while $P_1 = 1 - P_o$ is the probability for having at least one interstitial neighbor.

For the case that the interstitial carbon atoms are separated from each other due to mutual repulsion, the fraction of iron atoms neighboring an interstitial carbon atoms would be $6x$, and the fraction of iron atoms not neighboring a carbon atom would be $1 - 6x$. The fraction of iron atoms in Fe–C austenite not neighboring an interstitial carbon is given for the above models in Table I for a composition of 0.0875.

B. The Austenite Pattern

It has been shown by Roberts (1953) that a substantial fraction of the austenitic phase in high-carbon steels is retained upon quenching and that subsequent cooling to lower temperatures promotes the transformation of a portion of the retained austenite to martensite. The Fe–C austenite has been studied by means of the Mössbauer effect by a number of investigators (Gielen and Kaplow, 1967; Christ and Giles, 1967, 1968; Genin and Flinn, 1968; Lesoille and Gielen, 1972; Gridnyev et al., 1977; DeCristofaro and Kaplow, 1977a). Two problems were associated with the interpretation of the austenite pattern. One is the assignment of the peaks to the two iron configurations, nonequivalent in regard to the location of their interstitial carbon atoms. The second problem is the determination of the carbon atom distribution (i.e., random distribution versus separated C atom) due to a repulsive interaction between them.

A room-temperature Mössbauer spectrum of a 9 at.% Fe–C austenite is shown in Fig. 3. It is common practice to separate the room-temperature Mössbauer spectrum of retained austenite into two components. One is a single peak related to iron atoms having no nearest carbon neighbors, the other component is a quadrupole-split doublet with a positive isomer shift relative to the single peak, arising from iron atoms having one nearest

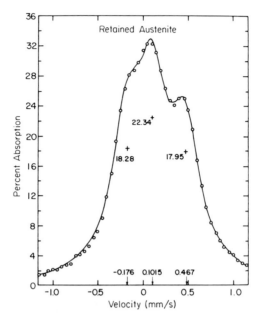

Fig. 3. Room-temperature Mössbauer spectrum of 8.5 at.% C retained austenite quenched to room temperature (Genin and Flinn, 1968).

carbon neighbor. There is reasonably good agreement about the values of the isomer shifts and of the quadrupole interaction as well as the assignment of the peaks to the two types of atoms in Fe–C austenite given by the various investigators, as Table II shows. Recently Williamson et al. (1979) have analyzed the Mössbauer pattern of retained austenite in a 5.43 at.% C alloy, in terms of two quadrupole split lines.

TABLE II

Mössbauer Parameters of High-Carbon Austenite[a]

Q.S.	IS_{Fe_0} [b]	$IS_{Fe_1} - IS_{Fe_0}$ [c]	Reference
0.625	− 0.1	0.043	Gielen and Kaplow (1967)
0.643	− 0.029	0.043	Genin and Flinn (1968)
0.620	− 0.08	0.05	Lesoille and Gielen (1972)
0.630	~ −0.1	0.04	DeCristofaro and Kaplow (1977a)

[a] Units are in mm/s.
[b] IS (isomer shift) with respect to α-Fe.
[c] IS (isomer shift): Fe_0—iron atoms without carbon nn in austenite. Fe_1—iron atoms with one carbon nn in austenite.

After carefully accounting for the effect of thickness, Williamson *et al.* (1979) obtained a substantially improved computer fit, interpreting the results as being in agreement with a random distribution of carbon atoms in austenite. Also Lesoille and Gielen (1972) reported that experimental results fit well the assumption of a random distribution. However, Genin and Flinn (1968), investigating a Fe–C austenite containing 9.5 at.% C, found that the measured relative central peak intensity of 0.43 corresponds exactly to $(1 - 6x)$, viz., the fraction of iron atoms having no nearest carbon neighbors as calculated in accordance with the C–C repulsion model. The conclusion from the other Mössbauer studies was also that the C–C repulsion model, as suggested on thermodynamic grounds, appears to be the appropriate one (e.g., DeCristofaro and Kaplow 1977a).

The conversion of the austenite pattern into one line at high temperature was observed by Lewis and Flinn (1968) in a 1.6 wt% C alloy. At a temperature of 895°C the average time between jumps of carbon atoms is about 10^{-8} s, which is more than one order of magnitude shorter than the mean lifetime of the Mössbauer transition. Under these circumstances, the spectra will show the time-average effect of the carbon atoms, and all iron atoms will appear equivalent. The resulting single line has an isomeric shift relatively close to that of the single line of retained austenite at room temperature if the temperature shift is taken into account. These experimental results support the assumption that the quadrupole interaction observed at room temperature originates from a local distortion of the cubic symmetry about some iron atoms due to the presence of a neighboring carbon atom.

The Fe–C austenite spectra reveals a comparatively large positive quadrupole interaction of about 0.63 ± 0.02 mm/s for iron atoms having a nn carbon atom. This large quadrupole effect, together with the small difference in isomeric shift (0.045 mm/s) between the two iron atom types in Fe–C austenite, has been interpreted by Gielen and Kaplow (1967) in the following way. The carbon atom becomes positively ionized by giving some electronic charge to the 4s conduction band of iron, producing a positive shift (between the iron atoms with one nn carbon atom and those with none) via long-range, conduction electron charge density oscillations around an impurity center in a metal. The screening s-electron charge density concentrates on the impurity and causes a depletion of 4s charge density on the neighboring iron atom. The quadrupole interaction was interpreted as being due to the ionization of carbon in austenite. Lesoille and Gielen (1972) used a correlation between the isomeric shift and quadrupole interaction suggested by Remy and Pollak (1965) in order to derive the sign of the quadrupole interaction. They found it to be negative and suggested that the efg tensor is of the "shielding" type. The approximation that the efg is axially symmetric was used by all investigators.

No antiferromagnetic transition has been observed upon cooling retained austenite contained in Fe–C alloys to liquid nitrogen temperatures. The austenite lines are readily discernible and may be unambiguously separated from the martensite pattern. The relative spectral areas (corrected for differences in f factors, etc.) can thus be used in order to determine, with high precision, the amounts of retained austenite in technically important cases (Swartzenbdruber et al., 1974). The method was recently described and summarized by Schwartz (1976) in Chapter 2 of Volume I of this treatise.

IV. The Martensitic Transformation and the Structure of Fe–C Martensite

A. The Fe–C Martensite

A thorough understanding of the mechanism of the martensitic transformation and of the details of the martensitic phase structure is essential to analyze the phase transformations and the metastable states encountered in the heat treatment of steels. The martensitic transformation in Fe–C alloys has been investigated for more than 100 years (Tschernov, 1868; Martens, 1880; Bain, 1924; Cohen, 1951; Kurdjumov, 1960). A number of characteristic features are commonly ascribed to a martensitic transformation. It is a diffusionless transformation exhibiting a macroscopic shear. No atomic movements over distances larger than the interatomic spacing are involved. The transformation occurs at a velocity close to that of the propagation of sound waves in crystals. Usually a fraction of the original phase transforms rapidly and no further change takes place if the temperature remains constant. This is the so-called "athermal" behavior, which is, however, occasionally followed by a small amount of isothermal transformation. During rapid and continuous cooling the martensitic transformation begins at a temperature M_s and continues until a temperature M_f is reached below which no transformation occurs. The martensitic transformation is in principle reversible, but some hysteresis occurs upon reversal.

The crystallography of Fe–C martensite has been widely studied by x-ray diffraction, electron microscopy, and other techniques. A considerable number of Mössbauer-effect investigations have also been concerned with the Fe–C martensite and its crystal structure, and a brief discussion of the iron–carbon martensite crystallography is in order at this point. The subject has been extensively described in several monographs (Wayman, 1964; Kaufman and Cohen, 1955; Kurdjumov et al., 1977). Bain (1924) was the first to show that the lattice transformation from fcc austenite to bct martensite proceeds by only slight relative movements of atoms. For this he used the correspondence of the two lattices, considering the fcc lattice as a bct lattice with a c/a ratio of $\sqrt{2}$. Subsequently, a homogeneous strain was used to convert the lattice parameters of the bct cell into those of martensite. More realistic models were suggested later, the one commonly accepted being that suggested by Kurdjumov and Sachs (1930). In a 1.4 wt% C steel they observed the $(111)_A/(001)_M$ and $[1\bar{1}0]/[1\bar{1}2]$ lattice relationships, the subscripts A and M referring to austenite and martensite lattices, respectively. A habit plane of about (225) was observed. From these relationships it was concluded that the Bain stress was insufficient to account for the coherent fit of the two structures while allowing growth of the bct phase. To explain the existence of an invariant plane, it has been suggested that simple shear deformation (lattice invariant strain) on a certain plane in a certain direction, which can occur by slip or by twinning, would decrease the lattice deformation (Bain strain) (Geniger and Trojano, 1949; Bowles and Mackenzie, 1954; Wechsler *et al.*, 1953). Since the intermediate transformations are linear, a matrix analysis was used (Christian, 1965) in order to obtain the total transformation: $F = SRB$, where B is the Bain distortion, S the twinning or slip, and R the invariant plane strain. For Fe–C martensite the orientation relationships and habit planes depend on the carbon concentration.

For plain carbon steel with less than 1.4 wt% C, the Kurdjumov–Sachs relationships hold, while for higher carbon contents the relationships of Nishiyama (1934) were observed: $[112]_A/[001]_M$ and (259) habit plane. Twinning has been observed by electron microscopy to be the mode of lattice invariant strain for high-carbon steel (Baker and Kelly, 1963; Kelly and Nutting, 1960). The addition of alloying elements strongly affects the internal substructure of the martensitic phase. Sachdev (1977); Sachdev *et al.* (1979) studied the substructure of high-carbon Fe–C martensites containing Mn. The Mn was added in order to provide a series of alloys with M_s temperatures varying from 320 to 150 K. The martensite habit plane was seen to change from $(259)_\gamma$ for a 1.85 wt% C alloy to $(225)_\gamma$ with the addition of 3 wt% Mn. The $(259)_\gamma$ martensites have a midrib containing $(112)_\alpha$ twins, but the rest of each such martensitic plate forms by faulting (twinning) on $\{011\}_\alpha$ planes as an essential part of the martensitic transformation.

The position of the carbon atoms in austenite and martensite has been derived based on the models for a migration of the iron atoms during the martensitic transformation and other relevant experimental data. The average unit cell volume in both lattices expands with the carbon content (Fig. 2). While the fcc lattice of austenite expands uniformly, in the tetragonal martensite the a parameter contracts and the c parameter increases with the carbon content (Roberts, 1953; Kurdjumov et al., 1977). The dilations, when averaged over the entire martensite crystal, lead to the following lattice dimensions:

$$C \text{ (in } kX) = 2.861 + 0.0116 \times (\text{wt\% C}),$$
$$a \text{ (in } kX) = 2.861 - 0.013 \times (\text{wt\% C}), \tag{6}$$
$$c/a = 1.000 + 0.045 \times (\text{wt\% C}).$$

The change of the lattice parameters of austenite and martensite with the carbon content is shown in Fig. 2. After much controversy it was shown by Winchell and Cohen (1962) that c/a extrapolates to unity when the carbon content tends to zero. In the fcc phase the carbon atom, occupying an octahedral interstice with six nearest neighbors, leads to a symmetrical expansion of the lattice. Since the austenite $\rightarrow$ martensite transformation is nondiffusive and displacive, the trapped carbon atom may be expected to tend to preserve the octahedral configuration of its neighbors. However, in the bcc cell the octahedral voids at $(0, 0, \frac{1}{2})$ and $(0, \frac{1}{2}, \frac{1}{2})$ are smaller than the tetrahedral ones at $(\frac{1}{2}, \frac{1}{4}, 0)$. The presence of a carbon atom in an octahedral void distorts it by pushing apart two iron atoms in a direction that becomes the c axis of the bct martensitic cell and by drawing together the four atoms in the transversal direction. Cohen (1962) has calculated that the two iron atoms adjacent to the interstitial carbon are pushed apart by 10% along the c axis, forming a so-called "distortion dipole," while a distortion of 36% would be required in order to restore an octahedron of regular shape and a bcc lattice. There are three sets of octahedral interstices in the bcc lattice (before the bct martensite forms); but only one of them becomes occupied by the carbon atoms, and the newly formed c axes align in parallel. According to Kurdjumov and Khachaturyan (1972), the long-range order parameter that describes the preferential carbon-atom occupancy among the three sublattices in martensite is defined by

$$Z = \tfrac{1}{2}(2n_3 - 1) = 1 - 3n_2 = 1 - 3n_1, \tag{7}$$

where $n_i (i = 1, 2, 3)$ is the occupational probability of the octahedral site of the ith kind.

Further, $n_1 + n_2 + n_3 = 1$ and $n_1 = n_2$. The long-range order parameter Z has values $0 \leqslant Z \leqslant 1$. $Z = 1$ if all carbon atoms occupy the third sublattice, viz., $n_3 = 1$. $Z = 0$ if all carbon atoms are distributed among the three sublattices in a random manner, viz., $n_1 = n_2 = n_3 = \frac{1}{3}$. The axial ratio of the bct martensite can thus be expressed as

$$c/a = 1 + 0.045 \times (\text{wt\% C})Z. \tag{8}$$

Commonly values reported in the literature assume $Z = 1$, as in Eq. (6).

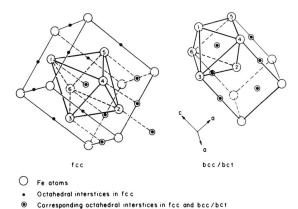

f c c   b c c / b c t

○   Fe atoms
•   Octahedral interstices in fcc
◉   Corresponding octahedral interstices in fcc and bcc/bct

Fig. 4. The crystallographic correspondence of fcc and bcc/bct lattices and octahedral interstitial sites. Note that only one of the three possible sets in martensite forms, thus defining the c direction of the unit cell and the alignment of the fourfold axes of the resulting nonregular octahedra (Cohen, 1962).

The crystallographic correspondence between the parent (fcc) and the product (bcc/bct) lattices, as well as the octahedral interstices, are shown in Fig. 4. The distribution of carbon in martensite has been found to be a random one as in a random solid solution, while in austenite the carbon atoms were found to be isolated (Aaronson *et al.*, 1966).

An anomalously low c/a ratio (see Fig. 5) was found in Fe–Mn–C martensite quenched to liquid nitrogen temperature (called "virgin martensite") by Lysak and Vovk (1965), Lysak *et al.*, (1967), Lysak and Nikolin (1966), but the normal c/a ratio is restored as the alloy is heated back to room temperature. Lysak and Vovk (1965) suggested that this behavior was due to the occupancy of the tetrahedral interstices by some of the carbon atoms, which migrated to the octahedral sites upon heating

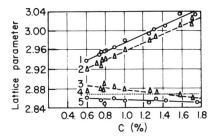

Fig. 5. Lattice parameters of virgin Fe–C martensite versus carbon concentration. Lines 1 and 5 are for a phase of normal tetragonality. Lines 2 and 3 are for a phase of low tetragonality. Line 4 is for pure iron (Lysak *et al.*, 1967).

to room temperature. Oshima and Wayman (1974) and Oshima *et al.* (1976) observed satellite spots in the diffraction patterns of virgin Fe–Mn–Cr–C martensites having anomalously low tetragonality. Fujita (1975) and Oshima *et al.* (1976) attribute the presence of these spots to the formation of an intermediate six-layer structure in the austenite–martensite transformation. This mechanism leads to a martensite with carbon atoms distributed between octahedral and tetrahedral sites in a 1:1 ratio. These suggestions were, however, refuted by theoretical calculations and by the results of neutron and diffuse electron scattering investigations (Khachaturyan and Onissimova, 1968; Etnin, 1973; Kurdjumov *et al.*, 1977). On the other hand, the partial occupation by carbon atoms of more than one set of octahedral sites was mentioned by Alshevsky and Kurdjumov (1968) and Kurdjumov *et al.* (1977) as the possible reason for the low value of the c/a ratio. The assumption that the c/a ratio as measured by x-ray diffraction is proportional to the degree of alignment of the c axes, i.e, the degree of occupancy of one of the three sets of octahedral sites, is very powerful. Using it, one can conclude, for example, that the increased values of the c/a ratio encountered in steels containing nickel are due to a high degree of occupancy of one set of octahedral sites (Kurdjumov *et al.*, 1977, Kurdjumov and Khachaturyan, 1975).

As seen from the above discussion, it is still not agreed upon whether carbon atoms do or do not occupy only octahedral sites in the bct lattice of Fe–C martensite.

B. Mössbauer Effect Investigations of Fe–C Martensite

1. The Crystal Structure and Atomic Configurations

The technical interest in this subject stimulated intensive Mössbauer investigations of the martensitic state and of interstitial alloys in general as soon as the Mössbauer technique became generally available. The most significant advantage of the Mössbauer spectroscopy with regard to the investigation of martensite is its very high sensitivity to changes in the atomic configuration of resonant atoms (Fe^{57}) in the immediate neighborhood of an interstitial atom. Mössbauer spectroscopy, via the hyperfine interactions, provides information on the magnetic state, electrical interaction, electric field gradients, chemical bonding, local crystal symmetry, lattice defects, dynamic parameters, elastic stresses, etc. Although the Mössbauer studies of martensite have provided a substantial amount of new information on atomic arrangements, there is still much controversy even on basic questions. For instance, the question of whether or not interstitial carbon atoms partially occupy tetrahedral interstices in virgin martensite has not yet been definitely resolved.

An interstitial carbon atom severely affects its iron neighbors by altering the electronic structure, magnetic interaction, crystal field symmetry, and even the position in the crystal lattice. The hyperfine interactions of Fe^{57} in the Fe–C martensite are perturbed by these effects. Thus, the Mössbauer spectra depend on the number and location of the carbon neighbors with respect to the iron atoms. The Mössbauer spectra have been analyzed as a superposition of components arising from the various types of iron atoms. No uniform definition or designation has been used by the various investigators for the types of iron atoms. The following notation will be used throughout this chapter:

Fe_0 iron atoms without nn or nnn carbon atoms,
Fe_1 iron atoms with one nn carbon atom,
Fe_2 iron atoms with two nn carbon atoms,
Fe_{nn} iron atoms with one nn carbon atom at an octahedral site,
Fe_{nnn} iron atoms with one nnn carbon atom at an octahedral site,
Fe_{nn}^t iron atoms with one nn carbon atom at a tetrahedral site.

The first Mössbauer investigations of the structure of martensite were carried out and published almost simultaneously by Genin and Flinn (1966), Christ and Giles (1967), Gielen and Kaplow (1967), Ron *et al.* (1967), Ino *et al.* (1967, 1968), and Moriya *et al.* (1968). The above studies may be considered at the "first phase" of Mössbauer investigations of the structure of martensite and will be so referred to. Their results, which were qualitatively similar, were all based on the assumption that the interstitial carbon atoms occupy an interstice of distorted octahedral symmetry. However, considerable disagreement arose over the interpretation and the assignment of various hyperfine interactions to specific types of iron atoms according to their location relative to an interstitial carbon atom.

In order to discuss these disagreements, a brief description of the various types of atomic groups, their designations, and the spectral components assigned to them, seems to be in order. It is generally agreed that in martensite the carbon atoms occupying an octahedral interstice produce a large local distortion (the "distortion dipole"). An iron atom at the [001] position (on the c axis) and four iron atoms at the (001) plane (perpendicular to the c axis) may therefore be designated as nearest, Fe_{nn}, and next nearest, Fe_{nnn}, neighbors, respectively, of an interstitial carbon at an octahedral site (see Fig. 6). Iron atoms remote from an interstitial carbon atom, i.e., such that have no interstitial nn or nnn may be designated as Fe_0. Table III lists the assignments given by the above-named investigators to the partially resolved Mössbauer patterns of martensite.

A Mössbauer absorption spectrum for a martensitic specimen containing 5.1 at.% C quenched from 850°C into ice water and immediately

TABLE III

Mössbauer Parameters for Various Groups of Iron Atoms in Fe–C Martensite[a]

Type of iron atom[b]	Hyperfine field (kOe)	Isomer shift[c] (mm/s)	Quadrupole splitting (mm/s)	Temperature measured	Carbon content (at.%)	References
$Fe_0 + Fe_{nnn}$	337 ± 3	−0.01 ± 0.05	0	rt	5.8	Gielen and Kaplow (1967)
Fe_{nn}	267 ± 3	−0.08 ± 0.05	0.19 ± 0.1	rt	5.8	
Fe_0	355	—	—	rt	4.5	Christ and Giles (1967, 1968)
Fe_{nn}	—	—	0.019 ± 0.009	rt	4.5	
Fe_0	334 ± 2	0.01 ± 0.05	0.10 ± 0.05	rt	4.2	Moriya et al. (1968)
Fe_{nnn}	342 ± 2	0.02 ± 0.05	−0.02 ± 0.05	rt	4.2	
Fe_{nn}	265 ± 2	−0.03 ± 0.05	0.13 ± 0.05	rt	4.2	Ino et al. (1968)
Fe_0	330	0	—	rt	1.76	Ron et al. (1968)
Fe_{nnn}	304	0.148 ± 0.02	—	rt	1.76	
Fe_{nn}	274	0.148 ± 0.02	—	rt	1.76	Ron and Schechter, (1968)
Fe_0	350	0.188	−0.06	77 K	9.0	Genin and Flinn (1966, 1968)
$Fe_{nn} + Fe_{nnn}$	340.74	0.544	+0.063	77 K	9.0	
Cluster	277.50	0.210	+0.159	77 K	9.0	

[a] Representative data from "phase-one" martensite studies (see text).

[b] According to the notation introduced in this paper (see Fig. 6): Fe_{nn}—iron nn to a carbon atom in an octahdral interstice, Fe_{nnn}—iron nnn to a carbon atom in an octahedral interstice, Fe_0—iron atom without nn or nnn carbon atom, and cluster—carbon-rich ordered region.

[c] With respect to α-Fe.

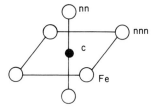

Fig. 6. Nearest (nn) and the next nearest (nnn) iron-atom neighbors to an interstitial carbon atom at an octahedral interstitial site ●—C, ○—Fe.

thereafter to liquid nitrogen temperature is shown in Fig. 7 (Ino *et al.*, 1968). The bar diagrams (a) and (b) symbolize the spectral components related to groups of iron atoms that are nearest, Fe_{nn}, and next nearest, Fe_{nnn}, neighbors, respectively, to a carbon atom in an octahedral interstice. The small absorption peak (c) at the center of the spectrum is produced by retained austenite.

As far as the hyperfine magnetic field attributed to the Fe_{nn} iron atoms is concerned, the values shown in Table III are all confined to the comparatively narrow range of 267–274 kOe, with the exception of the results obtained by Genin and Flinn (1966, 1967). Comparisons with their

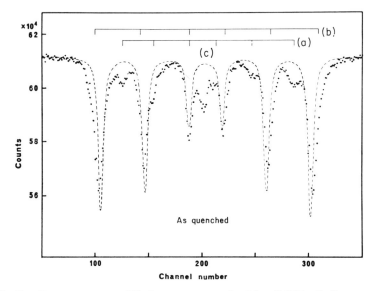

Fig. 7. Room-temperature Mössbauer spectrum of a 5.1 at.% C Fe–C alloy quenched into ice water and immediately thereafter to L. N. temperature. (a) As quenched (low-temperature phase); (b) aged at 20°C for 3 hr; (c) aged at 20°C for 76 hr; (d) aged at 80°C for 1 hr. (Measured at 77 K.) Dotted line, α-iron spectrum (Ino *et al.*, 1968).

results were impossible because they used different groupings of iron atom types for the interpretation of their spectra, as shown in Fig. 8. The values listed for the largest hyperfine magnetic field, related to the Fe_0 or $Fe_0 + Fe_{nnn}$ groups, appear to be in reasonable agreement if their dependence on the carbon concentration of the alloy, as well as on the different measurement temperatures, are allowed for. The isomeric shifts and the quadrupole splitting will be discussed later.

The above-noted Mössbauer investigations agreed in regard to the distorted octahedral symmetry of the carbon atom interstitial site. Also, the hyperfine characteristics of the iron atoms (Fe_{nn}) that are nearest neighbors to the carbon atom at the distortion dipole position were recognized to be the most influenced and thus showed the smallest magnetic hyperfine field. However, the opinions of the investigators were divided with regard to the group (or groups) of atoms to which the components having intermediate and largest magnetic hyperfine fields should be related.

To resolve this point, a number of studies were carried out subsequently. From a chronological point of view, they may be considered as the "second phase" of Mössbauer studies of the martensite structure. During this period a question arose as to whether or not at low temperature the tetrahedral interstice is occupied by some fraction of the carbon atoms. The nearest neighbor configurations of an interstitial carbon atom in octahedral and tetrahedral sites are shown in Fig. 9.

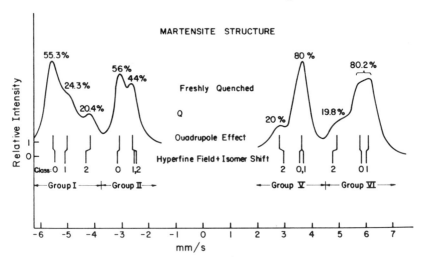

Fig. 8. A Mössbauer spectrum of a 1.96 wt% C martensite, taken at 77 K, and interpreted as a superposition of three iron-atom classes, 0, 1, and 2 as explained in text (Genin and Flinn, 1967, 1968).

(a)   (b)

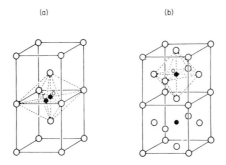

Fig. 9. Nearest-neighbor configuration of an interstitial carbon atom, 0; in the bcc structure (a) and in the fcc structure (b) O–Fe.

Fujita *et al.* (1971) investigated a 5.6 at.% Fe–C martensite at low temperature and interpreted the Mössbauer spectrum as consisting of three components designated a, b, and c in Fig. 10. They identified the a-component as related to type Fe_0 iron atoms and the c-component to the Fe_{nn} iron atoms in an octahedral interstice. However, they related the b-component to iron atoms in a tetrahedral interstice in accordance with the structure of the low-temperature K' martensite, as suggested by Lysak

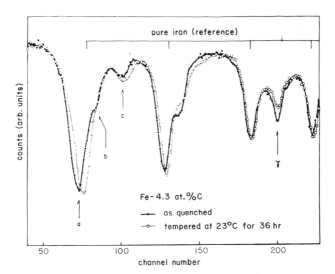

Fig. 10. Mössbauer spectra of 4.3 at.% C steel: —●—, as quenched to −196°C and —○—, aged at 23°C for 36 hr. The spectral components were related to (a) Fe atoms without C neighbors, (b) iron atoms nn to a C atom at tetrahedral interstice (Fe_{nn}^t in the present notation) (c) iron atoms nn to a C atom at an octahedral interstice (Fe_{nn} in the present notation) (Fujita *et al.*, 1971; Fujita, 1977).

and Vovk (1965). The intensity of the b-component was found to decrease while that of the c-component to increase during aging of the samples at 20°C for 30 min. The decrease in the b-component was twice as large as the increase of the c-component; this behavior corresponds to the change in the number of nn-iron atoms from 4 in a tetrahedral interstice to 2 in the octahedral one. Results of resistivity decay measurements were in accord with this interpretation. Moreover, the internal hyperfine magnetic field of 318 kOe fits a curve of the internal field versus distance from a carbon atom if the tetrahedral configuration is assumed. This curve is shown in Fig. 11. Table IV is a compilation of representative hyperfine parameters of various types of iron atom configurations taken from "second-phase" Mössbauer martensite studies.

Lesoille and Gielen (1972) followed Lysak's hypothesis that the tetrahedral and octahedral interstitial sites are equally populated and that a 9 at.% C virgin martensite was an ideal random solid solution. They unfolded the spectrum into essentially the following four types of iron atoms:

type I   remote from interstitial carbon atoms (Fe_0),
type IIa   nn to a carbon atom in an octahedral interstitial site (Fe_{nn}),
type IIb   nnn to a carbon atom in an octahedral interstitial site (Fe_{nnn}),
type III   nn to an interstitial carbon in a tetrahedral site (Fe_{nn}^t).

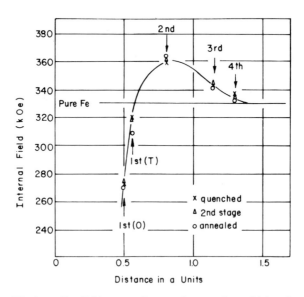

Fig. 11. The hyperfine field versus distance from an interstitial carbon atom in 1.8 at.% C Fe–C martensite. First (O) and first (T) correspond to Fe_{nn} and Fe_{nn}^t, respectively, in the present notation (Fujita, 1977).

TABLE IV

Mössbauer Hyperfine Parameters for Various Types of Iron Atoms in Fe–C Martensite[a]

Investigators	at.%	Interstitial site symmetry[b]	Distribution of C atoms[c]	Type of Fe atoms[d]	Hyperfine field (KOe)	$T_{meas.}$ (K)	Isomer Shift[e] (mm/s)	Quadrupole splitting (mm/s)	Remarks
DeCristofaro[f] and Kaplow (1977a)	8.2	OIS	Separated C atoms	Fe_{nn}	279	80	0.05	−0.15	
				Fe_{nnn}	331		0.04	−0.04	
				Fe_0	361		−0.04	0.0	(Third nearest n) + Hyperfine Field
Gridyew et al. (1977)	9	OIS	SS + clusters	Fe_{nn}	316	65	0.15	−0.21	
				Fe_{nnn}	335		0.12	0.08	
				Fe_0	357		0.22	0.01	(Third nearest n) + Second coordination sphere component
Cadeville et al. (1977)	5	OIS	—	Fe_{nn}	269		−0.03	0.22	
				Fe_{nnn}	349		−0.01	0.0	
				Fe_0	331		0.0	0.0	
Lesoille and Gielen (1972)	8.2	OIS + TIS	Ideal RSS	Fe_{nn}	322	L.N.	0.29	−1.23	
				Fe_{nnn}^t	274		−0.18	0.62	
				Fe_{nn}^t	322		0.02	−0.885	
								0.625	
				Fe_0	353		−0.09	—	
Fujita (1977) Fujita et al. (1971)	7.9	OIS + TIS	—	Fe_{nn}	275	L.N.	−0.06	0.10	
				Fe_{nnn}	360		0.06	−0.12	
				Fe_{nn}^t	320		0.12	−0.06	
Ino and Ito (1978)	2.7–7	OIS + TIS	—	Fe_{nn}		77			Value of hyperfine field not specified
				Fe_{nnn}^t			—	—	
				Fe_{nn}^t					

[a] Representative data from "phase-two" martensite studies (see text).
[b] OIS—octahedral interstitial site; TIS—tetrahedral interstitial site.
[c] SS—solid solution; RSS—random solid solution.
[d] Notation introduced in this chapter (see text).
[e] With respect to α-Fe.
[f] Also Choo and Kaplow (1973).

As opposed to other investigators, they related the smallest hyperfine magnetic field of 274 kOe (see Table IV) to the next nearest-neighbors iron atoms to an octahedral interstitial carbon. The iron atoms in the type III configuration are divided into two groups with respect to the value and sign of the quadrupole interaction.

In a recent study Fujita (1977) has concluded that a 1 : 1 occupancy of octahedral and tetrahedral sites at low temperature fits a new and elaborated model for the martensitic transformation and room-temperature aging behavior (Shiga et al., 1974, 1975; Fujita et al., 1974).

Ino and Ito (1979) have investigated a martensitic Fe–C alloy of about 5 at.% C. Virgin martensite was obtained by quenching carburized foils into ice water and immediately thereafter into liquid nitrogen. Their spectrum taken at liquid nitrogen temperature (shown in Fig. 12) was resolved into three ferromagnetic components, which were related to the following groups of iron atoms:

(a) remote from a carbon interstitial (Fe_0),
(b) nn to a carbon interstitial in a tetrahedral site (Fe_{nn}^t), and
(c) nn to a carbon interstitial in an octahedral site (Fe_{nn}).

They estimated the value of the magnetic hyperfine field by using a relation given by Ino et al. (1968), which correlates the number of nn iron atoms surrounding a carbon atom (in a given configuration) and the magnitude of the magnetic hyperfine field. The reduction of the internal magnetic field of pure iron H_{io} due to the presence of a nn carbon atom approximately obeys a simple rule: $(\Delta H_i / H_{io}) \, 1/n = 0.2$, where $\Delta H_i = H_{io} - H_i$, H_i is the internal magnetic field of an iron atom in an Fe–C alloy, and n is the number of carbon atoms, which are nn to an iron atom in a given configuration. The (b) component has been found to have a ratio of $H_i / H_{io} \sim 0.95$ and has therefore been identified as the contribution arising from the four iron atoms (Fe_{nn}^t) surrounding the carbon in a tetrahedral site.

Up to this point, we have discussed the results of those investigations that support the assumption that a part of the carbon atoms occupy tetrahedral interstitial sites. However, other investigators attempt to refute the concept of tetrahedral interstice occupancy (Choo and Kaplow, 1973; Gridnyev et al., 1977; DeCristofaro and Kaplow, 1977a; Cadeville et al., 1977).

Choo and Kaplow (1973) have investigated a 9 at.% C martensite and concluded that the virgin martensite Mössbauer spectrum can be resolved in terms of first-, second-, and third-nearest-neighbor iron atoms of a carbon interstitial, with octahedral symmetry of the interstitial site.

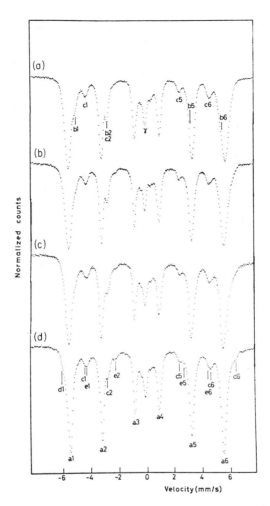

Fig. 12. Mössbauer spectra of a Fe 1.2 wt% C alloy: (a) as quenched (low temperature phase), (b) aged at 20°C for 3 hr, (c) aged at 20°C for 76 hr, (d) aged at 80°C for 1 hr (measured at 77 K). Treatments and measurement temperature are indicated in the figure. Note that new peaks appear on aging, e.g., peak e_2 in spectrum (d), related to ordered Fe_4C structure (Ino and Ito, 1978).

DeCristofaro and Kaplow (1977a) differentiated their virgin martensite spectrum into components related to the same group of iron neighbors as in the work of Choo and Kaplow (1973). From their analysis of the various configurations, they concluded that the model assuming that carbon atoms are initially distributed between tetrahedral and octahedral interstitial sites was incorrect.

Gridnyev *et al.* (1977) used six components in unfolding the Mössbauer spectrum of a 9 at.% C virgin martensite. Three of the components were related to the first-, second-, and third-neighbor iron atom to a carbon atom in an octahedral interstitial site, i.e., Fe_{nn}, Fe_{nnn}, and Fe_0, respectively. The remaining three components were related to iron atoms in carbon clusters and ordered Fe_4C regions. Analyzing the change upon room-temperature aging of the intensities of the various components, they came to conclusions that refute the assumption that carbon atoms occupy tetrahedral sites. Cadeville *et al.* (1977) have studied, by means of combined Mössbauer, x-ray, and electrical resistivity measurements, splat-quenched Fe–C alloys containing 2–5 at.% carbon. The high quenching speed involved prevented autotempering. The assignment of the spectral components to the various iron atom configurations (Table IV) was deduced from the relative spectral intensities and their dependence on the carbon content. They concluded that the interstitial carbon atoms occupy octahedral sites, and they were able to distinguish four components in the virgin martensite spectra. Three of the components correspond to Fe_0, Fe_{nn}, and Fe_{nnn} in the present notation, whereas the fourth component was attributed to iron atoms in the second coordination sphere of the interstitial carbon. It is seen in Fig. 13 that the relative intensities of the Fe_{nn} (F_1 in their notation) and Fe_{nnn} (Fe_3 in their notation) components are nearly linearly dependent on the carbon concentration of the alloy. They assigned the largest hyperfine field ($H_n \sim 349$ kOe, in a 4 at.% C alloy) to the Fe_{nnn} iron atoms, in agreement with the interpretation of Moriya *et al.* (1968).

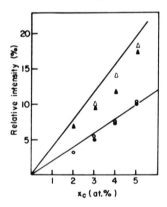

Fig. 13. Relative intensities of the subspectra 1 (O, ●) and 3 (△, ▲) attributed to axial and equatorial nearest Fe neighbors of a carbon atom as a function of the carbon content. The open points (O, △) correspond to a fit with three magnetic sites, the filled points (●, ▲) to four magnetic sites. The straight lines correspond to slopes of $2x_c$ and $4x_c$. Subspectra (1) and (3) correspond to Fe_{nn} and Fe_{nnn}, in the present notation, respectively. x_c is the carbon concentration of the alloy (Cadeville *et al.*, 1977).

From the above discussion it may be concluded that the question of the distribution of the interstitial carbons between the tetrahedral and octahedral sites in Fe–C martensite is not yet definitely solved by the Mössbauer investigations and further investigation of this problem is desirable.

The quadrupole interaction in Fe–C martensite has been interpreted in different ways depending on the symmetry of the interstitial site, which is still debated. When the absorber is magnetically ordered, as in martensite, the degeneracy of the nuclear magnetic sublevels is lifted by the Zeeman effect due to the interaction between the nuclear magnetic moment and the hyperfine field. The quadrupole interaction further changes the energy levels of the 14-keV excited state (see Chapter 1 in Volume 1 of this treatise for a more complete discussion). For the case of axial symmetry of the electric field gradient (efg), the quadrupole contribution to the hyperfine interaction is given by (Wetheim, 1964)

$$\epsilon = \pm \tfrac{1}{8} e^2 q Q (3 \cos^2 \theta - 1), \tag{9}$$

where-eq is the principal component of the efg, Q is the quadrupole moment of the nucleus, and θ is the angle between the axis of symmetry of the efg tensor and the direction of the hyperfine field, which is normally assumed to lie along the easy magnetization axis (ema). As pointed out by Wertheim (1964), if the efg tensor is axially symmetric, q and θ cannot be unambiguously determined from the hyperfine energies alone. Generally, experimenters have assumed a particular value for θ and analyzed the spectra under that assumption. As can be seen from Eq. (9), different values of q can be obtained, depending on the value assumed for θ. This complicates the interpretation of the spectra, and accounts for some of the discrepancies that appear in Tables III and IV.

Genin and Flinn (1968) have concluded from their Mössbauer results that the hyperfine field is largely oriented along the c axis. Consistent with this result, Lesoille and Gielen (1972) have assumed that the easy axis of magnetization is along the [001] direction, as determined by Izotov and Utevskii (1968). Lesoille and Gielen then conclude that in Fe–C martensite, q, the principal component of the efg, is approximately the same for all classes of iron atom having one nn carbon atom. They assign a negative sign to the quadrupole interaction of the iron atoms nn to a carbon atom at the distortion dipole position.

The variation of the isomeric shift and quadrupole interaction versus the distance from an interstitial carbon atom, given by Fujita (1977), is shown in Fig. 14. The systematic variation of the inversion in sign of the isomer shift and quadrupole splitting with the Fe–C distance were explained by a theory of Adachi and Imoto (see Fujita, 1977). Fujita's

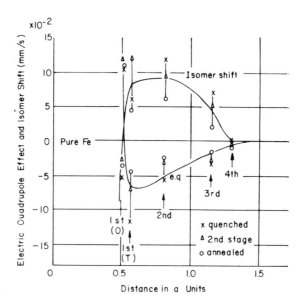

Fig. 14. The electric quadrupole interaction and isomeric shift versus distance from an interstitial carbon atom in Fe 1.8 wt% C martensite (Fujita, 1977).

interpretation is based on an oversimplifying approximation, using the distances of a bcc lattice. This disregards the changes that occur due to the presence of the interstitial carbon atom.

2. Aging of Fe–C Martensite

The process of segregation and redistribution of carbon atoms at room temperature is known to take place in low-carbon high M_s steels and also in steels of subambient M_s (Speich and Leslie, 1972; Cohen, 1962). Winchell and Cohen (1962) have shown that substantial increases in the electrical resistivity and hardness of Fe–C–Ni martensites occur during the early stages of aging at room temperature. Koval *et al.* (1969) have related the initial drop in resistivity of Fe–Mn–C martensite upon aging between −50 and 20°C to carbon ordering. Internal friction studies have also found evidence for structural changes that occur at room temperature in less than 1 hr (Marquis *et al.*, 1962; Schmidtmann *et al.*, 1965).

Genin and Flinn (1966) were the first to observe the considerable changes occurring in the Mössbauer spectra of Fe–C martensite due to aging at room temperature and to relate them to the clustering of carbon. As already mentioned, Genin and Flinn (1968) suggested that carbon atoms occupy an octahedral interstitial site, and interpreted the spectrum of

8.5 at.% martensite in terms of contributions from the following iron atom groups: "class zero"—corresponding to Fe_0—in the present notation, "class one"—corresponding to $Fe_{nn} + Fe_{nnn}$—in the present notation, and "class two"—corresponding to Fe_2—in the present notation. They suggested that "class two" environment exists in the vicinity of a cluster of carbon atoms. Upon aging at 0, 10, and 20°C, they found that the relative spectral intensity related to "class one" iron atoms decreased, while those related to the other two classes increased. No new component and no changes in the hyperfine parameters were observed. The results were interpreted as a two-stage process, the first being the clustering of carbon atoms in the martensite matrix and the second the formation of ϵ-carbide. They described the kinetics of the clustering of carbon atoms in martensite by the relationship $C - C_0 = kt^n$, where C is a parameter related to the clusters, C_0 the initial value of C, t time, k a constant, exponentially dependent on temperature, and n an exponent to be determined experimentally. Their results yielded $n = \frac{1}{3}$ and an activation energy of about 15.5 kcal/mole, a value close to the activation energy for diffusion of carbon in ferrite. The kinetics of the clustering match the kinetics determined by the electrical resistance and internal friction experiments mentioned above. The absence of any new hyperfine component in the magnetic hyperfine pattern at stage two was explained as due to the superparamagnetic state of the ϵ-carbide.

Choo and Kaplow (1973) have made a systematic study of the aging of a 8.2 at.% Fe–C martensite in the temperature range 20–82°C. They unfolded the spectrum of virgin martensite into three components related to groups of iron atoms having first-, second-, and third-nearest-neighbor carbon atoms with hyperfine fields of 280, 334, and 364 kOe, respectively. In martensite, freshly quenched and cooled to low temperature, they found the carbon atoms to be widely separated, rather than statistically distributed. Upon room temperature aging for periods as long as three months, the carbon atoms agglomerate into an ordered structure, Fe_4C (isomorph with Fe_4N). By means of difference spectra (shown in Fig. 15), they have identified the following three components related to the Fe_4C structure: iron atoms at the cube center, Fe^f; iron atoms at the corners, Fe^c, and iron atoms without carbon atoms in their neighborhood, Fe^p (Fe_0—in the present notation). The Mössbauer hyperfine parameters are shown in Table V. The hyperfine field of 280 kOe attributed to the first neighbors (Fe_1—in the present notation) is seen to be unchanged and the corresponding intensity to have increased. The ratio of the magnitudes of the hyperfine field of Fe^c to that of Fe^f is approximately $\frac{3}{2}$. This and the relation of the hyperfine fields to that of α-iron suggest that magnetic

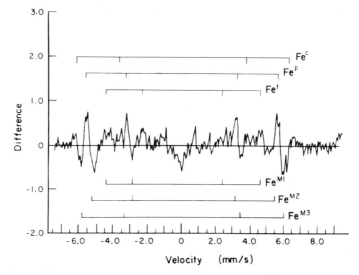

Fig. 15. Difference spectrum of 8.2 at.% Fe–C martensite. This spectrum was obtained by subtracting a spectrum of martensite aged at room temperature for $14\frac{1}{2}$ days from virgin martensite spectrum. A difference spectrum is obtained by subtracting two sets of data after normalizing them relative to one another (Choo and Kaplow, 1973). [Reprinted with permission from *Acta Met.* **21**, 725. Copyright 1973, Pergamon Press, Ltd.]

moments of about 3 μ_B and 2 μ_B are associated with Fe^c and Fe^t groups, respectively. The relative change in the number of each type of iron atoms is shown in the last column of Table V. The calculated and experimentally derived data seem to fit quite well. After tempering at 82°C, the Fe_4C formed at room temperature was replaced by a new structure, which was identified as ϵ-carbide of approximate composition Fe_2C, yielding a hyperfine field of 240 kOe. Other investigators have also reported on the appearance of additional components in the spectra of martensite upon aging. (Gridnyev *et al.*, 1977; DeCristofaro and Kaplow, 1977a; Ino and Ito, 1978). Their interpretations were qualitatively similar to those discussed above, which involve the clustering of carbon at room temperature.

Recently DeCristofaro *et. al.* (1978) have studied the rate of clustering of carbon atoms into ordered regions of Fe_4C in a 8.2 at.% C alloy at two temperatures, 295 and 323 K. They used the following rate equation: $\ln(R_t) = \text{const} - Q/RT$, where R_t is the rate of clustering of carbon as determined from the change in relative spectral areas of particular components of the martensite Mössbauer spectra before and after the aging, Q an activation energy, R the gas constant, and T the aging temperature in degrees Kelvin. The activation energy was found to be 21.4 kcal/mole.

TABLE V

Mössbauer Hyperfine Parameters Associated with Iron-Atom Types
in Martensite and Fe_4C[a]

Type of iron atom	Hyper-fine field (kOe)	Isomer shift[b] (mm/s)	Quadrupole splitting (mm/s)	Change in intensity (relative)	
				Theory	Experiment
Virgin martensite (measured at $-193°C$)[c]					
1st neighbors: Fe^{M1}	280	-0.046	0.168	1.0	1.3
2nd neighbors: Fe^{M2}	334	$+0.148$	0.002	-4.0	-3.3
3rd neighbors: Fe^{M3}	364	$+0.066$	0.030	-5.8	-6.3
After RT aging (measured at $-193°C$)					
1st and 2nd neighbors (cube-face atoms); Fe^f	280	$+0.096$	0.011	1.0	1.3
3rd neighbors (cube-corner atoms); Fe^c	386	$+0.097$	0.008	1.0	1.6
Interstitial-free atom	348	$+0.014$	0.002	7.8	7.8
After 82°C aging (measured at room temperature)					
X spectrum (ϵ carbide?)	240	0.00	0.03		

[a] Taken from Choo and Kaplow (1973).
[b] With respect to α-Fe.
[c] Fe^{M1}, Fe^{M2}, and Fe^{M3} correspond to Fe_{nn}, Fe_{nnn}, and Fe_0 in the present notation.

On the other hand, investigators using the hypothesis of partial occupancy of tetrahedral interstitial sites in the low-temperature martensite phase K' were also able to explain changes in the spectrum. Fujita *et al.* (1971) and Fujita (1977) have attributed the smallest hyperfine field ($H_{eff} = 270$ kOe) to the two iron atoms (Fe_{nn}) at the [001] position near the carbon in the octahedral-interstitial site and the intermediate component ($H_{eff} = 310$ kOe) to nn iron atoms (Fe_{nn}^t) near a tetrahedral-interstitial carbon. The increase in the relative intensity of the peak arising from the Fe_{nn} atoms during room temperature aging was explained as indicating the diffusion of carbon atoms from tetrahedral to octahedral interstitial sites, in accordance with the observed $K' \rightarrow \alpha_m$ transition. This interpretation is in agreement with a comprehensive model for the crystallography of the martensitic transformation given by Fujita (1977). Ino and Ito (1978) have interpreted their virgin martensite spectrum as consisting of three components resulting from iron atoms Fe_{nn}^t, Fe_{nn}, and Fe_0 (see Fig. 12). Upon aging at 20°C, new peaks that were related to the ordered Fe_4C clusters appeared in the spectra.

V. Iron Carbides

A. The Precipitation of Carbides in Fe–C Alloys and Steels

The carbides formed in the course of the tempering process in steels can be described by the general scheme:

$$\text{carbon clusters} \rightarrow \text{ordered structures} \, \epsilon \rightarrow \chi \rightarrow \theta,$$

where ϵ and χ are transition carbides, and θ is the metastable-phase cementite. The precipitation process depends on the composition and the details of the heat treatment. The addition of Si retards the precipitation of carbide, while other elements may enhance it. In some alloys this sequence may not be followed or some stages may overlap. Also, the stoichiometry of the transition carbides may vary (Dünner and Müller, 1965; Palatnik et al., 1966). The morphology of precipitated carbides and their kinetics of precipitation are important factors for the control of the properties of steel by means of heat treatment.

In Fe–C alloys the process of segregation of carbon may start during the quenching from austenite. For alloys with high M_s temperature, auto-tempering is known to occur. In Fe–C alloys and steels of low M_s temperature, the segregation of carbon starts at about room temperature. Carbon-rich clusters (Johnson, 1964, 1967), as well as very small coherent precipitates having the $\{100\}_\alpha$ habit and approximately isomorphous with $Fe_{16}N_2$, have been found to form at the very early stages of tempering of steels (Leslie, 1961; Keh and Leslie, 1963; Doremus and Koch, 1960). The first-ordered structure that forms upon room-temperature aging of Fe–C martensite is Fe_4C (isomorph with Fe_4N), as has been already mentioned. A transition η-carbide having an approximate composition of Fe_2C and orthorhombic structure has been found to form in a Fe 5.43 at.% C alloy after aging at 150°C for 16 hr (Hirotsu and Nagakura, 1972, 1974; Williamson et al., 1979).

The transition ϵ carbide has a hexagonal structure and a composition approximately between Fe_2C and $Fe_{2.2}C$ (Gridnev and Petrov, 1962; Ruhl and Cohen, 1969; Jack, 1951a). It precipitates in the form of thin platelets lying parallel to $\{110\}_\alpha$ planes. It has been found to form in high carbon Fe–C martensite at temperatures close to room temperature (Genin and Flinn, 1968; Choo and Kaplow, 1973). However, the upper temperature limit of the ϵ-carbide precipitation has not yet been unambiguously established. There is little agreement in the literature on the Mössbauer hyperfine parameters of the ϵ carbide precipitated in Fe–C martensite as measured in situ, i.e., within the martensitic matrix which dominates the spectrum. The results of Mössbauer investigations of iron carbides are summarized in Table VI. It has been found by magnetic measurements that ε carbide is ferromagnetic and has a Curie temperature of 380°C. A magnetic hyperfine field related to ε carbide has been found in Mössbauer spectra of Fe–C martensite tempered in the range 80–150°C (Ino et al., 1968; Choo and Kaplow, 1973; Ino and Ito, 1978; Williamson et al., 1979) and in carbides extracted from a commercial steel (Mathalone et al., 1971b). On the other hand, Genin and Flinn (1968) and LeCaer et al. (1971) have reported the presence of a superparamagnetic component related to the ε carbide in the spectra of tempered martensite and extracted carbides, respectively. The addition of more than 1% of Si stabilized the ε carbide and caused the phase to persist at higher tempering temperatures (Gordine and Codd, 1969; Mathalone et al., 1971b).

Ruhl and Cohen (1969) have revealed the existence of an hcp ε phase (different from ε carbide) in splat-quenched Fe–C alloys. Upon heating at 140–200°C for 1hr, the ε phase decomposes into ε carbide and martensite, while upon heating at 333–460°C for 1 hr, it decomposes into ferrite and cementite. A recent Mössbauer investigation has used splat-quenched Fe–C alloys in which the ε phase was of an extended carbon concentration range Fe_6C–Fe_3C (Dubois and LeCaer, 1977). The ε phase was found to be magnetically ordered. The dependence of the hyperfine field and isomer shift on the carbon content is shown in Fig. 16. The ε phase is stabilized by increasing content of Si, see Fig. 17. However, in alloys containing more than 5 wt% Si, an orthorhombic structure appears, the lattice parameters of which vary with the silicon content.

The transition χ carbide, also known as the Hägg carbide, has a monoclinic crystal structure and composition close to Fe_5C_2. It precipitates in steels at aging temperatures above 200°C. However, for χ carbide precipitated in steels, the orthorhombic structure and deviations from stoichiometry have also been reported (Hofer and Cohen, 1959; Jack, 1948; Arentz et al., 1973). At 250°C and higher temperatures the χ carbide transforms into the metastable θ carbide known as cementite.

Cementite has the composition of Fe_3C and the orthorhombic crystal structure with space group P_{nma} (Fasisca and Jeffrey, 1965). The carbon atoms are located in interstitial positions of a close-packed structure of iron atoms having a pleated layer hexagonal arrangement. Two types of iron atoms are distinguished in this structure: an iron atom Fe_G at a general G site, which has 3 C and 11 Fe nearest neighbors, and an iron atom Fe_S at a special S site, which has 2 C and 12 Fe nearest neighbors.

The kinetics of carbide precipitation in steels have been widely studied (Leslie, 1961; Keh and Leslie, 1963; Butler, 1966; Wells and Butler, 1966; Duggin, 1968). A Mössbauer study of carbon precipitation kinetics has been performed on a Fe–C alloy of nearly eutectoid composition containing 0.7 wt% C (3.2 at.%) and on a low alloy eutectoid steel containing 0.92

TABLE VI Results of Mössbauer Investigations of Iron Carbides [a]

Carbide	Preparation method	Structure method	H (kOe)	Isomer shift[b] (mm/s)	(mm/s)	Comments	Reference
ε-Fe₂C	Extracted from Fe–C–Si alloy tempered at 300°C for 1 hr	Hex.—x-ray and electron diffraction	0	0.36(2)	0.76(2)	Superparamagnetic	LeCaer et al. (1971)
ε-Fe₂C	Extracted from Fe–C–Si alloy tempered at 300°C for 1 hr	Hex.—x-ray and electron diffraction	230(4) 174(5)	0.244(4) 0.22(4)	0.10		Mathalone et al. (1971b)
ε-Fe₂C	Extracted from Fe–C–Si–Mn alloy tempered at 300°C for 2 hr	Hex.—x-ray	162(5)	0.18(2)			
ε-Fe₂C	Splat-quenched Fe-1.86°C alloy tempered at 82°C for 62 h	Not identified	240(5)	0(.05)	0.03(5)		Choo and Kaplow (1973)
ε-Fe₂C	Carburized and quenched Fe–C alloy tempered at 140°C for 1 hr	Not identified	– 265			ε-carbide persists up to tempers at 340°C	DeCristofaro and Kaplow (1978)
ε-Fe₂C	Synthesized in catalysis reaction at 160°C	Hex.—x-ray, thermo-magnetic analysis	237(5) 170(3) 130(6)	0.35(5) 0.20(5) 0.30(5)		Suggests alternate structure similar to χ-carbide	Maksimov et al. (1974b)
ε'-Fe₂.₂C	Synthesized in catalysis reaction at 115°C	Not identified	– 170 0	0.25(10)	0.9(1)	T_c clearly different from that of ε-Fe₂C Superparamagnetic component	
ε'-Fe₂.₂C	Synthesized in catalysis reaction at 255°C	Hex.—x-ray	173(1)	0.22(1) 0.23(1)	– 0.13(1) 0.96(1)	Superparamagnetic component	Amelse et al. (1978)
ε'-Fe₂.₂C	Synthesized in catalysis reaction at 400°C	Hex.—x-ray	173(1)	0.25(1) 0.23(1)	– 0.29(1) 0.90(1)	Superparamagnetic component	
χ-Fe₅C₂	Extracted from Fe–C–Si–Mn alloy tempered at 400°C for 2 hr	Monoclinic—x-ray	179(5)	0.28(2)			Mathalone et al. (1971b)

			H (kOe)[b]	IS	QS	Remarks	Reference
χ-Fe₅C₂	Carburized and quenched Fe–C alloys tempered at 220°C for 1 hr	Not identified	194(5)	0.28(2)			Ino et al. (1968)
χ-Fe₅C₂	Synthesized		218(2)	0.26(2)			Genin et al. (1975)
			186(2)	0.23(2)			
			111(3)	0.17(2)			Bernas et al. (1967)
χ-Fe₅C₂	Synthesized		222(3)	0.35(4)			
			184(3)	0.30(4)			
			110(6)	0.30(8)			LeCaer et al. (1976)
χ-Fe₅Ĉ₂	Synthesized		216(2)	0.23(2)	0.08(4)		
			185(2)	0.20(2)	0.10(2)		
			185(2)		−0.24(4)		
			124(4)				Genin et al. (1975)
θ-Fe₃C	Extracted from Fe–C–Si alloy after tempering at 580°C	Orthorhombic—x-ray	208(1)	0.19(1)			
Fe₃C	Extracted from Fe–C–Si alloy after tempering at 400°C	"Highly faulted" cementite (θ)—TEM	215(1)	0.23(2)		Suggested structure intermediate between χ- and θ-carbides	Mathalone et al. (1971b)
			204(1)	0.21(2)			
			192(3)	0.20(2)			
θ-Fe₃C	Extracted from Fe–C–Si–Mn alloy after tempering at 500°C for 2 hr	Orthorhombic—x-ray	179(3)	0.19(1)	0.04(2)	Suggested H low because of presence of Mn	Ron et al. (1968)
θ-Fe₃C	Extracted from Fe–C–Si–Mn–Cr alloy after tempering from 250 to 500°C for 1 and 5 hr		195–201	0.17(3)			
θ-Fe₃C	Carburized and quenched Fe–C alloys tempered at 520°C for 1 hr	Not identified	208(2)	0.18(5)	0.01(3)		Ino et al. (1968)
θ-Fe₃C	Synthesized	Orthorhombic—x-ray	208(2)	0.29(2)			Bernas et al. (1967)
θ-Fe₃C	Extracted from Fe–C alloy		207	0.17(2)	− 0.58		Ron and Mathalone (1971)
			205	0.17(2)	0.32		
θ-Fe₃C	Synthesized		208(2)	0.18(1)	0.02(1)		LeCaer et al. (1976)
			206(2)				

[a] Taken from Williamson et al. (1979); all values obtained at room temperature; significant figure in parentheses—uncertain. See also Shinjo et al. (1964).
[b] With respect to α-Fe.

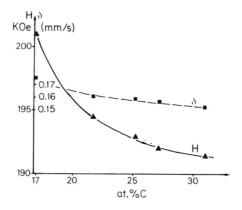

Fig. 16. Variation of the hyperfine field H and isomeric shift δ with the carbon content of the Fe–C ϵ-phase obtained in splat-quenched Fe–C alloys (Dubois and Le Caer, 1977). [Reprinted with permission from *Acta Met.* **25**, 609. Copyright 1977, Pergamon Press, Ltd.]

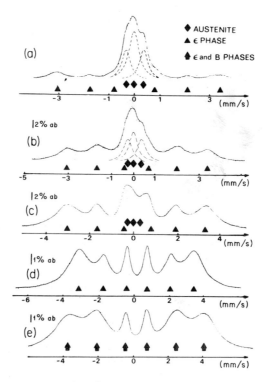

Fig. 17. Room-temperature Mössbauer spectra of splat-cooled Fe–C–Si alloys of following compositions, in weight percent: (a) Fe-2.56Si-3.22C, (b) Fe-2.87Si-3.56C, (c) Fe-2.91Si-4.18C, (d) Fe-3.6Si-4.17C, (e) Fe-5.27Si-4.08C. The alloys contain prevailing amounts of ϵ-phase. Spectrum D shows only ϵ-phase (Dubois and LeCaer, 1977). [Reprinted with permission from *Acta Met.* **25**, 609. Copyright 1977, Pergamon Press, Ltd.]

wt% Mn as the main alloying element (Mathalone *et al.*, 1971a). The fractional amount of cementite $R(tt)$ formed upon tempering the alloys for constant periods of time is shown as a function of the anneal temperature tt in Fig. 18a. The changes in the hyperfine field of Fe–C cementite and the amount of Mn in the $(Fe_{1-x}Mn_x)_3C$ cementite precipitated from the above alloyed steel as a function of tt are shown in Fig. 18b and c. The dependence of R on tt for an austeniting temperature of $T_A = 950°C$ yields a rate equation with a single activation energy Q,

$$R(T) = C\exp(-Q/kT), \tag{10}$$

where C is a constant, T the tempering temperature in degrees Kelvin, and k the Boltzmann constant. A logarithmic plot of $R(T)$ versus $1/kT$ is shown in Fig. 19. The value derived for Q was 0.3 eV. The dependence of the precipitation rate $R_F(T_A)$ on the austeniting temperature T_A was found

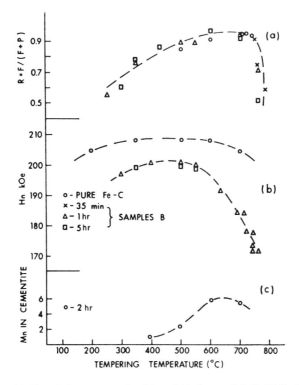

Fig. 18. (a) The relative amount of stable carbide (cementite) R, (b) the hyperfine field at 300°K on the Fe^{57} nucleus in cementite H_n, (c) the concentration of Mn wt% in $(Fe_{1-x}Mn_x)_3C$ cementite (calculated), as a function of tempering temperature. The carbides were chemically extracted from a Fe-0.7 wt% C alloy and an eutectoid steel, containing 0.92 wt% Mn as the main alloying element (Mathalone *et al.*, 1971a).

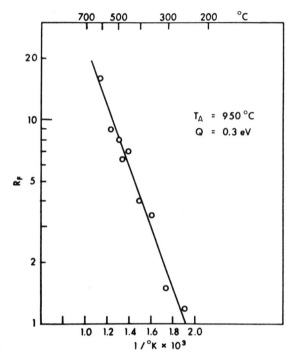

Fig. 19. A logarithmic plot of the fraction of stable (ferromagnetic) carbide, R_F versus $1/kT$, where T is the tempering temperature (Mathalone *et al.*, 1971a).

to yield

$$R_F(T_A) = 1 - A \exp\left[-(n/n_0)^\gamma\right] \quad \text{with} \quad A > 1, \qquad (11)$$

where $n = n(T_A)$ is the equilibrium vacancy concentration and n_0 and γ are constants. The value of 0.3 eV found for Q is rather low and was explained as representing the energy required for the transformation of a transition carbide into cementite. The dependence of the precipitation rate on T_A implies the possibility of formation of vacancy-carbon couples, which may facilitate the nucleation process. The increase in the hyperfine field with *tt* for temperatures up to 400°C was thought to be related to a slight decrease in the carbon content of the Fe–C carbides; for *tt* above 500°C, the decrease in the hyperfine field of the alloyed carbides was related to the increase of Mn concentration in them.

The effective Debye temperature θ of carbides extracted from the above alloys was determined through the Mössbauer effect (Mathalone *et al.*, 1970). According to the Debye model for a homogeneous and harmonic lattice (Goldanskii and Herber, 1968), the f factor was determined from

$$f(T,\theta) = \exp(-2W) = \exp\left\{ -\frac{6E_r}{k\theta}\left[\frac{1}{4} + \left(\frac{T}{\theta}\right)^2 \int_0^{\theta/T} \frac{x\,dx}{e^x - 1} \right] \right\},$$

(12)

where θ is the effective Debye temperature, T the absolute temperature, k the Boltzmann constant, m the nuclear mass, W the Debye–Waller factor, and $E_r = E_\gamma/2mc^2$ the recoil energy of a free atom. The $f(T,\theta)$ curves plotted in Fig. 20 were calculated from Eq. (12) with θ as a parameter. For $tt > 500°C$ the effective Debye temperature θ was found to increase with tt. This was related to the change in carbon content of the cementites.

Single crystals are usually used for the determination of the easy magnetization axis (ema). However, it is very difficult, if at all possible, to prepare single crystals of Fe_3C (Fasiska and Jeffrey, 1965). The ema of Fe_3C was determined by Gonser et al. (1972) by using Mössbauer spectroscopy on a polycrystalline sample having a strong preferred orientation. The Fe_3C powder was chemically extracted by a procedure described by Mathalone et al. (1971b). The platelike particles obtained had the shape of

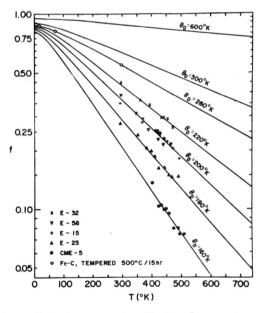

Fig. 20. The recoil-free factor on a logarithmic scale versus temperature of measurement for carbides extracted from steel after various heat treatments. The solid curves are calculated for various values of θ according to the Debye model. The experimental data are fitted to the calculated lines (Mathalone et al., 1970).

flakes with diameters of 50–100 μ and a diameter-to-thickness ratio of about 20:1. The powder was lightly pressed within a sample holder, and the crystalline preferred orientation (texture) was determined by x-ray diffraction. The Mössbauer spectrum of this sample is seen in Fig. 21. The $\Delta m = 0$ transition lines (lines 2 and 5) are considerably weaker than for a random orientation of spins. The pertinent intensity ratio, ρ for thin absorbers, is

$$\rho = \frac{I_2}{I_3} = \frac{I_5}{I_4} = \frac{4\sin^2\gamma}{1 + \cos^2\gamma}, \tag{13}$$

where I_i is the intensity of the ith line ($i = 2, 3, 4, 5$) and γ is the angle between the gamma-ray propagation axis and the spin direction. The value measured from the spectrum was $\rho = 0.65$ as compared with $\rho = 2$ for a random spin orientation. Correlating the x-ray texture and Mössbauer results, the spins in Fe_3C were found to be oriented within a solid angle of 20°C along the crystallographic c axes. Formulas were derived to make this correlation method generally applicable.

An x-ray diffraction and Mössbauer spectroscopy investigation on carbides precipitated throughout the precipitation process for a wide range of tempering temperatures in a commercial steel containing 0.38 wt% C, 1.33 wt% Si, and 1.13 wt% Mn, and balance Fe was carried out by

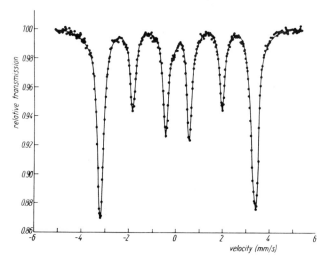

Fig. 21. The room-temperature Mössbauer spectrum of flake-shaped Fe_3C powder shows that lines 2 and 5 ($\Delta m = 0$) are weaker. From the line intensity ratio and correlation with preferred orientation, determined by x-ray diffraction, the easy magnetization axis of cementite was determined (Gonser *et al.*, 1972).

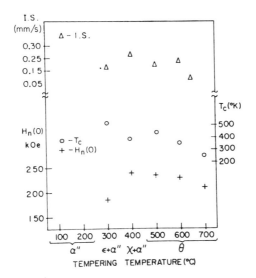

Fig. 22. Curie temperatures, hyperfine fields (extrapolated to 0°K), and isomeric shifts at room temperature are plotted versus tempering temperatures for carbides extracted from steel containing 0.38 wt% C, 1.33 wt% Si, 1.13 wt% Mn, and balance Fe (Mathalone *et al.*, 1971b).

Mathalone *et al.* (1971b). The carbides were chemically extracted from the tempered steel. The ϵ, χ, and θ carbides were identified by x-ray diffraction in accordance with Schenck *et al.* (1966), and ASTM cards, and their Mössbauer characteristics, were determined. The ϵ carbide was found to persist at a relatively high temperature of 300°C. This was attributed to the effect of Si on the carbide precipitation kinetics. For the same reason, the χ carbide also persists to 400°C. The hyperfine fields, the Curie points, and isomer shifts of the θ carbide were found to vary with the tempering temperature. The variation in these characteristics was related to changes in the Mn content of the $(Fe_{1-x}Mn_x)_3C$ carbide. The isomer shifts, hyperfine fields, and Curie points for the ϵ, χ, and θ carbides versus the tempering temperature are shown in Fig. 22.

B. The Structure and Properties of Iron Carbides

The use of Mössbauer spectroscopy to characterize binary and alloyed iron carbides is now quite common. The information derived from Mössbauer spectroscopy studies reveals details that complement the information on structures provided by x-ray diffraction and electron microscopy. Carbides formed during the process of precipitation in steels normally have an imperfect structure and varying properties, while synthesized carbides can be produced with controlled stoichiometry and structural defects. The properties and structure of both types of iron carbides will be briefly discussed here.

Synthesized ϵ-iron carbide has been studied by means of Mössbauer spectroscopy, thermomagnetic, and x-ray measurements [Arentz *et al.* (1973); see also Table VI]. A Curie point of 380°C was found by using Mössbauer measurements in correspondence with the magnetic measurements of Kagan (1959). The ϵ carbide has the hcp structure with interstitial carbon atoms in octahedral interstices. The magnetic hyperfine field in carbides is merely determined by the Fe–C distance and the number of nearest carbon neighbors, whereas the quadrupole interaction normally depends on the crystalline direction. Thus the spectrum of ϵ carbide should consist of either a single hyperfine field or two hyperfine field components differing in their quadrupole interaction. However, as seen from Table VI, most investigators found for synthetic ϵ carbide a single hyperfine field along with a central part related to a superparamagnetic component.

The synthetic χ carbide (Hägg carbide) has been studied using Mössbauer spectroscopy (Bernas *et al.*, 1967; Genin *et al.*, 1975; LeCaer *et al.*, 1976). A Mössbauer spectrum of Fe_5C_2, along with a spectrum of Fe_3C for comparison, is shown in Fig. 23. The Hägg carbide has a monoclinic structure and the composition Fe_5C_2. The lattice parameters are $a = 11.563$ Å, $b = 4.573$ Å, $c = 5.058$ Å, and $\beta = 97°44$, as summarized by Foct *et al.* (1978). The structure of the χ carbide is similar to that of cementite and can be described as having the carbon atoms in interstitial positions in a

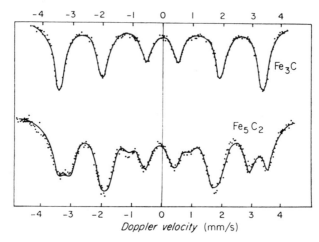

Fig. 23. Room-temperature Mössbauer spectra of Fe_3C (a) and Fe_5C_2 (b) (Bernas *et al.*, 1967). [Reprinted with permission from *J. Phys. Chem. Solids* **28**, 17. Copyright 1967, Pergamon Press, Ltd.]

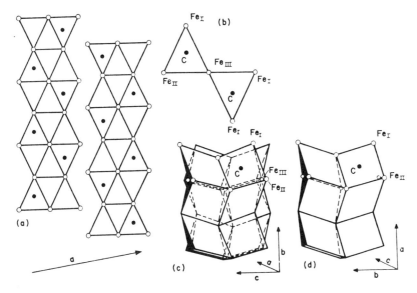

Fig. 24. The structure of χ-Fe_5C_2 and Fe_3C: (a) and (b) schematic representation of Fe_5C_2, (c) Fe_5C_2, (d) Fe_3C. Fe_I, Fe_{II}, and Fe_3 are inequivalent iron sites (Bernas *et al.*, 1967). [Reprinted with permission from *J. Phys. Chem. Solids* **28**, 17. Copyright 1967, Pergamon Press, Ltd.]

structure of iron atoms in pleated layers of prismatic sheets. The structure of χ carbide can be obtained by introducing stacking faults into the structure of cementite (Fig. 24). Above the Curie temperature $T_c \sim 521$ K, the spectrum consists of three doublets, the hyperfine parameters of which are given in Table VII. Below T_c the spectrum, as interpreted by Bernas *et*

TABLE VII

Mössbauer Parameters and Magnetic Moments of χ carbide[a]

	$\delta(mm/s)$ relative to Fe	H (kOe)	μ^b $(T = 18°C)$	μ^c $(T = 18°C)$	H/μ (kOe/μ_B)
Fe_3C^d	0.29 ± 0.02	208 ± 2	1.60		130
Fe_5C_2	0.35 ± 0.04	222 ± 3	1.48	1.71	130
	0.30 ± 0.04	184 ± 3		1.42	
	0.30 ± 0.08	110 ± 6		0.85	
(site Fe_{III})					

[a] Taken from Bernas *et al.* (1967).
[b] Average magnetic moment.
[c] Magnetic moment on individual Fe sites from ME supposing $H/\mu = 130$ kOe/μ_B.
[d] Assuming only one hyperfine field in Fe_3C.

al. (1967), consists of three hyperfine field components corresponding to three distinct Fe-atom sites Fe_I, Fe_{II}, and Fe_{III} (see Table VII). The atoms of site Fe_{II} split into two groups Fe'_{II} and Fe_{II} having different values of the quadrupole interaction (LeCaer *et al.*, 1976).

The Fe_3C cementite has the orthorhombic crystal structure (Fasiska and Jeffrey, 1965) and the following lattice parameters: $a = 4.5248$ Å, $b = 5.0896$ Å, and $c = 6.7443$ Å. There are two crystallographically inequivalent iron sites Fe_G and Fe_S, as described in Section A. At room temperature, two slightly different hyperfine fields of 205 and 207 kOe were attributed to the Fe_G and Fe_S iron sites, respectively (Ron and Mathalone, 1971). The sites have the same isomer shifts, and the small difference in the hyperfine fields would not be distinguishable in the Mössbauer spectrum. However, due to the different quadrupole interaction at the two iron sites, the room-temperature spectrum shows a broadened and considerably shortened sixth line ($+ \frac{3}{2} \rightarrow + \frac{1}{2}$ transition), as seen in Fig. 25. Above the Curie temperature of cementite, which is about 210°C (Apayev *et al.*, 1961; LeCaer *et al.*, 1976), the spectrum consists of two doublets, with quadrupole splittings of 0.32 and -0.58 mm/s for the Fe_G and Fe_S sites, respectively (Ron and Mathalone, 1971). Different values for the quadrupole interaction in cementite, above T_c and for the efg asymmetry parameter η below T_c, have been found by LeCaer *et al.* (1976) and Foct *et al.*, (1978).

An interpretation of the changes in the electronic structure of interstitial compounds of C, N, and B in iron has been given by Bernas *et al.* (1967).

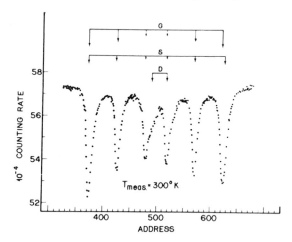

Fig. 25. Room-temperature Mössbauer spectrum of Fe_3C below the Curie temperature. The arrows G and S indicate the position and relative intensities of Fe_G and Fe_S components. The sixth line of the spectrum is broaden and shortened, mainly due to the different quadrupole interaction of the two components (Ron and Mathalone, 1971).

A "donor model" was used, which is essentially a rigid-band model in which the number of d-band electrons in the metal is increased by the valence electrons of the metalloid. In compounds where only iron atoms at a single site exist, the change in saturation magnetization reflects a change in the number of unpaired d electrons (e.g., FeB–Fe$_2$B). However, in compounds where the iron atoms occupy more than one site, the situation becomes more complex. The isomer shift is not only sensitive to changes in the 4s-electron density but also to changes in the 3d-electron configuration, both of which change the screening. The experimental results for Fe$_5$C$_2$ of Bernas *et al.* (1967), (see Table VII), according to which the value of $\overline{H}/\overline{\mu}$ was nearly constant, suggest that the conduction electron polarization was either small or proportional to the magnetization. It was also concluded that through the localized hybridization in a common band of the metal d electrons and the metalloid sp electrons, the total number of d electrons on the iron site remains practically constant and that these electrons carry the magnetic moment.

It has been mentioned that Ino *et al.* (1968) have made an attempt, in Fe–C alloys, to correlate the number n of interstitial near-neighbor carbon atoms to a given Fe site and the hyperfine field related to it. Figure 26 shows, as suggested by Foct *et al.* (1978), a correlation between n, the number of interstitial carbon atoms, the Fe–C distance, and the corre-

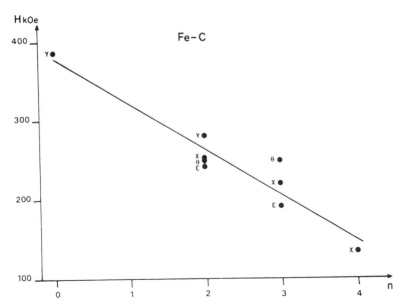

Fig. 26. Hyperfine fields (at various sites) in iron carbides as a function of the number of interstitial neighbors n (Foct *et al.*, 1978).

sponding hyperfine fields of various atomic configurations in clusters and γ, ϵ, χ and θ iron carbides.

Huffman *et al.* (1967) have studied alloyed carbides having the cementite structure $(Fe_{1-x}Mn_x)_3C$ with $0 < x < 0.6$ and a Hägg carbide $(Fe_{1.1}Mn_{3.9})C_2$, over the temperature range 12–500°K. The hyperfine field and the Curie temperature were found to decrease very rapidly with Mn content, extrapolating to zero at 34% Mn and 25% Mn, respectively. The results were interpreted as consistent with the donation of 2p electrons of the carbon to the d band, giving Fe in cementite an average configuration of approximately $3d^{7.66}4S^1$.

VI. Fe–N Alloys

A. Phases in the Fe–N Binary System

Nitrogen atoms occupy interstitial sites in the lattice of iron. In α-iron, the solubility of nitrogen is limited and depends on the phase with which the α-iron is in equilibrium. The binary phase diagram shown in Fig. 27 was given by Paranjpe *et al.* (1950). The maximum solubility of N in α-Fe is 0.1 wt% at 590°C, which decreases to ~ 0.004 wt% when the temperature decreases to 200°C. The γ-phase, nitrogen austenite, is stable above

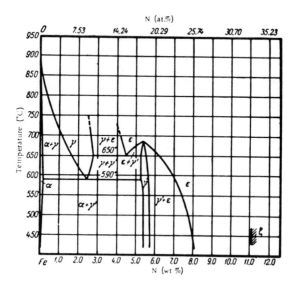

Fig. 27.   Fe–N phase diagram (Vol., 1967b).

590°C and has a structure isomorphous with fcc Fe–C austenite (Jack, 1951b). The Fe–N austenite is an interstitial solid solution in which the N atoms are randomly distributed on octahedral sites. The nitrogen concentration reaches a maximum of 2.8 wt% at 650°C. The lattice parameter of the Fe–N austenite increases linearly with nitrogen concentration (Paranjpe *et al.*, 1950). At 590°C the γ-phase undergoes an eutectoid decomposition $\gamma \to \alpha + \gamma'$ at a nitrogen concentration of 2.30 wt% N.

The γ'-Fe$_4$ has a range of stoichiometry between 5.30 and 5.75 wt% N. The iron atoms in γ'-Fe$_4$N form an fcc lattice in which the N atoms occupy $\frac{1}{4}$ of the octahedral interstices at $\frac{1}{2}, \frac{1}{2}, \frac{1}{2}$ in an ordered manner. The hexagonal ϵ nitrides exist over a wide range of composition, between ϵ-Fe$_4$N and Fe$_2$N. The ϵ-lattice parameters increase with nitrogen concentration, while the c/a ratio changes for 1.6 to 1.65. The nitrogen atoms occupy octahedral interstices in the hcp-iron lattice. Superlattice reflections have been reported for the Fe$_3$N and Fe$_2$N compositions and other superlattices for compositions in between these (Hendricks and Kosting, 1930). The ξ-Fe$_2$N structure, which has an orthorhombic unit cell, exists at a nitrogen concentration of 1 wt% N (Jack, 1948).

Metastable phases and states in the Fe–N system are of technological and scientific importance. Fe–N austenite having a concentration higher than 2.4 wt% can be retained at room temperature. Below this concentration a partial transformation to martensite occurs. In bct Fe–N martensite, the nitrogen atoms occupy one set of octahedral interstices carried over from the austenite by the Bain transformation. The c axes of the Fe–N martensite unit cells are aligned in parallel by a mechanism similar to that mentioned previously for the Fe–C martensite.

Unlike C in Fe–C martensite, nitrogen is negatively charged, and the interaction between nitrogen atoms (in Fe–N martensite) is perhaps slightly attractive (DeCristofaro and Kaplow, 1977a).

In virgin Fe–N martensite at room temperature, nitrogen atoms agglomerate and form local regions of ordered α''-Fe$_{16}$N$_2$. The α''-Fe$_{16}$N$_2$ structure consists of a few expanded bct unit cells, where the N atoms occupy $\frac{1}{24}$ of the octahedral interstices, in a completely ordered state. Aging at 100°C leads to decomposition of martensite to α-iron and α''-Fe$_{16}$N$_2$. Aging above 160°C leads to the precipitation of the stable γ''-Fe$_4$N precipitates.

B. Fe–N Austenite and Martensite

Fe–N austenite is retained at room temperature upon quenching alloys having a sufficient concentration of nitrogen. The austenite lines appear in the central part of the Mössbauer spectra because austenite is para-

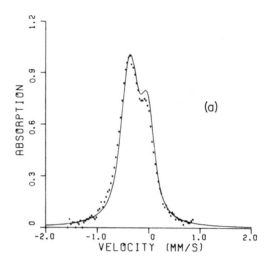

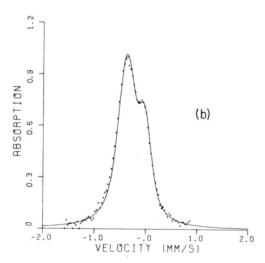

Fig. 28. Room-temperature Mössbauer spectra of 2.34 wt% N austenite: (a) fitted by computer (line) using a model based on separated nitrogen atoms; (b) fitted by computer (line) using a model of randomly distributed nitrogen atoms (DeCristofaro and Kaplow, 1977a).

magnetic. This facilitates their differentiation from other, mostly ferromagnetic, Fe–N phases.

By analogy with the Mössbauer spectrum of Fe–C austenite, the spectrum of retained Fe–N austenite has been interpreted as a superposition of two components (Ino et al., 1967; Moriya et al., 1968; Gielen and Kaplow, 1967; DeCristofaro and Kaplow, 1977a). The room-temperature spectra, shown in Figs. 28a and b, for 2.34 wt% N austenite consist of a single line and a doublet. These were related to Fe atoms having no nitrogen nn Fe_0 and one nitrogen nn Fe_1 (DeCristofaro and Kaplow, 1977a). The value of the quadrupole splitting, 0.29 mm/s (see Table VII), is much smaller than in the Fe–C austenite. A nitrogen nn atom produces a positive shift of 0.21 mm/s with respect to the Fe_0 iron-atom group. The relative intensities of the doublet and the single peak indicate that a fraction of 0.495 ± 0.008 of iron atoms neighbor an N atom. This is close to the value obtained for a random distribution model given by $1 - (1 - y)^6$, where y is the atomic fraction of N, with $y = 0.0956$ (see Table I).

Comparing the fits of computer-generated spectra and experimental data points, it is seen that the random distribution model (in Fig. 28b) provides a better fit than the one based on the model of separated N atoms (in Fig. 28a). Evidence exists that nitrogen atoms in Fe–N austenite are negatively charged and have a slightly attractive interaction (Elliot, 1963; Nagakura, 1968; Nagakura and Tenehashi, 1968). This is in contrast to carbon interstitial atoms in austenite, which are heavily positively charged and have a repulsive interaction. It may therefore be expected that the perturbation of the electronic structure of the iron in austenite, caused by an interstitial nn atom, is smaller for nitrogen than for carbon (Jack and Jack 1973).

Fe–N martensite is metastable, has (in average) the bct structure, and is ferromagnetically ordered. In analogy to the Fe–C martensite, the low symmetry, lattice distortion, and the effect of interstitial atoms lead to a number of magnetically inequivalent iron-atom sites in the Fe–N martensite. A spectrum for virgin Fe 2.83 wt% N martensite is shown in Figs. 29a and b. The computer fit of Fig. 29b is a superposition of four hyperfine field components related to Fe atoms having first, second, and third nitrogen neighbors (Fe_1, Fe_2, and Fe_3 in the present notation) and one having further than third-neighbor nitrogen (see Table VIII). The relative spectral intensities, calculated with the assumption of a random distribution of nitrogen on one of the three interstitial sublattices, were 0.221, 0.306, 0.298, and 0.175 for the iron-atom groups with first, second, third, and further nitrogen neighbor, respectively.

Aging of Fe–N Martensite below 200°C results in the formation of α-iron and α''-$Fe_{16}N_2$, an ordered arrangement of N atoms in the bct matrix. Garwood and Thomas (1973) have found, by an electron micro-

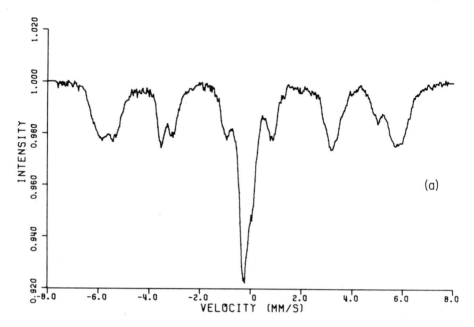

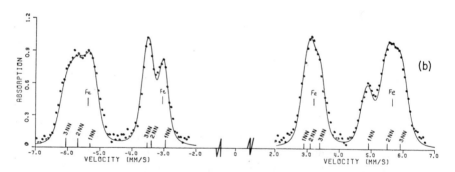

Fig. 29. (a) Mössbauer spectrum of Fe-2.83 wt% N martensite. (b) Comparison of a spectrum for virgin martensite, with a computer-generated one based on a model of randomly distributed nitrogen atoms (line). Positions labeled 1NN, 2NN, and 3NN correspond to peaks from the first- , second- , and third-nearest-neighbor iron atoms of a nitrogen atom. Positions labeled Fe correspond to peaks from iron atoms that are further than third neighbor (DeCristofaro and Kaplow, 1977a).

scopic investigation, that the α''-Fe$_{16}$N$_2$ precipitates homogeneously (similar to spinodal decomposition), and with continued aging it acquires a lamellar structure. The α''-Fe$_{16}$N$_2$ formed in nitrided steel has been measured by Mössbauer spectroscopy (Moriya *et al.*, 1973; Bainbridge *et al.*, 1973). Recently DeCristofaro and Kaplow (1977a) have used the Mössbauer effect to study the aging of Fe–N martensite. A Mössbauer spectrum taken at $-193°$C of 2.83 wt% N martensite, aged 24 hr at 100°C, is shown in Fig. 30. The bars on the top of the spectrum show the positions of the outer lines from three groups of iron atoms having first-, second-, and third-neighbor N atoms in the α''-Fe$_{16}$N$_2$ structure. The relative spectral intensities were taken as $1:2:1$, corresponding to the relative number of atoms in the iron-atom groups in this structure.

Figure 31b shows a difference spectrum, viz., a plot of differences between the spectrum for an alloy aged at room temperature (Fig. 31a) and a spectrum for virgin martensite (Fig. 29). The peaks in the difference spectrum are seen to correspond to the hyperfine fields of α''-Fe$_{16}$N$_2$, as they were determined for martensite aged at 100°C (Fig. 30). It was therefore concluded that changes observed in the Mössbauer spectra of Fe–N martensite due to room-temperature aging are a result of agglomeration and ordering of nitrogen into the α''-Fe$_{16}$N$_2$ structure.

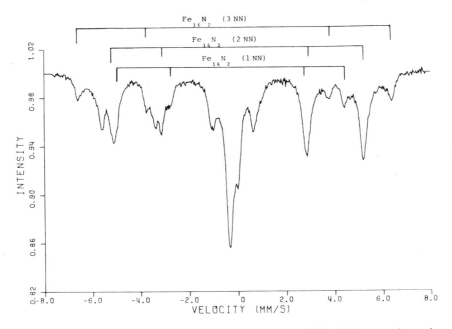

Fig. 30. Mössbauer spectrum measured at $-193°$C of Fe–2.83 wt% N martensite aged 24 hr at 100°C. The bar diagrams at the top of the spectrum indicate the α''-Fe$_{16}$N$_2$ components (DeCristofaro and Kaplow, 1977a).

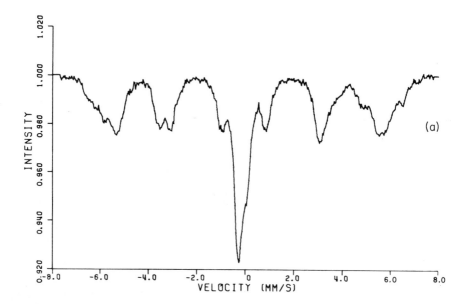

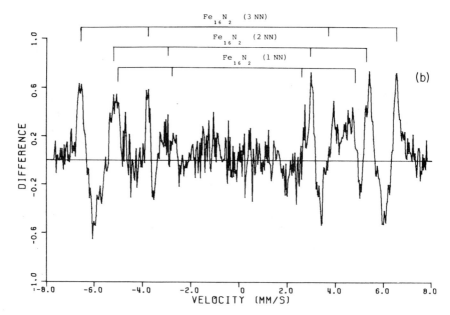

Fig. 31. (a) Mössbauer spectrum of Fe-2.83 wt% N martensite aged one month at room temperature. (b) Difference spectrum obtained by subtracting the experimental spectrum of virgin martensite (Fig. 29) from the spectrum of the martensite aged at room temperature (Fig. 31a). The bar diagrams indicate the peaks related to the appearance of $Fe_{16}N_2$ (DeCristofaro and Kaplow, 1977a).

C. Iron Nitrides

γ'-Fe_4N is a stable precipitate that forms on aging Fe–N martensite above 160°C (Jack, 1948). Stoichiometric Fe_4N nitride can be synthesized by a method described by Arnott and Wold (1960). The nitride Fe_4N has been studied by x-ray (Jack, 1948), neutron (Frazer, 1958), and electron diffraction (Guilland and Creveaux, 1946) and by Mössbauer effect (Shirane et al., 1962; Nozik et al., 1969; Clauser, 1970; Wood and Nozik, 1971). The γ'-Fe_4N has a simple crystal structure, which may be visualized as an fcc-iron lattice with nitrogen in the body center octahedral interstice (Jack, 1948). The corner atom Fe^c is surrounded by 12 Fe nn at 2.96 Å, while the face center atom Fe^f has two nitrogens at 1.90 Å as the nn. The Fe^c and Fe^f are two inequivalent iron sites in the ratio of 1 : 3. Correspondingly, the Mössbauer spectrum was unfolded into two components related to these two iron sites (Shirane et al., 1962). Magnetic moments of $\sim 3\mu_B$ and $2\mu_B$ have been assigned to the Fe^c and Fe^f sites, respectively, by Wiener and Berger (1955). Later, Nozik et al. (1969) have shown that a Mössbauer spectrum for γ'-Fe_4N taken with high resolution consists of three six-line spectral components (see Fig. 32) of relative spectral intensities of 1 : 2 : 1.

Clauser (1970) and Wood and Nozik (1971) have related the component having the largest hyperfine field and zero quadrupole interaction to the Fe^c atoms, while the two other components have been related to Fe^f atoms split into two groups having different orientations of the efg with respect to the easy magnetization axis of the crystal. The Mössbauer parameters assigned to the three iron-atom groups are given in Table VIII. The easy magnetization axis has been found to be parallel to the $\langle 100 \rangle$ direction (Wood and Nozik, 1971). In the γ'-Fe_4N structure, the principal component of the efg is perpendicular to the cube face and therefore will be parallel to the easy magnetization axis for one face and perpendicular for the two others. Therefore, the group of Fe^f atoms with the principal component of the efg perpendicular to the easy magnetization axis has an intensity double to that of other groups. Shirane et al. (1962) have found a positive isomeric shift of 0.15 mm/s between the Fe^c and Fe^f atoms in the γ'-Fe_4N. Assuming that nitrogen acts as a "donor" giving one electron to each of the nearest iron neighbors, they suggest the electronic configuration to be $3d^74s^1$ for Fe^c and $3d^84s^1$ for the Fe^f atoms. However, Nagakura (1968), on the basis of electron diffraction measurements, has refuted this interpretation suggesting a configuration of γ'-$Fe^c(Fe^{f+1/3})_3N^{-1}$.

Eichel and Pitsch (1970) have pointed out that the relationship between the 3d and 4s states and the isomeric shift (Walker et al., 1961) applied to the electronic configuration above would predict a more positive shift than obtained by experiment (Shirane et al., 1962; Wood and Nozik, 1971, Table VII).

TABLE VIII

Mössbauer Parameters Associated with Iron-Atom Types in Fe–N Alloys

Iron atom type	Relative intensity	T (°C)	Isomer[a] shift, (mm/s)	Quadrupole splitting (mm/s)	Hyperfine magnetic Field (KOe)	Reference
Fe 2.34 wt% N austenite		20				DeCristofaro and Kaplow (1977a)
N neighbor	0.45		−0.3	0.29	—	
No N neighbor	0.55		−0.39	—	—	
Fe 2.83 wt% N versus martensite[b]		−193				DeCristofaro and Kaplow (1977a)
1 nn of N	0.221		−0.11	−0.32	316	
2 nn	0.306		−0.12	0.12	346	
3 nn	0.298		−0.06	0.0	370	
No N neighbor	0.175				340	
$Fe_{16}N_2$		20				DeCristofaro and Kaplow (1977a)
1 nn of N	0.25		−0.15	−0.17	288	
2 nn	0.50		−0.06	0.09	314	
3 nn	0.25		−0.09	−0.09	399	
γ'-Fe_4N		rt				Nozik et al. (1969)
Fe^c	1		0.24	0.0	340.6	
Fe^f_I	2		0.52	−0.22	215.5	
Fe^f_{II}	1		−0.15	+0.43	219.2	
ε-$Fe_{3.2}N$		22				Eickel and Pitch (1970)
Fe_ε^{-1}	7.33		0.33	—	238	
$Fe_\varepsilon^{-0.5}$	1		0.24	—	298	

[a] with respect to α-iron.
[b] 1 nn, 2 nn, and 3 nn correspond to Fe_{nn}, Fe_{nnn}, and Fe_0 in the present notation.

The ϵ-$Fe_{3.2}N$ nitride has been investigated by Eickel and Pitsch (1970). The Fe_xN nitrides with $3.3 \leqslant x \leqslant 2$ form a hcp lattice at 550° (Pearson, 1958). These structures were described as sequences of Fe atom arrays parallel to the closed-packed plane of a hexagonal lattice with octahedral interstices. The interstices are only partially occupied by N atoms (Jack, 1952; Hendricks and Kosting, 1930). It can be shown that the environment of Fe atoms in ϵ-Fe_3N and Fe^f in γ'-Fe_4N is similar insofar as they both have the same number of neighbors of different kinds, viz., Fe atoms, N atoms, and unoccupied octahedral interstices. The Mössbauer spectrum of ϵ-$Fe_{3.2}N$ (Eickel and Pitsch, 1970) has been decomposed into two components (see Table VIII). One of 0.88 relative spectral intensity, was related to the group of iron atoms having two nitrogen neighbors; the other was related to iron atoms having one nitrogen neighbor. Magnetic moments of $2.4 = \mu_B$ and $1.9 = \mu_B$ were derived for the groups having one and two N neighbors, respectively. The electronic configuration ϵ-$(Fe^{+3.2})_2N^{-3}$, as derived by Nagakura and Tanehashi (1968) from electron diffraction, has been used by Moriya *et al.* (1973) to discuss the Mössbauer pattern of Fe–N alloys of ϵ and other structures. Recently DeCristofaro and Kaplow (1977b) have investigated ϵ-iron–nitrogen alloys having a wide range of composition. By quenching from 700°C, the hcp phase could be completely retained at room temperature for compositions as low as 17.1 at.% nitrogen. Mössbauer spectra representative for quenched alloys in the concentration range of 17.1–23.8 at.% nitrogen are shown in Fig. 33. The

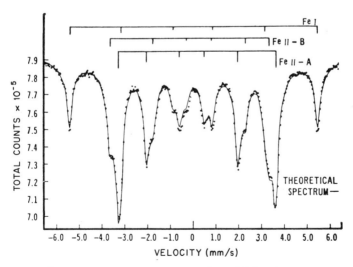

Fig. 32. Mössbauer spectrum of γ'-Fe_4N. The solid line is a computer fit of a superposition of three components related to inequivalent iron sites. In the present notation the sites are Fe^c corner atoms and Fe^f face center atoms, which subdivide into two groups of different quadrupole interaction (Nozik *et al.*, 1969). [Reprinted with permission from *Solid State Commun.* 7, 1677. Copyright 1969, Pergamon Press, Ltd.]

ferromagnetic spectra were interpreted as a superposition of three components related to Fe atom groups having one- , two- , and three-N-atom neighbors. The spectrum for the 17.1 at.% N is seen to be paramagnetic at room temperature. The Curie temperature and the hyperfine magnetic field are seen in Fig. 34 to depend strongly on the nitrogen concentration of the alloy. The Curie point has a maximum of 294°C at

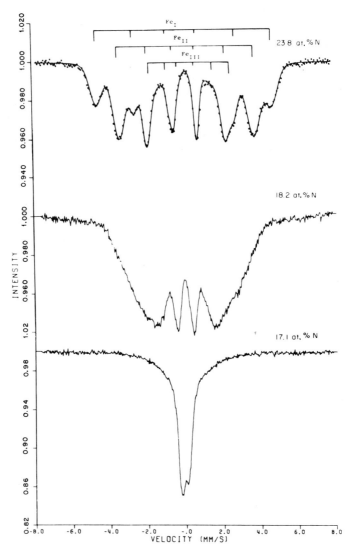

Fig. 33. Room-temperature Mössbauer spectra of ε-nitrides quenched from 700°C. Compositions are shown in the figure (DeCristofaro and Kaplow, 1977b).

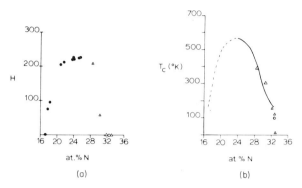

Fig. 34.   Curie temperatures and hyperfine magnetic fields (of Fe$_{II}$ site) of ϵ nitrides plotted as a function of nitrogen content.   —, Bridelle (1955); $\triangle$, Mekata *et al.* (1972); $\circ$, Chabanel *et al.* (1968); ---, $\bullet$, DeCristofaro and Kaplow (1977b).

$\sim$ 24 at.% N. Mössbauer measurements indicate that the hexagonal ϵ-phase undergoes ordering of nitrogen atoms on interstitial sites, confirming the suggestion by Hendricks and Kosting (1930).

Finally, an attempt to systemize the hyperfine fields in iron nitrides and to represent them as a function of the number of interstitial nearest neighbors n is shown in Fig. 35 (Foct *et al.*, 1978).

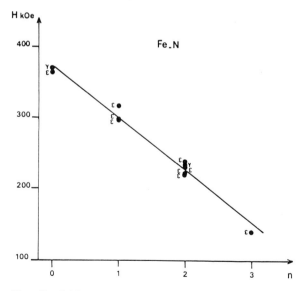

Fig. 35.   Hyperfine fields at various sites in iron nitrides as a function of the number of interstitial neighbors n (Foct *et al.*, 1978).

VII.   Miscellaneous Applications

This closing section demonstrates the potential of the Mössbauer spectroscopy as an analytical tool in the metallurgy of iron-base alloys. A
number of technical problems and the contribution of the Mössbauer
spectroscopy of their understanding are briefly discussed. The problems
that are presented were not selected to provide a comprehensive review of
all possible applications; rather they are representative examples for the
use of this technique.

Bogachev et al. (1975) have investigated the segregation of the main
alloying elements Mn ($\sim$ 8 wt%) in a low-carbon (C $\sim$ 0.07 wt%) steel in
the temperature range of maximum embrittlement. Processes of short-range
segregation at the temperature of maximum embrittlement were traced by
Mössbauer spectroscopy. They suggested that the short-range segregation
may be one of the reasons for the irreversible temper embrittlement of this
steel. The redistribution and ordering of carbon at room and lower
temperatures in Fe–C, Fe–Mn–C, and Ce–Ni–C steels in the austenitic
and martensitic states have been studied by Mössbuer spectroscopy by
several authors (Narkhov et al., 1975; Gruzin et al., 1975). A thorough
analysis of Fe–Cr carbides appearing in alloyed steels during their production and heat treatment has been carried out by Kuzmann et al. (1976)
using the Mössbauer method.

The basic Mössbuer data on iron nitrides have been used to study
problems of technical importance by several investigators. Maksimov et al.
(1974a) have studied the process of decomposition of ϵ-iron nitride in the
temperature range of 380°C–480°C. They found that the ϵ-phase decomposes into regions having different Curie temperatures while retaining the
unchanged crystal structure. This was related to local variations in the
nitrogen concentration. The final decomposition product was found to be
γ'-Fe$_4$N. Yamakawa et al. (1978) and Ujihira and Handa (1978) have
studied surface layers of nitrided steels using CEMS (Conversion Electron
Mössbauer Spectroscopy) and x-ray Mössbauer reflection. They were able
to identify the iron nitrides progressing layer by layer in from the surface.
A thorough analysis of the iron nitrides and intermediate states that form
at the surface is given by Ujihira and Handa (1978).

Mössbauer spectroscopy has been employed by Huffman and Podgurski
(1975) in a study of the nitride phases generated in Fe–Mo and Fe–Ti
alloys by gasphase nitrogeneration at 460°C–600°C. Satellite peak intensities were used to determine the extent to which the solute atoms have been
removed from the solid solution. It was also possible to determine the
excess of N in the ferrite matrix, the formation of secondary phases, and in
some cases the approximate dimensions of dispersed particles. Salomon
and Levinson (1977) have investigated the redistribution of Ni and Cr

during the phenomenon of "475°C embrittlement" of a duplex stainless steel and single-phase ferritic alloys. The alloys studied included two binary Fe–Cr alloys and multicomponent alloys containing Ni, Mo, and Cu, as well as Fe and Cr. In all alloys there is a hyperfine field increase with aging at 475°C, while paramagnetic peaks (and precipitates) are observed only in alloys containing Ni. These results were interpreted as due to spinodal decomposition. Swartzendruber *et al.* (1974) have used the Mössbauer technique for the examination of welds and castings and have demonstrated the importance of the backscattering mode of Mössbauer measurements for such examinations. Albritton and Lewis (1971) have deviced computer programs that permit quantitative evaluation of amounts of minor phases in stainless steel. Using this technique, they were able to analyze the 885°C embrittlement in ferritic stainless steels. Mössbauer spectroscopy (Vyunnik, 1975) has been used to study the influence on the precipitation process of Co^{60} γ-ray irradiation $(3 \times 10^{12} - 4 \times 10^{16}$ quanta$/cm^2 s)$ during tempering at 720 and 800°C of an EL69 steel containing Cr. The carbide phase $Cr_{23}C_6$ in which Cr has replaced iron was found to form in the irradiated samples after treatments shorter than in the nonirradiated ones.

The effect on the Mössbauer patterns of plastic deformation of Fe–C alloys and alloyed steels has been subject to theoretical and experimental investigations (Kvashina and Krivoglaz, 1967; Lagunov *et al.*, 1971; Gridnyev *et al.*, 1975; Yurchikov *et al.*, 1968; Abu Elnasr and Reid, 1974). Plastic deformation alters severely the ferromagnetic structure, viz., domains orientation and distribution of the ferromagnetic α-iron and martensite. The dependence of the Mössbauer parameters of the magnetic hyperfine splitting on the magnetic structure of the alloy makes the interpretation difficult. Discrepancies between theoretical prediction and experimental measurements lead to nonconclusive results.

VIII.   Suggestions for Further Research

About 15 years of Mössbauer research have provided much detailed information regarding the crystal structure, atomic configurations, magnetic structure, etc., for the equilibrium phases and metastable states of the Fe–C and Fe–N systems. Although valuable structural, kinetic, and thermodynamic conclusions have been derived, a number of important problems have remained unresolved. Their reinvestigation can be expected in the near future, stimulated by currently developed stronger sources, higher resolution, and more effective methods of analysis. The emerging backscattering technique is of particular value as it provides a method that does not require thinning of specimens, which often leads to the formation of artifacts.

The following problems are among those demanding further investigation:

1.   (a) Do carbon atoms partially occupy tetrahedral interstices in virgin martensite at low temperature?

(b) Do the changes observed at room and lower temperatures indicate a redistribution of tetrahedral to octahedral interstices, thereby increasing the c/a ratio?

(c) Do the above changes indicate the agglomeration of carbon atoms into clusters or ordered regions?

These studies should be carried out in a wide range of carbon concentrations and under conditions which eliminate autotempering and formation of artifacts.

2.   Room-temperature ordering and clustering processes of carbon atoms in martensite must be studied in a wide range of carbon concentrations.

3.   The investigation of structural details of carbides and nitrides should be continued.

4.   The utilization of the Mössbauer technique for routine testing in industrial quality control is feasible in many cases. It would be especially useful to improve backscatter techniques for this purpose since sample preparation would be minimized.

Acknowledgments

The author is indebted to Professor J. Reseck for his stimulating discussions and critical reading. The educating and stimulating discussions with Professor M. Cohen, Professor R. Kaplow, and Professor G. Olson are greatly appreciated.

References

Aaronson, H. *et al*. (1966). *Trans AIME* **236**, 753.
Abu Elnasr, T. Z. H., and Reid, C. N. (1974). *Met. Sci*. **8**, 33.
Albritton, O. W., and Lewis, J. M. (1971). *Weld. Res. Suppl*. 327-S.
Alshewsky, Yu. L., and Kurdjumov, G. V. (1968). *Fiz. Met. Metallov*. **25**, 172.
Amelse, J. A., Butt, J. B., and Schwartz, L. H. (1978). *J. Phys. Chem*. **82**, 558.
Apayev, B. A., Yakovlev, M. B., and Tikhonov, G. F. (1961). *Fiz. Met. Metallov*. **12**, 208.

Arentz, R. A., Maksimov, Yu. B., Suzdalyev, I. P., Imshennik, V. K., and Krupiansky, Yu. F. (1973). *Fiz. Met. Metallov.* **36**, 277.

Arnott, R. J., and Wold, A. (1960). *Phys. Chem. Solids* **15**, 152.

Bogachev, I. N., Charushnikova, G. A., Ovchinnikov, V. V., and Litvinov, V. S. (1975). *Fiz. Met. Metallov.* **39**, 129.

Bain, E. C. (1924). *Trans AIME* **70**, 25.

Bainbridge, J., Channing, R. A., Whitlow, W. H., and Pend, R. E. (1973). *J. Phys. Chem. Solids* **34**, 1579.

Baker, A. I., and Kelly, P. M. (1963). "Electron Microscopy and Strength of Crystals," Wiley (Interscience), New York.

Bernas, H., Campbell, I. A., and Fruchart, R. (1967). *J. Phys. Chem. Solids* **28**, 17.

Bowles, T. S., and MacKenzie, J. K. (1954). *Acta Metall.* **2**, 138.

Bridell, R. (1955). *Ann Chim. Ser.* 12 **10**, 824.

Butler, J. F. (1966). *J. Iron Steel Inst.* **204**, 127.

Cadeville, M. C., Friedt, J. M., and Lerner, C. (1977). *J. Phys. F* **7**, 123.

Chabanel, M., Janot, C., and Motte, J. P. (1968). *Co. R. Acad. Sci. Paris* **266**, B419.

Choo, W. K., and Kaplow, R. (1973). *Acta Metall.* **21**, 725.

Christ, B. W., and Giles, P. M. (1967). "Mössbauer Effect Methodology," Vol. 13, p. 37. Plenum Press, New York.

Christ, B. W., and Giles, P. M. (1968). *Trans. AIME* **242**, 1915.

Christian, J. W. (1965). "The Theory of Transformations in Metals." Pergamon, Oxford.

Clauser, M. J. (1970). *Solid State Commun.* **8**, 781.

Cohen, M. (1951). "Phase Transformations," in p. 588. Wiley, New York.

Cohen, M. (1962). *Trans. TMS-AIME* **224**, 638.

DeCristofaro, N., and Kaplow, R. (1977a). *Metall. Trans.* **8A**, 35.

DeCristofaro, N., and Kaplow, R. (1977b). *Metall. Trans.* **8A**, 425.

DeCristofaro, N., Kaplow, R., and Owen, W. S. (1978). *Metall. Trans.* **9A**, 821.

Doremus, R. H., and Koch, E. F. (1960). *Trans. Metall. Soc. AIME* **218**, 591.

Dubois, J. M., and Le Caer, G. (1977). *Acta Metall.* **25**, 609.

Duggin, M. J. (1968). *Trans. MS-AIME* **242**, 1091.

Dünner, Ph., and Müller, S. (1965). *Acta Metall.* **13**, 25.

Eickel, K. E., and Pitsch, W. (1970). *Phys. Status Solidi* **39**, 121.

Elliott, N. (1963). *Phys. Rev.* **129**, 1120.

Etnin, I. R. (1973). *Dok. Akad. Nauk SSSR* **17**, 1021.

Fasiska, E. J., and Jeffrey, G. A. (1965). *Acta Crystallogr.* **19**, 469.

Foct, J., Le Caer, G., Dubois, J. M., and Faivre, R. (1978). *Int. Conf. Carbides, Borides, and Nitrides in Steels, Kolobrzeg, Poland.* Private communication.

Frauenfelder, H., Nagle, D. E., Taylor, R. D., Cochran, R. F., and Vischer, W. M. (1962). *Phys. Rev.* **126**, 1065.

Frazer, B. C. (1958). *Phys. Rev.* **112**, 751.

Fujita, F. E. (1975). *In* "Topics in Applied Physics: Mössbauer Spectroscopy" (U. Gonser, ed.), Vol. 5, p. 220. Springer-Verlag, Berlin and New York.

Fujita, F. (1977). *Metall. Trans.* **8A**, 1727.

Fujita, F. E., Moriya, T., and Ino, H. (1971). *Proc. Int. Conf. Sci. Technol. Iron Steel Part II. Trans Iron Steel Inst. Jpn.* **11**, 1273.

Fujita, F., Shiga, C., Moriya, T., and Ino, H. (1974). *Jpn. Inst. Met.* **38**, 1030.

Garwood, R. D., and Thomas G. (1973). *Metall. Trans.* **4**, 225.

Geniger, A. B., and Trojano, A. R. (1949). *Trans AIME* **185**, 590.

Genin, J. M., and Flinn, P. A. (1966). *Phys. Lett.* **22**, 392.

Genin, J. M., and Flinn, P. A. (1967). Ph.D. Thesis, Carnegie Institute of Technology.

Genin, J. M., and Flinn, P. A. (1968). *Trans. AIME* **242**, 1419.

Genin, J. M., Le Caer, G., and Simon, A. (1975). *Proc. Int. Conf. Mössbauer Spectrosc.*, 5th, Czechoslovakia p. 318.

Gielen, P. M., and Kaplow, R. (1967). *Acta Metall.* **15**, 49.

Goldanskii, V. I., and Herber, R. H. (1968). "Chemical Applications of Mössbauer Spectroscopy." Academic Press, New York.

Gonser, U. (1968/9). *Mater. Sci. Eng.* **3**, 1.

Gonser, U. (1971). *In* "Introduction to Mössbauer Spectroscopy" (L. May, ed.), Chapter 8. Plenum Press, New York.

Gonser, U., (ed.) (1975). "Topics in Applied Physics," Vol. 8, Mössbauer Spectroscopy. Springer-Verlag, Berlin and New York.

Gonser, U., Ron, M., Ruppersberg, H., Keune, W., and Trautwein, A. (1972). *Phys. Status Solidi (a)* **10**, 493.

Gordine, J., and Codd, I. (1969). *J. Iron Steel Inst. (London)* **207**, 461.

Greenwood, N. N., and Gibb, T. G. (1971). "Mössbauer Spectroscopy," Chapman and Hall, London.

Gridnyev, W. N., and Petrov, Yu (1962). *Fiz. Met. Metallov.* **13**, 686.

Gridnyev, W. N., Gavrolyukh, J. G., Diechtyar, I. Ya., Meshkov, Yu. Ya., Nizin, P. C., and Prokopyenko, V. G. (1975). *Dokl. Acad. Nauk Ukr. S. S. R.* 75.

Gridnyev, W. N., Gavrelyuk, W. G., Nemoshkalyenko, W. W., Polyushkin, Yu. A., and Rasumov, O. N. (1977). *Fiz. Met. Metallov.* **43**, 582.

Gruzin, P. L., Radionov, Yu. L., and Li, Yu. A. (1975). *Fiz. Met. Metallov.* **39**, 77.

Guilland, C., and Cerveaux, H. (1946). *C. R. Acad. Sci. Paris* **222**, 1170.

Hanna, S., and Preston, R. (1965). *Phys. Rev.* **139**, 722.

Hendricks, S. S., and Kosting, P. B. (1930). *Z. Kristallogr.* **74**, 571.

Hirotsu, Y., and Nagakura, S. (1972). *Acta Metall.* **20**, 645.

Hirotsu, Y., and Nagakura, S. (1974). *Trans. Jpn. Inst. Met.* **15**, 129.

Hofer, L. J., and Cohn, E. M. (1959). *J. Am. Chem. Soc.* **81**, 1576.

Huffman, G. P., and Podgurski, M. M. (1975). *Acta Metall.* **23**, 1367.

Ino, H., and Ito, T. (1978). *J. Phys.* **40**, 644.

Ino, H., Moriya, T., Fujita, F. E., and Maeda, Y. (1967). *J. Phys. Soc. Jpn.* **22**, 346.

Ino, H., Moriya, T., Fujita, F. E., Maeda, Y., Ono, Y., and Inokuti, Y. (1968). *J. Phys. Soc. Jpn.* **25**, 88.

Isotov, V., and Utevskii, L. (1968). *Phys. Met. Metallov.* **25**, 751.

Jack, K. H. (1948). *Proc. R. Soc. London Ser.* 34, *A* **195**, 56.

Jack, K. H. (1951a). *J. Iron Steel Inst.* **169**, 26.

Jack, K. H. (1951b). *Proc. R. Soc. London Ser. A* **208**, 216.

Jack, K. H. (1952). *Acta Crystallogr.* **5**, 404.

Jack, D. H., and Jack, K. H. (1973). *Mater. Sci. Eng.* **11**, 1.

Johns, R. D. (1973) *Iron and Steel* **46**, 33, 137.

Johnson, R. A. (1964). *Acta Metall.* **12**, 1215.

Johnson, R. A. (1967). *Acta Metall.* **15**, 513.

Kagan, Yu. I. (1959). *Fiz. Met. Metallov.* **8**, 535.

Kaufman, L., and Cohen, M. (1955). The mechanism of phase transformation in metals. *Inst. Met. Monogr. Rep.* **18**, 187.

Keh, R. S., and Leslie, W. C. (1963). *Mat. Sci. Res.* **1**, 208.

Kelly, P. M., and Nutting, J. (1960). *Proc. R. Soc. London Ser. A* **259**, 45.

Khachaturyan, A. G., and Onissimova, T. A. (1968). *Fiz. Met. Metallov.* **26**, 12.

Koval, Y., Titov, I., and Khandros, L. (1969). *Fiz. Met. Metallov.* **27**, 65.

Kurdjumov, G. V. (1960). *J. Iron. Steel Inst.* **195**, 26.

Kurdjumov, G. V., and Khachaturyan, A. G. (1972). *Metall. Trans.* **3**, 1069.

Kurdjumov, G. V., and Khachaturyan, A. G. (1975). *Acta Metall.* **23**, 1077.
Kurdjumov, G. V., Utyovskii, L. M., and Entin, R. I. M. (1977). "Transformations in Iron and Steel." "Nauka," Moscow (in Russian)
Kurdjumov, G. V., and Sachs, G. (1930). *Z. Phys.* **63**, 325.
Kuzmann, E., Bene, E., Domonkos, L., Hegedus, Z., Nagy, S., and Vertes, A. (1976). *J. Phys.* **37**, 66–409.
Kvashina, L. B., and Krivoglaz, M. A. (1967). *Fiz. Met. Metallov.* **23**, 3.
Lagunov, V. A., Polozerko, V. I., and Stepanov, V. A. (1971). *Sov. Phys. Solid State* **12**, 2480.
LeCaer, G., Simon, A., Lorenzo, A., and Genin, J. M. (1971). *Phys. Status Solidi (a)* **6**, K97.
LeCaer, G., Dubois, J. M., and Senateur, J. P. (1976). *J. Solid State Chem.* **19**, 19.
Leslie, W. C. (1961). *Acta Metall.* **9**, 1004.
Lesoille, M., and Gielen, P. M. (1972). *Met. Trans.* **3**, 2681.
Lewis, S. L., and Flinn, P. A. (1968). *Phys. Status Solidi* **26**, K51.
Lysak, L., and Vovk, Y. (1965). *Fiz. Met. Metallov.* **20**, 540.
Lysak, L., and Nikolin, B. (1966). *Fiz. Met. Metallov.* **22**, 730.
Lysak, L., Vovk, Y., and Polishchuk, Y. (1967). *Fiz. Met. Metallov.* **23**, 898.
Maksimov, Yu. W., Susdalev, I. P., Kushnerov, M. Ya., and Arentz, P. A. (1974a). *Fiz. Met. Metallov.* **37**, 268.
Maksimov, Yu. V., Suzdalev, I. P., Arents, R. A., and Loktev, S. M. (1974b). *Kinet. Catal.* **15**, 1144.
Marquis, H., Dube, A., and Letender, G. (1962). *Mem. Sci. Rev. Metall.* **59**, 119.
Martens, A. (1180). *Z. Deut. Ing.* **24**, 398.
Mathalone, Z., Ron, M., and Shechter, H. (1970). *Appl. Phys. Lett.* **17**, 32.
Mathalone, Z., Ron, M., and Niedzwiedz, S. (1971a). *J. Met. Sci.* **6**, 957. Published with permission of Chapman and Hall, London.
Mathalone, Z., Ron, M., Pipman, J., and Nadiv, S. (1971b). *J. Appl. Phys.* **42**, 687.
Mekata, M., Yoshimura, H., and Takaki, H. (1972). *J. Phys. Soc. Jpn.* **33**, 62.
Moon, K. (1963). *Trans. TMS-AIME* **227**, 1116.
Moriya, T., Ino, H., Fujita, F. E., and Maeda, Y. (1968). *J. Phys. Soc. Jpn.* **24**, 60.
Moriya, T., Sumitomo, Ino, H., Fujita, E., and Maeda, Y. (1973). *J. Phys. Soc. Jpn.* **35**, 1378.
Mössbauer R. L., (1958). *Z. Phys.* **151**, 124.
Nagakura, S. J. (1968). *J. Phys. Soc. Jpn.* **25**, 488.
Nagakura, S., and Tanehashi, K. (1968). *J. Phys. Soc. Jpn.* **25**, 840.
Narkhov, A. V., Makarov, V. A., and Kozlova, O. S. (1975). *Fiz. Met. Metallov.* **40**, 93.
Nishiyama, Z. (1934). *Sci. Rep. Tohoku* **23**, 638.
Nozik, A. J., Wood, J. C. Jr., and Haacke, G. (1969). *Solid State Commun.* **7**, 1677.
Okabe, T., and Guy, A. G. (1970). *Metall. Trans* **1**, 2705.
Okabe, T., and Guy, A. G. (1973). *Metall. Trans* **4**, 2673.
Oshima, R., and Wayman, C. M. (1974). *Scripta Metall.* **8**, 223.
Oshima, R., Azuma, H., and Fujita, F. E. (1976). *Proc. Int. Symp. Martens. Trans., 1st, Suppl. Trans JIM* **17**, 293.
Palatnik, L. S., and Boronin, S. W. (1966). *Fiz. Met. Metallov.* **21**, 217.
Paranjpe, V. G., Cohen, M., Bever, M. B., and Floe, C. F. (1950). *J. Met.* **2**, 261; *Trans. AIME* **188**, 261.
Pearson, W. B. (1958). "Handbook of Lattice Spacing and Structure of Metals and Alloys," pp. 984, 1425. Pergamon, Oxford.
Remy, P. H., and Pollak, H. (1965). *J. Appl. Phys.* **36**, 860.
Roberts, C. S. (1953). *Trans. AIME* **197**, 203.
Ron, M. (1973). The Application and Potential of Mössbauer Spectroscopy, Review, Symp. Proc. International Atomic Energy Agency, SM-159/32, Vienna.
Ron, M., and Mathalone, Z. (1971). *Phys. Rev. B* **4**, 774.

Ron, M., Niedzwiedz, S., Kidron, A., and Shechter, H. (1967). *J. Appl. Phys.* **38**, 590.

Ron, M., Shechter, H., and Niedzwiedz, S. (1968). *J. Appl. Phys.* **39**, 265.

Ruhl, R., and Cohen, M. (1969). *Trans. Metall. Soc. AIME* **245**, 241.

Sachdev, A. K. (1977). Ph.D. Thesis, MIT.

Sachdev, A. K., Cohen, M., and Vandersande, J. B. (1979). To be published.

Salomon, H. D., and Levinson, L. M. (1977). General Electric Rep. No. 76CRD295.

Schenck, H., Macken, M., Butenuth, G., and Potthast, E. (1966). "Untersuchugen über der Egistenzbereiche de Eisencarbide." Westdeutscherverag, Köln.

Schmidtmann, V. E., Hongardy, L. M., and Schenck, H. (1965). *Arch. Eisenhuttenwes.* **36**, 191.

Schwartz, L. H., (1976). *In* "Applications of Mössbauer Spectroscopy" (R. L. Cohen, ed.), Vol. 1, p. 37. Academic Press, New York.

Shiga, C., Kimura, M., and Fujita, F. (1974). *Jpn. Inst. Met.* **38**, 1037.

Shiga, C., Fujita, F., and Kimura, M. (1975). *Jpn. Inst. Met.* **39**, 1205.

Shinjo, T., Itoh, F., Takei, H., Nakayama, Y., and Shikazono, N. (1964). *J. Phys. Soc. Jpn.* **19**, 1252.

Shirane, G., Takei, W. J., and Ruby, S. (1962). *Phys. Rev.* **126**, 49.

Speich, G. R., and Lieslie, W. C. (1972). *Metall. Trans.* **3**, 1043.

Swartzendruber, L. J., Bennett, L. H., Schoefer, E. A., DeLong, W. T., and Campbell H. C. (1974). *J. Weld.* **53**, I-S.

Tschernov, D. K. (1868). *Russ. Tech. Soc.* **7**, 427.

Ujihira, Y., and Handa, A. (1978). *J. Phys.* **40**, C2-886.

Vol, A. E. (1967a). "Handbook of Binary Metallic Systems, Structure and Properties," Vol. II, p. 401. Israel Program for Scientific Translations, Jerusalem.

Vol, A. E. (1967b). "Handbook of Binary Metallic Systems, Structure and Properties," Vol. I, p. 558. Israel Program for Scientific Translations, Jerusalem.

Vyunnik, I. M. (1975). *Fiz. Met. Metallov.* **39**, 142.

Walker, L. R., Vertheim, G. K., and Jaccarino, V. (1961). *Phys. Rev. Lett.* **6**, 98.

Wayman, C. M. (1964). "Introduction to Crystallography of Martensitic Transformation," MacMillan, New York.

Wechsler, M. S., Lieberman, D. S., and Read, T. A. (1953). *Trans AIME* **197**, 1503.

Wertheim, G. K. (1964). "Mössbauer Effect: Principles and Applications." Academic Press, New York.

Westgren, A., and Phragmen, C. (1922). *J. Iron Steel Inst London* **105**, 241.

Wells, M. G. H., and Butler, J. F. (1966). *Trans. Am. Soc. Met.* **59**, 727.

Wever, F. (1924). *Z. Electrochem.* **30**, 376.

Wiener, C. W., and Berger, J. A. (1955), *J. Met.* **7**, 360.

Williamson, D. L., Nakazawa, K., and Krauss, G. (1979). *Met. Trans.*, **10A**, 1351.

Winchell, P. G., and Cohen, M. (1962). *Trans. Am. Soc. Met.* **55**, 347.

Wood, J. C. Jr., and Nozik, A. J. (1971). *Phys. Rev. B* **4**, 2224.

Yamakawa, K., and Fujita, F. E. (1978). *Met. Trans.* **9A**, 91.

Yurchikov, Ye. Ye., Menshikov, A. Z., Tsurin, V. A., and Kalashnikova, L. V. (1968). *Fiz. Metal. Metallov.* **26**, 108.

Diffusion in Solids and Liquids

Paul A. Flinn

Intel Corporation
Santa Clara, California

I. Introduction

Mössbauer spectroscopy can be used to observe diffusive effects in solids under three different sets of circumstances: (1) when the diffusing atom itself has a Mössbauer resonant nucleus, (2) when a Mössbauer active atom, itself quasi-stationary, experiences a changing hyperfine interaction as a result of the motion of a more rapidly diffusing species in the material, and (3) when the gamma radiation is scattered by some material, not necessarily Mössbauer active, and the Mössbauer effect is used for energy analysis of the scattered radiation. The qualitative features of these various situations can readily be understood on the basis of simple approximate arguments. In the first case, if the atom containing the resonant nucleus moves rapidly enough to travel a distance that is large relative to the gamma-ray wavelength in a time that is short relative to the nuclear lifetime, we may expect a line broadening since the coherence of the gamma ray is destroyed by the motion. It is as if the lifetime of the state were reduced to approximately the time required for the atom to travel a distance of one wavelength. In the case of diffusion in crystalline solids, where the jump distance is normally quite large relative to the gamma-ray wavelength, a single jump is sufficient to destroy coherence

APPLICATIONS OF MÖSSBAUER
SPECTROSCOPY, VOL. II

almost completely; if the time between jumps is short relative to the Mössbauer lifetime, the linewidth is increased over the natural value by the ratio of the natural lifetime to the time between jumps. For example, for the case of ^{57}Fe in an iron–silicon alloy, the jump distance is about 2.5 Å, while the gamma-ray wavelength is 0.86 Å. At 1350 K the time between jumps for the iron atoms is about 3.6×10^{-8} s, and the mean lifetime of the 14-keV level is 140×10^{-9} s, so that we expect a line broadened by about a factor of 4 over the natural linewidth. Broadening of roughly this order is observed; we will discuss the detailed theory and quantitative results in Section II.

If some hyperfine interaction, such as an electric field gradient acting on a nuclear electric quadrupole moment, varies in time as a result of diffusion, the Mössbauer spectrum may be changed in shape. If the diffusion is sufficiently rapid, so that the EFG or other perturbation changes at a rate much higher than the Larmor frequency associated with the interaction, only the average value of the perturbation is observed. Such time-dependent hyperfine effects may occur either as a result of diffusive motion of the Mössbauer atom, where the line broadening effect discussed above must also be considered, or as a result of some other motion, such as that of a rapidly diffusing point defect. For example, carbon in solid solution in iron produces a quadrupole splitting of the levels of the neighboring nuclei of 0.64 mm/s; the corresponding Larmor frequency is 7.4 MHz. At 895°C, the jump rate of the carbon atoms is $\sim 10^{9}$ s^{-1}; consequently, no quadrupole splitting is expected or observed. This is analogous to diffusional narrowing in NMR (nuclear magnetic resonance). Again, details will be discussed in Section III.

We see then that the range of time scales over which the Mössbauer effect can be used to observe diffusion is determined by the characteristic times associated with the resonance: the natural lifetime and the Larmor precession times of the hyperfine interactions. An additional restriction is imposed by the need for a reasonably large resonant fraction; since rapid diffusive motion normally occurs only at elevated temperatures, it is necessary to use relatively low energy resonances for which the resonant fraction is still large at moderately high temperatures. The choice of available isotopes is, therefore, quite limited; almost all of the work has been done with ^{57}Fe, some with ^{119}Sn, and a little with ^{181}Ta.

Experimental observations of these effects of diffusive motion on Mössbauer spectra were reported quite early in the history of Mössbauer spectroscopy: for diffusion in liquids, the broadening of the resonance for iron dissolved in glycerol was observed by Bunbury et al. in 1963, and for diffusion in solids, the case of iron in copper and gold was reported by Knauer and Mullen in 1968. An example of the relaxation of the hyperfine interaction due to diffusive motion was seen for iron dissolved in silver chloride: the quadrupole splitting arising from the EFG due to a vacancy associated with the iron collapsed when the motion of the vacancy became sufficiently rapid (Lindley and Debrunner, 1966).

II. General Theory: Line Broadening

The general theory for the effect of the motion of the resonant atom on the Mössbauer lineshape is similar to that for neutron diffraction; an early treatment based on the Van Hove space–time correlation function was given by Singwi and Sjolander (1960); the analysis has since been extended by many authors, e.g., Boyle and Hall (1962) and Karyagin (1974, 1975).

Since the usual Mössbauer experiments involve only one resonant atom interacting at a time, we need consider only the self-correlation function $G_s(\mathbf{r}, t)$. In the classical limit this function has a simple interpretation: if we choose our origin of coordinates at the location of the particle of interest at time $t = 0$, then $G_s(\mathbf{r}, t)$ is simply the probability that the particle will be at position $\mathbf{r}$ at a later time t. In the quantum case, such a simple description violates the uncertainty principle, and a more complicated definition is needed:

$$G_s(\mathbf{r}, t) = \left\langle \int \delta\{\mathbf{r} - \mathbf{R}(0) - \mathbf{r}^1\}\delta\{\mathbf{r}^1 - \mathbf{R}(t)\} \, dr^1 \right\rangle, \qquad (1)$$

where $\mathbf{R}(t)$ is the Heisenberg operator

$$\mathbf{R}(t) = \exp[(iH/\hbar)t]\mathbf{R}\exp[(-iH/\hbar)t]. \qquad (2)$$

The Mössbauer line shape is given by the Fourier transform with respect to space and time of $G(\mathbf{r}, t)$:

$$\sigma(\omega) = \frac{\Gamma}{4\hbar} \int \exp\left(\frac{-\Gamma}{2\hbar}t\right)\exp(-i\omega t)\exp(i\mathbf{k}\cdot\mathbf{r})G(\mathbf{r}, t)d\mathbf{r}\, dt, \qquad (3)$$

where $\hbar\omega = E - E_0$, κ is the propagation vector of the gamma ray ($|\kappa| = 2\pi/\lambda$), and Γ is the linewidth.

From a semiclassical viewpoint, the interpretation of this equation can be seen as follows: we can view the active nucleus as a transmitter (or receiver) with a resonant energy E_0 and a sharpness of tuning characterized by the exponential decay factor $\exp(-\Gamma t/2\hbar)$. We choose the origin of space coordinates at the position of the active nucleus at $t = 0$. If the nucleus is displaced by a vector $\mathbf{r}$ at time t, there are two phase shifts: $\exp(-i\omega t)$, due to the difference between the resonant frequency and the observed frequency, and $\exp(i\kappa\cdot\mathbf{r})$, due to the spatial displacement of the nucleus. $G(\mathbf{r}, t)$ provides the proper weighting factor when we sum over space and time and average over all nuclei.

It is sometimes convenient to introduce the complex variable $p = -i\omega + \Gamma/2$ and write Eq. (3) in the form

$$F(p) = \text{Re} \int_{t=0}^{\infty} \exp(-pt) \, dt \int \exp(i\boldsymbol{\kappa} \cdot \mathbf{r}) G(\mathbf{r}, t) \, d\mathbf{r}. \tag{3a}$$

Often it is useful to consider an intermediate step in the transformation; after carrying out the integration over space but not time, we have the "intermediate scattering function":

$$F_s(\boldsymbol{\kappa}, t) = \int \exp(i\boldsymbol{\kappa} \cdot \mathbf{r}) G(\mathbf{r}, t) \, d\mathbf{r}, \tag{4}$$

so that

$$F(p) = \text{Re} \int_0^{\infty} \exp(-pt) F_s(\boldsymbol{\kappa}, t) \, dt. \tag{5}$$

In order to obtain a simple explicit expression for the line shape, Singwi and Sjolander made several approximations to the form of $G(\mathbf{r}, t)$: they assumed the diffusive jumps to be long relative to the gamma-ray wavelength, the jump directions to be distributed isotropically, so that it was only necessary to consider the magnitude of $\mathbf{r}$, the directions of successive jumps to be uncorrelated, and the vibrational motions of the nuclei to be uncorrelated with their diffusive motions. With these approximations, the evaluation of Eq. (3a) is straightforward.

$G_d(\mathbf{r}, t)$ is governed by the following relatively simple integrodifferential equation:

$$dG(\mathbf{r}, t)/dt = -\nu G(\mathbf{r}, t) + \nu \int p(\mathbf{r}_1) G(\mathbf{r} - \mathbf{r}_1, t) \, d\mathbf{r}_1, \tag{6}$$

where ν is the jump rate and $P(\mathbf{r}_1)$ is the probability that a given jump will correspond to a translation $\mathbf{r}_1$. The first term on the right-hand side of Eq. (6) is the decrease in $G(\mathbf{r}, t)$ due to jumps away from $\mathbf{r}$; the second term is the decrease in $G(\mathbf{r}, t)$ due to jumps to the point $\mathbf{r}$ from elsewhere. The product $P(\mathbf{r}_1) G(\mathbf{r} - \mathbf{r}_1)$ is simply the combined probability that an atom is at the point $(\mathbf{r} - \mathbf{r}_1)$ and the jump translation will be $\mathbf{r}_1$, so that after the jump the atom will be at $\mathbf{r}$.

The convolution in the second term can be converted to a product by Fourier transformation with respect to space; we thus obtain

$$(d/dt)F(\boldsymbol{\kappa}, t) = -\nu F(\boldsymbol{\kappa}, t) + \nu p(\boldsymbol{\kappa}) F(\boldsymbol{\kappa}, t), \tag{7}$$

where

$$p(\boldsymbol{\kappa}) = \int e^{i\boldsymbol{\kappa} \cdot \mathbf{r}} p(\mathbf{r}) \, d\mathbf{r}. \tag{8}$$

This can easily be integrated to give

$$F(\kappa, t) = \exp(-\nu\{1 - p(\kappa)\}t). \tag{9}$$

If we now insert Eq. (9) in Eq. (5), we observe that the two time-dependent exponents combine to give a term $\exp\{-(\Gamma + \nu(1 - p\{\kappa\})t\}$; the line shape is still Lorentzian (*for a specific* κ), but the width is increased by an amount $\nu\{1 - p(\kappa)\}$, just as if the lifetime had been correspondingly reduced.

If we take $P(\mathbf{r})$ as isotropic and of fixed length

$$P(\mathbf{r}) = \delta(r_0), \tag{10}$$

then

$$p(\kappa) = \sin(\kappa r_0)/\kappa r_0 \tag{11}$$

and the broadening is given by

$$\nu\left(1 - \frac{\sin \kappa r_0}{\kappa r_0}\right) \simeq \nu \tag{12}$$

since $\kappa r_0 \gg 1$ for typical jump distances in crystals.

As we shall see, this result is a reasonable first approximation for the case of diffusion in crystalline solids, but a greatly improved result can be obtained by the use of a more realistic model. In a single crystal, the directions in which an atom can jump are not, of course, distributed isotropically; they are restricted to a small, discrete set by the requirements of the crystal structure. The integral in Eq. (6) is replaced by a sum over nearest-neighbor translations; thus we have

$$\frac{dG(\mathbf{r}, t)}{dt} = -\nu G(\mathbf{r}, t) + \frac{\nu}{n} \sum_{\mathbf{l}} G(\mathbf{r} - \mathbf{l}, t) \tag{13}$$

or

$$\frac{dG(\mathbf{r}, t)}{dt} = \frac{\nu}{n}\left(\sum_{\mathbf{l}} \hat{g}(\mathbf{l}) - n\right) G(\mathbf{r}, t), \tag{14}$$

where $\hat{g}(\mathbf{l})$ is the displacement operator for a nearest-neighbor translation $\mathbf{l}$. The values of $p(\kappa)$ have been worked out for the following simple structures (Chudley and Elliot, 1961):

$$\text{sc } p(\kappa) = \tfrac{1}{3}\{\cos(\kappa_x a) + \cos(\kappa_y a) + \cos(\kappa_z a)\}, \tag{15a}$$

$$\text{bcc } p(\kappa) = \cos\left(\tfrac{1}{2}\kappa_x a\right)\cos\left(\tfrac{1}{2}\kappa_y a\right)\cos\left(\tfrac{1}{2}\kappa_z a\right) \tag{15b}$$

$$\text{fcc } p(\kappa) = \tfrac{1}{3}\left\{\cos\left(\tfrac{1}{2}\kappa_x a\right)\cos\left(\tfrac{1}{2}\kappa_y a\right) + \cos\left(\tfrac{1}{2}\kappa_y a\right)\cos\left(\tfrac{1}{2}\kappa_z a\right)\right.$$
$$\left. + \cos\left(\tfrac{1}{2}\kappa_z a\right)\cos\left(\tfrac{1}{2}\kappa_x a\right)\right\}, \tag{15c}$$

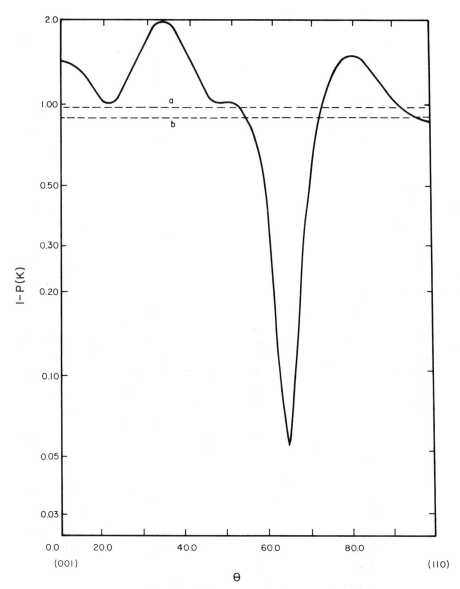

Fig. 1. Anisotropy of line broadening in the body-centered-cubic case. The function $1 - p(\kappa)$, evaluated along the $\{1\bar{1}0\}$ zone, according to Eq. (15b), with $\kappa a = 21.1$. Line a corresponds to the isotropic approximation [Eq. (11)], and line b corresponds to the broadening for a randomly oriented polycrystalline aggregate [Eq. (16)].

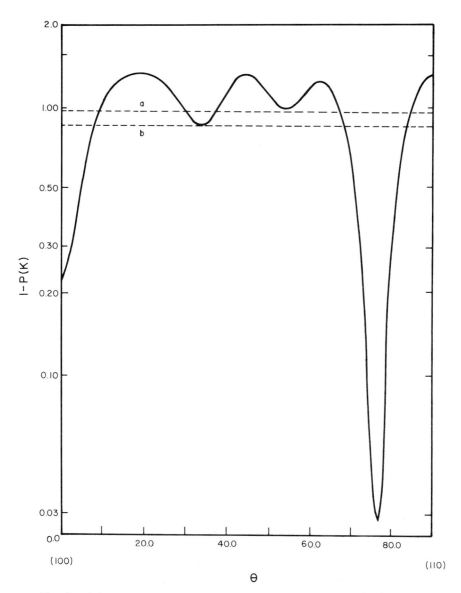

Fig. 2. Anisotropy of line broadening in the face-centered-cubic case. The function $1 - p(\mathbf{\kappa})$, evaluated along the $\{1\bar{1}0\}$ zone, according to Eq. (15c), with $\kappa a = 26.8$. Line a corresponds to the isotropic approximation [Eq. (11)], and line b corresponds to the broadening for a randomly oriented polycrystalline aggregate [Eq. (16)].

where κ_x, κ_y, and κ_z are the components of κ referred to the crystal axes and a is the lattice parameter.

This anisotropy is not a small effect. The anisotropy factor $1 - p(\kappa)$ can vary by more than a factor of 10 in a modest angular range, as may be seen in Figs. 1 and 2. This rapid angular variation is a consequence of the large value of the jump distance relative to the wavelength and the correspondingly large value of the product κa.

This explicit dependence on κ is, of course, simply a consequence of the fact that only the component of motion of the atom along the gamma-ray direction has any effect on the phase of the radiation; the component of motion normal to the gamma-ray direction produces no phase shift and, therefore, does not contribute to the broadening. Unfortunately, for reasons which will be discussed later, direct observation of this κ dependence for the single crystal case is extremely difficult, and most observations are carried out on polycrystalline aggregates of more or less random orientations. To obtain the predicted line shape for such a case, we must average the appropriate case of Eq. (15) over all orientations, so that

$$\sigma(\omega) \alpha \int_{4\pi} \frac{\{1 - p(\kappa)\}}{\omega^2 + \nu^2 \{1 - p(\kappa)\}^2} d\Omega. \tag{16}$$

We note several features of this result: it is not equivalent to the assumption of isotropic jumps, the predicted line shape is not Lorentzian, and the linewidth is somewhat less than that given by the Singwi–Sjolander approximation (Eq. (12)). The effect on line shape is shown in Figs. 3 and 4.

Howe and Morgan (1976) have carried out a detailed analysis for the more complicated case of the $NiAs$–CdI_2 structure (that of iron telluride and selenide). The lower symmetry results in highly non-Lorentzian line shapes for this case.

A more difficult approximation to remove concerns the correlation of successive jumps. Diffusion in solids normally occurs by the vacancy mechanism; a given atom can move only into a vacant site. There is a substantial probability that the next jump of the same atom will involve the same vacancy and a jump direction clearly related to the direction of the original jump. For example, it is quite likely that a second interchange of position between the particular atom and the vacancy will occur, so that the second jump vector is the negative of the first. This effect is well known in the theory of bulk diffusion, and the appropriate correction, the Bardeen–Herring correlation factor, has been calculated to high accuracy for many cases. These results are, unfortunately, inadequate for the Mössbauer line shape since they concern only the long time average of the

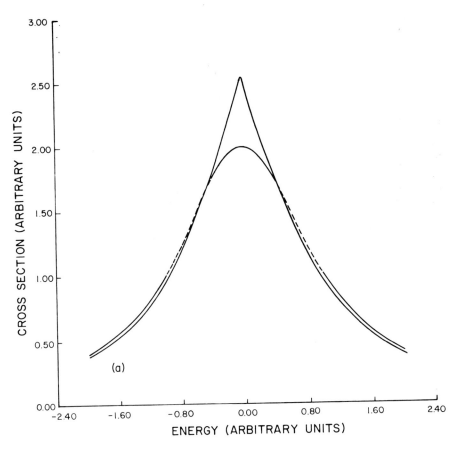

CROSS SECTION (ARBITRARY UNITS)

(a)

ENERGY (ARBITRARY UNITS)

Fig. 3a. Absorption line shapes as calculated for polycrystalline cubic materials by Dibar Ure and Flinn (1973), according to Eq. (16). The dotted curves show the isotropic approximation for comparison. Face-centered-cubic case, $\kappa a = 30.1$ as for Fe in Au.

correlation, while we need more detailed information about the correlation as a function of time.

This problem has been treated by several workers, both by analytical approximation (Krivoglaz and Repetskii, 1971; Wolf, 1977; Bender and Schroeder, 1979) and by direct numeral calculation (Dibar Ure and Flinn, 1977). Equation (14) must be replaced by an equation governing the joint probability distribution of active atom and vacancy $P(\mathbf{r}_a, \mathbf{r}_v, t)$.

$$(d/dt)P(\mathbf{r}_a, \mathbf{r}_v, t) = \nu\hat{\theta}P(\mathbf{r}_a, \mathbf{r}_v, t), \tag{17}$$

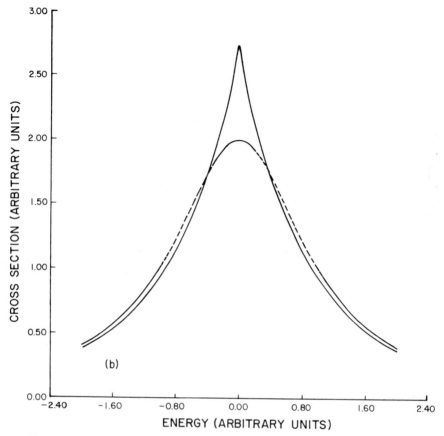

Fig. 3b. Absorption line shapes as calculated for polycrystalline cubic materials by Dibar Ure and Flinn (1973), according to Eq. (16). The dotted curves show the isotropic approximation for comparison. Face-centered-cubic case, $\kappa a = 26.8$ as for Fe in Cu.

where

$$\hat{\theta} = \left\{ 1 - \delta(\mathbf{r}_a, \mathbf{r}_v) - \sum_{\mathbf{l}} \delta(\mathbf{r}_v - \mathbf{r}_a + \mathbf{l}) \right\} \left\{ \sum_{\mathbf{l}} \frac{\hat{g}_v(\mathbf{l})}{n} - 1 \right\}$$
$$+ \left\{ \sum_{\mathbf{l}} \delta(\mathbf{r}_v - \mathbf{r}_a + \mathbf{l}) \right\} \left\{ \frac{\hat{g}_v(-\mathbf{l})\hat{g}_a(\mathbf{l})}{n} + \sum_{\mathbf{l}^1} \frac{\hat{g}_v(\mathbf{l}^1)}{n} - 1 \right\}. \quad (18)$$

After solving this equation by numerical methods for a finite lattice, we can obtain $G(\mathbf{r}, t)$ by summing over all vacancy positions:

$$G(\mathbf{r}_a, t) = \frac{1}{N_1} \sum_{\mathbf{r}_v} P(\mathbf{r}_a, \mathbf{r}_v, t), \quad (19)$$

where N_1 is the number of lattice points.

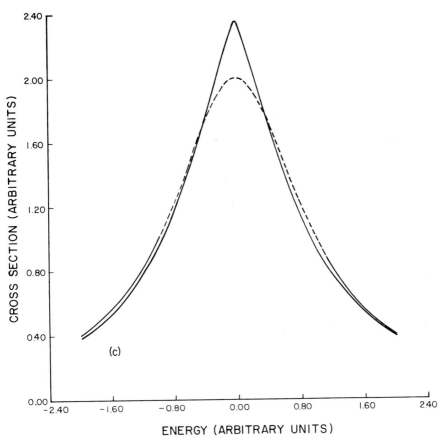

Fig. 3c. Absorption line shapes as calculated for polycrystalline cubic materials by Dibar Ure and Flinn (1973), according to Eq. (16). The dotted curves show the isotropic approximation for comparison. Body-centered-cubic case as for Fe + 3% Si.

Some results of these calculations are shown in Figs. 5–7. Bender and Schroeder (1979), using an iterative analytic technique, have recently obtained quite similar results. We see that the effect is rather similar to that for the bulk diffusion correction: just as the rate of bulk transport is reduced by the correlation factor, the line broadening is less than that given by the approximation of uncorrelated motion. Again, we see that a non-Lorentzian line shape appears when we leave the simple theory. The quantitative effect on the linewidth is quite close to that of the correlation factor for bulk diffusion, so that it can be used as a reasonable approximation if detailed calculations are not available.

The case of diffusion in liquids presents somewhat different problems with approximations: most liquids provide the simplification of isotropy,

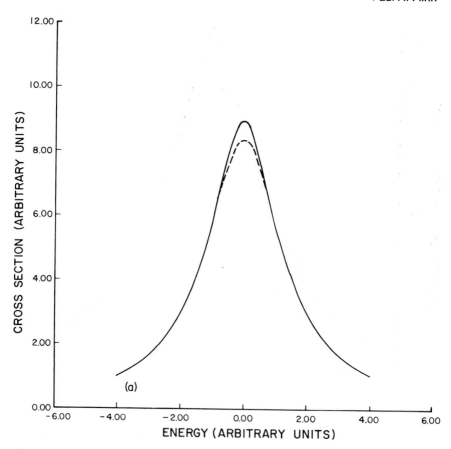

Fig. 4a. Mössbauer spectra for polycrystalline cubic materials as calculated by Dibar Ure and Flinn (1973). The broadening functions of Fig. 3, based on Eq. (16), have been convoluted with Lorentzians of natural width to show the effort of combined source and absorber broadening. The dotted curves show the isotropic approximation. Face-centered-cubic case, $\kappa a = 30.1$ as for Fe in Au.

and we do not expect the complications of the correlations associated with the vacancy mechanism, but we lack a well-established and generally accepted model for the details of diffusion in liquids. Many early treatments of diffusion in liquids were based on a quasi-solid model in which diffusive jumps were large; some Mössbauer data were interpreted in terms of such a picture. Such a model, however, is inconsistent with molecular dynamics simulations of liquids, in which quasi-continuous, rather than jump, motion is seen (Rahman, 1971). Since these simulations are quite successful in predicting both equations of state and transport properties for liquids, they provide a highly credible picture for liquids.

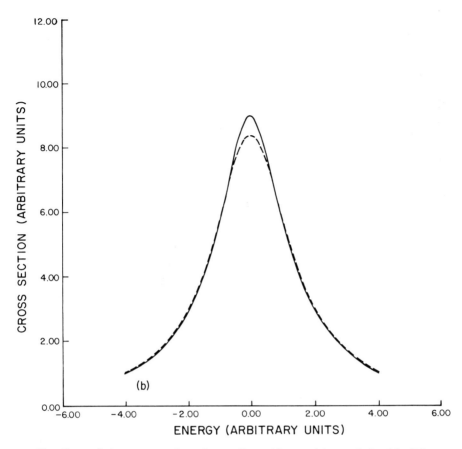

Fig. 4b. Mössbauer spectra for polycrystalline cubic materials as calculated by Dibar Ure and Flinn (1973). The broadening functions of Fig. 3, based on Eq. (16), have been convoluted with Lorentzians of natural width to show the effort of combined source and absorber broadening. The dotted curves show the isotropic approximation. Face-centered-cubic case, $\kappa a = 26.8$ as for Fe in Cu.

It seems appropriate then (Flinn, 1977) to construct an approximate theory for the liquid case based on quasi-continuous, rather than large jump, motion. We can take advantage of the simplicity of the classical limit since, under normal circumstances, the mean-square amplitude of motion is large relative to the zero point motion.

Molecular dynamics calculations have shown that the self-correlation function for a liquid is given, except for times less than about 10 ps, reasonably well by the "Gaussian approximation":

$$G_s(\mathbf{r}, t) = \{2\pi w(t)\}^{-3/2} \exp\{-r^2/2w(t)\} \qquad (20)$$

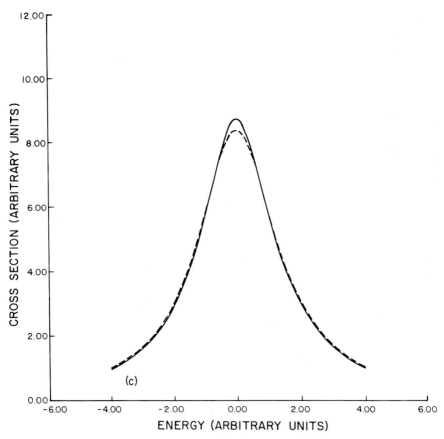

Fig. 4c. Mössbauer spectra for polycrystalline cubic materials as calculated by Dibar Ure and Flinn (1973). The broadening functions of Fig. 3, based on Eq. (16), have been convoluted with Lorentzians of natural width to show the effort of combined source and absorber broadening. The dotted curves show the isotropic approximation. Body-centered-cubic case as for Fe + 3% Si.

so that

$$F_s(\kappa, t) = \exp\{-\kappa^2 w(t)/2\}, \qquad (21)$$

with $w(t)$ being the component along the gamma-ray direction of the mean-square displacement of the active atom. Our problem now is to obtain an expression for $w(t)$ in terms of the spectral density function $f(\omega)$, which characterizes the overall dynamic behavior of the liquid. Its Fourier transform is the velocity autocorrelation function

$$\{v_\kappa(0)v_\kappa(t)\} = (kT/M)\int_0^\infty f(\omega)\cos(\omega t)\, d\omega, \qquad (22)$$

where $v_\kappa(t)$ is the component of velocity along the gamma-ray direction at

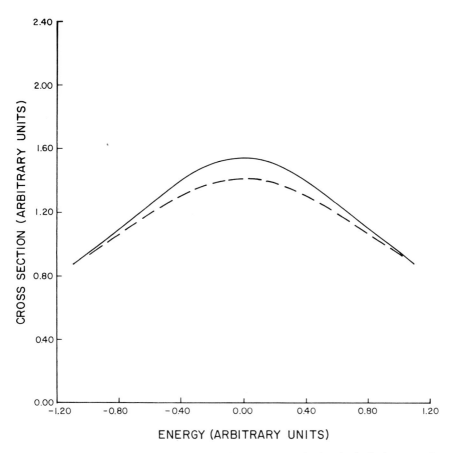

Fig. 5. Broadening due to diffusion by the vacancy mechanism in the body-centered-cubic case as calculated by Dibar Ure and Flinn (1977) for observation in the [710] direction. The dashed curve shows the result for the empty lattice approximation (scaled by the Bardeen–Herring correlation factor f). This is one of the directions where the correction to the scaled empty lattice approximation is largest (about 10%).

time t. It has been shown (Egelstaff, 1967) that $w(t)$ is given by

$$w(t) = 2\int_0^t (t - \tau)\{v_\kappa(0)v_\kappa(\tau)\} \, d\tau. \tag{23}$$

After substituting (22) in (23) and carrying out the integration over τ, we have

$$w(t) = \frac{2kT}{M} \int_0^\infty \left\{ \frac{1 - \cos(\omega t)}{\omega^2} \right\} f(\omega) \, d\omega. \tag{24}$$

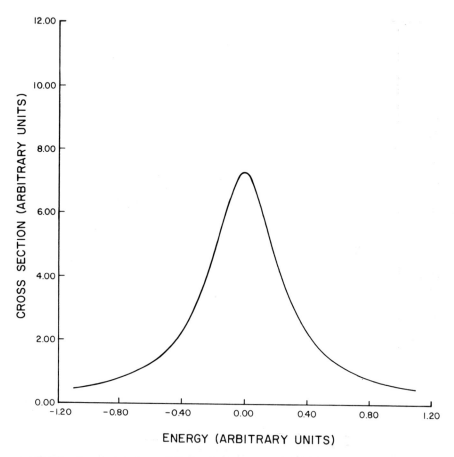

Fig. 6. Broadening due to diffusion by the vacancy mechanism in the body-centered-cubic case as calculated by Dibar Ure and Flinn (1977) for observation in the [0, 3, 1] direction, averaged over a 5° solid angle about this direction. In this direction the difference between this calculation and the scaled empty lattice approximation is too small to see on the graph.

We now introduce an explicit, approximate form for $f(\omega)$:

$$f(\omega) \simeq \frac{2MD}{kT\pi} + b|\omega| \exp(-|\omega|/\omega_1) + f_2(\omega), \qquad (25)$$

where D is the diffusion coefficient for the liquid and ω_1 is a cutoff frequency.

The constant term is required by the condition that

$$f(0) = \frac{2MD}{kT\pi}. \qquad (26)$$

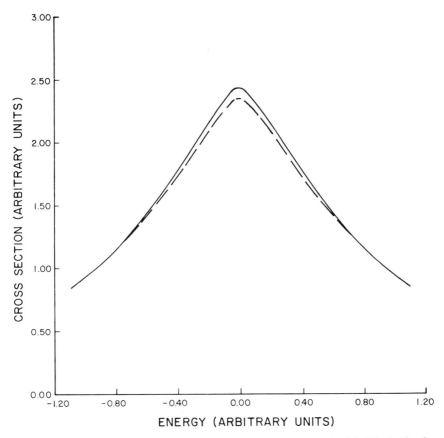

Fig. 7. Line shape for a randomly oriented polycrystalline sample with diffusion by the vacancy mechanism as calculated by Dibar Ure and Flinn (1977). The dashed curve shows the result for the scaled empty lattice approximation. The widths differ by about 5%.

The second term is introduced on the basis of Raman scattering measurements, which indicate a quasi-linear behavior of $f(\omega)$ for small ω. The third term $f_2(\omega)$ contains the bulk of the power spectrum; it must behave as ω^2 for small ω and vanish suitably at large ω; it is otherwise unrestricted. Next, we introduce (25) into (24) and integrate to obtain

$$w(t) = 2Dt + (2kT/M)\left\{b\ln\left(1 + \omega_1^2 t^2\right) + \langle\omega^{-2}\rangle\right\} \qquad (27)$$

as an approximation for all times longer than a few picoseconds.

For moderately short times, $\omega(t)$ has an approximately constant value $2kT\langle\omega^{-2}\rangle/M$, which is similar to the result for a crystalline solid; at long times the diffusion term $2Dt$ dominates; in the transition region the time dependence is logarithmic.

By using (5) and (21), we can now calculate the line shape

$$F(p) = \exp\left(-\frac{kT}{M}\langle\omega^{-2}\rangle\kappa^2\right)\text{Re}\left\{\int_0^\infty (1 + \omega_1 t^2)^{-B}e^{-p_1 t}\,dt\right\}, \quad (28)$$

where $p_1 = p + \kappa^2 D$ and $B = \kappa^2 kTb/M$. By expanding the integrand in power series, we obtain the approximate result

$$\sigma(\omega)\alpha\exp\left\{-\left(\frac{\kappa^2 kT}{M}\langle\omega^{-2}\rangle + B\ln\{1 + \omega_1^2 t_c^2\}\right)\right\}$$

$$\times \left\{\frac{\gamma}{\gamma^2 + \omega_M^2} + \frac{2B\gamma\omega^2 M}{\gamma^2 + \omega_M^2}\right\}, \quad (29)$$

where $\gamma = \Gamma/2 + \kappa^2 D$ and $t_c = \gamma^{-1}$.

The various parts of this expression have reasonably simple physical interpretations: the first term in the exponent is the usual vibrational contribution to the Debye–Waller factor; the second term in the exponent arises from the additional, characteristically liquid, motion associated with the quasi-linear term in the spectral density; the leading frequency-dependent term gives the usual Lorentz shape, with a width determined by the sum of the natural width and the diffusional broadening. The second frequency-dependent term, arising from the quasi-linear part of the spectral density, produces a deviation from the Lorentz shape; but it is numerically small and will be difficult to observe. We note that the diffusional broadening is given by $\kappa^2 D$, a result quite different from that of the large jump case of the quasi-crystalline model for liquids, where the corresponding term is $6D/\overline{x^2}$, where $\overline{x^2}$ is the mean-square jump length.

If the liquid undergoes a glass transition on cooling, we expect B to vanish below the glass-transition temperature. The line shape then becomes the same as that for a crystalline solid.

III. General Theory: Relaxation Effects

The theory for relaxation effects is somewhat simpler in the case that the Mössbauer active atom is quasi-stationary, and the changes in the hyperfine interaction are due to the motion of a more rapidly diffusing species. Such a situation occurs, for example, when a metallic material contains an interstitial solute. An interstitial solute, such as carbon or nitrogen in an iron-based alloy, moves about far more rapidly than the iron atoms since no vacancies are needed. The concentration of interstitial atoms is so low that virtually all interstitial sites are vacant; only the activation energy for motion is needed. Fortunately, the formal analysis for this case is sufficiently general to include other situations, such as relaxation due to a moving vacancy.

In principle, we must consider three types of hyperfine interactions: monopole (isomer shift), magnetic dipole, and electric quadrupole. An interstitial neighbor to a Mössbauer active atom will, in general, change the isomer shift and the electric quadrupole interaction. In particular, if the host material provides sites of cubic symmetry for the host atoms, the presence of the interstitial will destroy the cubic symmetry and produce a quadrupole splitting. If the material is magnetic, the presence of the interstitial may be expected to alter the magnetic hyperfine interaction; however, there do not appear to be any experimentally accessible cases for observation of this effect. Alloys that provide sufficient solubility for interstitials have Curie temperatures below the temperature where diffusion of interstitials occurs at a reasonable rate. We shall, therefore, consider in detail only the effects associated with monopole and electric quadrupole interactions.

An early treatment of these effects was given by Krivoglaz and Repetskii in 1966; several workers have later carried out more extensive analyses. Dattagupta (1976a) has given a detailed analysis, based on the work of Blume (1968), for the case of interstitials in a cubic metal (C or N in austenite). Since the austenite is not magnetic, the interactions treated are the monopole, taken as H_1 when the active atom does not have an interstitial as a nearest neighbor, and H_0 when it does have an interstitial nearest neighbor, and an electric quadrupole interaction, with a coupling constant q, present only when the interstitial nearest neighbor is present. When no neighboring interstitial is present, the interaction is simply H_1; when an interstitial is present, the interaction is H_0 plus a quadrupole term, which depends on the orientation of the interstitial. In this structure, if we take the active atom at the origin of coordinates, the possible interstitial sites lie in the x, y, and z directions. The corresponding values for the interactions are

$$
\begin{aligned}
\text{no interstitial nearest neighbor} \quad & H = H_0, \\
\text{interstitial in } z \text{ direction} \quad & H = H_0 + q\{3I_z^2 - I(I+1)\}, \\
\text{interstitial in } x \text{ direction} \quad & H = H_0 + q\{3I_x^2 - I(I+1)\}, \\
\text{interstitial in } y \text{ direction} \quad & H = H_0 + q\{3I_y^2 - I(I+1)\}.
\end{aligned} \tag{30}
$$

We now define a stochastic function $f(t)$, which is chosen to jump among the values 0, 1, 2, and -2 at random instants of time. No interstitial neighbor corresponds to $f = 0$, an interstitial in an X site to $f = 1$, and so on. The time-dependent interaction can then be written

$$H(t) = \tfrac{1}{4}(1 - f)(4 - f^2)H_1 + \tfrac{1}{3}f(4 - f^2)\{H_0 + q(3I_z^2 - I(I + 1))\}$$
$$+ \tfrac{1}{24}f(f + 2)(f^2 - 1)\{H_0 + q(3I_x^2 - I(I + 1))\}$$
$$+ \tfrac{1}{24}f(f - 2)(f^2 - 1)\{H_0 + q(3I_y^2 - I(I + 1))\}. \tag{31}$$

The Mössbauer line shape is given by

$$F(p) = \text{Re} \int_0^\infty dt \exp(-pt) \sum_{m_1 m_0 m_1' m_0'} \langle \tfrac{3}{2}m_1|A^+|\tfrac{1}{2}m_0\rangle\langle\tfrac{1}{2}m_0'|A|\tfrac{3}{2}m_1'\rangle$$

$$\times \left(\langle\tfrac{1}{2}m_0|\exp\left\{i\int_0^\tau H(t')\,dt'\right\}|\tfrac{1}{2}m_0'\rangle\right.$$

$$\times\langle\tfrac{3}{2}m_1'|\exp\left\{-i\int_0^\tau H(t')\,dt\right\}|\tfrac{3}{2}m_1\rangle\Big)_{\text{Av}}, \tag{32}$$

where A is the interaction between the nucleus and the electromagnetic field associated with the γ ray.

By using the results of Blume, this can be reduced to the form

$$F(p) = \text{Re} \sum_{m_1 m_0 m_1'} \langle\tfrac{3}{2}m_1|A^+|\tfrac{1}{2}m_0\rangle\langle\tfrac{1}{2}m_0|A|\tfrac{3}{2}m_1'\rangle$$

$$\times \langle\tfrac{3}{2}m_1'|\left\{\sum_{ab} P_a(a|\{p - W - i\omega_0 F_1 - iV_2 F_2 - iV_3 F_3\}^{-1}|b)\right\}|\tfrac{3}{2}m_1\rangle, \tag{33}$$

where the indices a and b specify the four different states corresponding to the values 0, 1, 2, and -2 for $f(t)$, P_a is the *a priori* probability of occurrence of the state a, V_2 and V_3 are given by

$$V_2 = q\{3I_z^2 - I(I + 1)\}, \qquad V_3 = q\{I_+^2 - I_-^2\},$$

$F_1, F_2, F_3,$ and W are matrices given by

$$F_1 = \begin{bmatrix} 1 & 0 & 0 & 0 \\ 0 & 0 & 0 & 0 \\ 0 & 0 & 0 & 0 \\ 0 & 0 & 0 & 0 \end{bmatrix}, \quad F_2 = \begin{bmatrix} 0 & 0 & 0 & 0 \\ 0 & 1 & 0 & 0 \\ 0 & 0 & -\tfrac{1}{2} & 0 \\ 0 & 0 & 0 & -\tfrac{1}{2} \end{bmatrix}, \quad F_3 = \begin{bmatrix} 0 & 0 & 0 & 0 \\ 0 & 0 & 0 & 0 \\ 0 & 0 & \tfrac{3}{4} & 0 \\ 0 & 0 & 0 & -\tfrac{3}{4} \end{bmatrix},$$

$$W = \begin{bmatrix} -3\lambda'' & \lambda'' & \lambda'' & \lambda'' \\ \lambda' & -2(\lambda + \lambda') & -(2\lambda + \lambda') & \lambda \\ \lambda' & \lambda & \lambda & \lambda \\ \lambda' & \lambda & \lambda & -(2\lambda + \lambda') \end{bmatrix}, \tag{34}$$

and the λs are the jump rates (probabilities per unit time for jumping): λ is the rate at which jumps occur between two nonequivalent nearest-neighbor sites; λ' is the rate at which jumps occur from nearest-neighbor sites away from the active atom to non-nearest-neighbor sites; and λ'' is the rate at which jumps occur into nearest-neighbor sites from non-nearest-neighbor sites.

Evaluation of the stochastic averages requires determination of the relative jump probabilities that appear in W. If an interstitial is in a nearest-neighbor site, it can jump to any of 12 neighboring sites, only four of which are also still nearest-neighbor sites of the active atom, and only two of these produce a new quadrupole interaction. Thus if the jump frequency is ν, we have

$$\lambda = \nu/6 \tag{35}$$

and

$$\lambda' = 8\nu/12 = 4\lambda. \tag{36}$$

To find λ'', we must take into account the fact that jumps into nearest-neighbor sites can only come from adjacent sites; the probability that such a site is occupied is x, the concentration of the interstitial atoms. Since there are eight such sites and two out of 12 possible jumps from such sites into the nearest-neighbor sites, we have

$$\lambda'' = x \times 8 \times \tfrac{2}{12}\nu = 8x\lambda. \tag{37}$$

After rather extensive algebra, Dattagupta obtains the final result for line shape:

$$F(p) = \mathrm{Re}\left\{ \frac{1}{\lambda' + 3\lambda''} \frac{\lambda' + \{\lambda'(2\lambda'' - \lambda) + \lambda''(p - i\omega_0 + 3\lambda'')\}g}{(p - i\omega_0 + 3\lambda'')(1 - \lambda g) - \lambda'\lambda''g} \right\}, \tag{38}$$

where

$$g = 3(p + \lambda' + 3\lambda)\{(p + \lambda' + 3\lambda)^2 + 9q^2\}^{-1}.$$

In the limiting case of low temperatures, this reduces to

$$F(p) = \mathrm{Re}\left\{ \frac{1}{1 + 6x}\left(\frac{1}{p - i\omega_0} + \frac{6xp}{p^2 + 9q^2} \right) \right\}. \tag{39}$$

This is the line shape of a triplet: a central peak arising from the iron atoms with no nearest neighbors and a quadrupole split doublet, displaced from the central triplet, arising from those iron atoms that have an interstitial nearest neighbor.

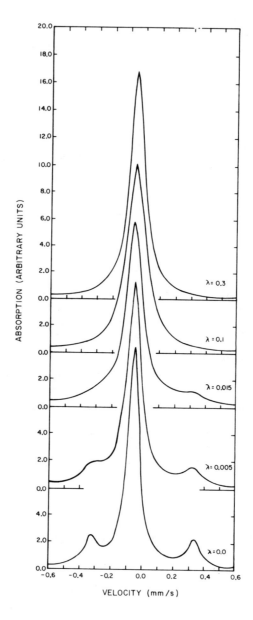

Fig. 8. Calculated line shape according to Eq. (38) for an alloy such as iron–carbon austenite with a 4 at.% interstitial content (Dattagupta, 1976a). For this case the temperatures corresponding to the values of λ in the figure are as follows:

λ	0.0	0.005	0.015	0.1	0.3
T(K)	− 273	552	567	659·	723

In the high-temperature limit, the line becomes a single sharp peak with no quadrupole splitting and an isomer shift corresponding to the average composition. The progressive change in line shape with temperature is shown in Fig. 8.

Two other limiting cases of Eq. (38) are applicable to situations of experimental importance. One is the case where the quadrupole splitting can be neglected so that q is zero, and we have

$$g = 3/(p + \lambda' + 3\lambda). \tag{40}$$

The change in line shape with temperature for this case, which is apparently appropriate for hydrogen in tantalum, is shown in Fig. 9.

The case of pure quadrupole interaction also occurs; e.g., a bound vacancy moving around an iron solute atom in silver chloride. In this situation, λ' and λ'' are zero and Eq. (38) simplifies to

$$F(p) = \text{Re} \frac{(p + 3\lambda)}{p(p + 3\lambda) + gq^2}. \tag{41}$$

This result was obtained earlier by Tjon and Blume (1968). The line shape at various temperatures for this situation is shown in Fig. 10.

A more complicated situation arises when the active atom itself moves in such a way that the hyperfine interaction changes as a result of the diffusive motion. Then we must consider the broadening that results directly from the translational motion and, as well, the potential narrowing due to the changing hyperfine interaction, as pointed out by Krivoglaz and Repetskii. Such a situation can arise in the case of substitutional diffusion in a crystal where the sites for the active atom are not cubic; this can occur locally, as in a concentrated alloy, even if the basic symmetry in the pure material is cubic. As we shall see later, however, it is unlikely that interesting effects would be observable in such a case.

Another situation that has been investigated in theory and experiment is that which occurs in liquids. The irregularity of the local environment in liquids normally results in a substantial quadrupole interaction. The orientation of the electric field gradient is subject to relatively rapid change, due to rotation of the active atom, as well as to its translation. Detailed analyses of the situation have been given by Dattagupta (1975, 1976b) for two cases: a "strong collision" model in which each jump of the atom results in the rotation of the EFG through an arbitrary angle and a model in the spirit of Stokes, where both translation and rotation are governed by related diffusion equations. The results of the second model have a greater *a priori* plausibility and are in better accord with experimental results. It is the only case we shall discuss here.

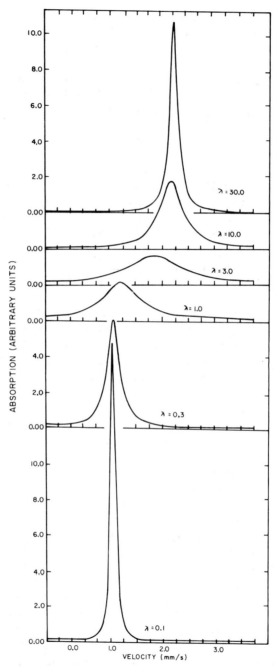

Fig. 9. Calculated line shape according to Eq. (40) for an alloy containing an interstitial that produces an isomer shift but negligible quadrupole splitting. At low temperatures a small peak with large isomer shift is also present, but off scale in the figure.

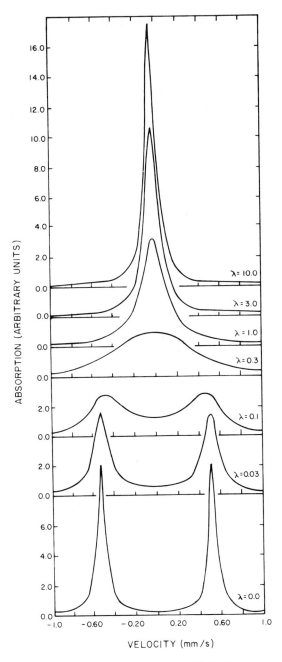

Fig. 10.   Calculated line shape according to Eq. (41) for the case of pure quadrupole interaction, e.g., a bound vacancy moving around an iron atom.

In this model, the rotational motion is described by the diffusion equation

$$\frac{\partial \Psi}{\partial t} = d \nabla_\Omega^2 \Psi, \tag{42}$$

where $\Psi(\Omega, t)$ is the probability that the nucleus has the orientation with respect to the EFG at time t and d is a rotational diffusion constant.

The time-varying hyperfine interaction is described by

$$H = \sum_{m=-2}^{2} V_{2m} Y_{2m}, \tag{43}$$

where

$$V_{20} = Q\left(\frac{4\pi}{5}\right)^{1/2} \{3I_z - I(I+1)\},$$

$$V_{2\pm 1} = \mp Q\left(\frac{6\pi}{5}\right)^{1/2} \{I_\pm I_z - I_z I_\pm\}, \tag{44}$$

$$V_{2\pm 1} = Q\left(\frac{6\pi}{5}\right)^{1/2} (I_\pm)^2,$$

and Y_{2m} is a diagonal stochastic matrix. The translational motion is assumed to occur in extremely small jumps and to be governed by the usual diffusion equation with a diffusion coefficient D, so that

$$D = \tfrac{1}{6}\lambda\langle r^2 \rangle, \tag{45}$$

where λ is the jump rate and $\langle r^2 \rangle$ is the mean-square jump distance. The line shape predicted by this model is of the form

$$F(p) = \text{Re}\left[\frac{p + D\kappa^2 + 6d}{(p + D\kappa^2 + 6d)(p + D\kappa^2) + 9Q^2}\right]. \tag{46}$$

If we now assume that the translational and rotational diffusion coefficients are related by the Stokes approximation

$$D = \frac{kT}{6\pi a\eta} \qquad d = \frac{D}{a^2}, \tag{47}$$

where a is the effective radius of the moving particle, we find

$$F(p) = \text{Re}\left[p + d\kappa^2 + \frac{9Q^2}{p + D\kappa^2 + \dfrac{\lambda\langle r^2 \rangle}{a^2}\left\{1 - \dfrac{\kappa^2\langle r^2 \rangle}{6}\right\}}\right]^{-1} \tag{48}$$

The relaxation effects are contained in the term

$$\frac{\lambda \langle r^2 \rangle}{a^2} \left\{ 1 - \frac{\kappa^2 \langle r^2 \rangle}{6} \right\}.$$

Under normal circumstances this is approximately equal to zero since $\langle r^2 \rangle$ is much smaller than a^2 and Eq. (46) reduces to

$$F(p) = \text{Re}\{ p + D\kappa^2 + (9Q^2/p + D\kappa^2) \}. \tag{49}$$

This is simply the expression for a quadrupole-split spectrum, broadened by diffusion and with no relaxation effects.

This rather puzzling disappearance of the expected collapse of the quadrupole splitting can be roughly understood on the following bases: relaxation effects depend on the coherence of the radiation during the time the interaction is changing; if the active atom moves a substantial distance during the time required for a substantial change in the hyperfine interaction, the necessary coherence is lost and no relaxation is seen. We can expect to see combined diffusive broadening and relaxation effects only if a case can be found in which a, the effective Stokes radius, is quite small, so that rotation occurs quite rapidly relative to translation. Similarly, we may expect the case of jump diffusion in a crystal with noncubic local environment to be relatively uninteresting, as mentioned previously, since the jumps necessary to change the hyperfine interaction also largely destroy the gamma-ray coherence.

IV. Experimental Results

Line broadening due to substitutional diffusion in simple metals is particularly amenable to analysis since the mechanism is reasonably well understood, and accurate measurements of the bulk diffusion coefficients are available for most cases. Mössbauer investigations have been carried out for iron in dilute solution in a variety of metals: copper (Knauer and Mullen, 1968a), gold (Knauer and Mullen, 1968b), aluminum (Sorensen and Trumpy, 1973), titanium (Lewis and Flinn, 1972), and yttrium (Carpenter and Cathey, 1977). There have also been measurements of iron-based alloys containing enough solute (silicon or vanadium) to close the gamma loop (Lewis and Flinn, 1969; Lindsey, 1976). Despite the apparent simplicity of the situation, the work has generated extensive controversy.

In their early work on the diffusion of iron in copper and gold, Mullen and co-workers reported a line broadening only about one-half as large as that they calculated on the basis of the Singwi–Sjolander approximation and bulk diffusion data. Many hypotheses have been offered to account for this discrepancy. Part of the difficulty lies in the treatment of correlation. In their analysis, Mullen *et al.* use the bulk diffusion data and the Bardeen–Herring correlation factor to calculate the jump rate of the active atoms, then insert the jump rate in the Singwi–Sjolander approximation to calculate the linewidth, with no correction for the effect of correlation in the Mössbauer case. As we saw earlier, the effect of correlation on the linewidth, although rather complicated, can be approximated reasonably well by using the Bardeen–Herring factor. This correction removes much of the discrepancy. Substantial additional narrowing is accounted for by including the effect of crystal structure and using the correct average over orientation, rather than the isotropic approximation of the Singwi–Sjolander formula. Whether a significant discrepancy still remains after these corrections is a matter of controversy.

An unambiguous resolution of the situation is made difficult by certain experimental problems. Several factors introduce uncertainties into the broadening measurements: the thin foil samples normally used are likely to have some texture, so the average over all orientations appropriate for a randomly oriented material will not be precisely correct (the large anisotropy of the broadening makes even modest deviations from randomness important); the measurement and control of sample temperature is much more difficult than for conventional bulk diffusion measurements; the thin foils used are vulnerable to composition change by evaporation or reaction during the measurement; and the "blackness effect" requires a correction, which is difficult to make accurately. This latter problem arises from the need to use rather "black," that is, highly absorbing, samples in order to obtain sufficient absorption in the upper part of the temperature range studied, where the line is not only broad but also weak, due to the small resonant fraction at high temperatures.

Methods of correcting for sample blackness are well known, but they are effective in giving reliable results only if such factors as unbroadened source linewidth, background intensity, nondiffusive broadening (such as unresolved hyperfine interactions) in the sample, sample thickness, and uniformity of sample cross section are all accurately known. Insufficient detail concerning these matters often makes it difficult to evaluate published results to the necessary accuracy.

In some alloy systems, bulk diffusion measurements alone have not led to unambiguous interpretation, and in two cases Mössbauer measurements have contributed to the understanding of the special mechanisms involved. These are both cases of unusually rapid diffusion: iron in titanium and iron in yttrium. In the case of iron in yttrium (Carpenter and Cathey,

1977), the bulk diffusion coefficient at 683 K is estimated to be $7 \times 10^{-13} m^2/s$, while at this temperature there is no observable broadening of the Mössbauer resonance, which places an upper bound on the value of D as seen by the Mössbauer effect of about $2 \times 10^{-16} m^2/s$. This discrepancy of more than three orders of magnitude is clear evidence that a simple vacancy mechanism is not appropriate here. The result is, however, quite consistent with a substitutional–interstitial mechanism already proposed for other systems where the solute is substantially smaller than the solvent atom. According to this model, most of the solute atoms are in substitutional sites at any given time and relatively immobile; while a small fraction are in interstitial sites and travel extremely rapidly. Since the Mössbauer effect sees all the atoms, the line shape is completely dominated by the vast majority of quasi-stationary atoms and no broadening is observed.

The case of iron in titanium (Lewis and Flinn, 1972), although less dramatic, is also one for which Mössbauer spectroscopy provides useful information about the diffusion mechanism in a somewhat complicated case. Both self- and solute diffusion in beta titanium have long been known to be unusually rapid and to show non-Arrhenius behavior. Various explanations have been offered, including that of rapid diffusion down dislocations. Since this would involve only a few of the iron atoms at any one time, such a mechanism would give rise to very little line broadening —a situation similar to the iron in yttrium case. In fact, the broadening observed is of the same order of magnitude as that predicted by simple theory, so that the dislocation mechanism and all other mechanisms requiring the participation of only a small fraction of the atoms at a given time are eliminated. The Mössbauer results are, however, consistent with the Kidson mechanism (Kidson, 1963), in which the rapid diffusion is a consequence of the presence of large numbers of extrinsic vacancies associated with impurities, such as oxygen, in the titanium.

Mössbauer studies of the diffusion of iron in more complicated systems have generally been less controversial since accurate predictions of expected results on the basis of bulk diffusion measurements and simple models have not been available. Howe and others (Howe et al., 1976; Howe and Morgan, 1976; Howe and Tsuji, 1977; Tsuji et al., 1977) have carried out a series of measurements on iron tellurides and selenides, which are highly anisotropic and have two different iron sites. In such cases the Mössbauer effect is particularly valuable since it can provide details unobtainable by bulk diffusion measurements. Quite strongly non-Lorentzian line shapes are observed, as shown in Fig. 11, and can be interpreted on the basis of a detailed model of the diffusive behavior.

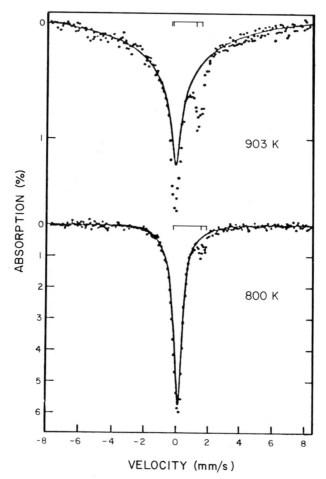

Fig. 11. Mössbauer spectrum of $Fe_{1.33}$ Te_2 (Howe and Morgan, 1976. *J. Phys.* C **9**, 4463. Copyright of The Institute of Physics). The solid curves show the non-Lorentzian broadened spectra of the $Fe_{1.33}$ Te_2, with contributions from 0.2% Fe_2SiO_4 impurity subtracted.

The diffusion of iron in wustite (FeO) has been studied by several investigators (Valov *et al.*, 1970; Greenwood and Howe, 1972; Anand and Mullen, 1973). This case is of particular interest in that the oxygen content and corresponding number of iron vacancies can be changed at fixed temperature by varying the oxygen pressure. Since the Mössbauer method is nondestructive, measurements as a function of composition and temperature can be made on a single sample. The effects of composition (and vacancy concentration) were demonstrated by the work of Greenwood and

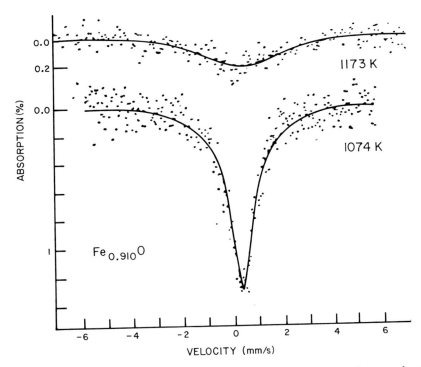

Fig. 12a.   Line broadening due to diffusion in $Fe_{0.910}O$ as observed by Greenwood and Howe (1972).

Howe, as shown in Fig. 12. Since the vacancies in Fe_xO are known to form clustered arrays, the relatively weak dependence of the broadening on composition is understandable. Mössbauer measurements of line broadening are roughly consistent with bulk diffusion measurements, but experimental disagreements about both sets of measurements, the relation between gas pressure and composition, and the complexity of the mechanism preclude a rigorous comparison.

Diffusion broadening has also been observed in CoO, but the deviations from stoichiometry are much smaller, and the diffusion much less rapid than in FeO, so that the effect is small. Song and Mullen (1975) observed a broadening of about 0.37mm/s at 1150 K, as compared with 15.2 mm/s for FeO at the same temperature.

Relaxation effects due to the motion of point defects in crystalline solids have also been observed. In the case of interstitials in metals, the predicted disappearance of quadrupole splitting for iron–carbon austenite at high

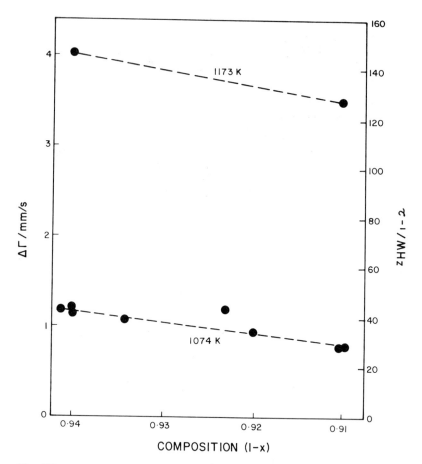

Fig. 12b. Composition dependence of the line broadening in $Fe_{(1-x)}O$ as observed by Greenwood and Howe (1972).

temperatures has been seen as shown in Fig. 13 (Lewis and Flinn, 1968), although the intermediate transition region has not yet been studied, due to the instability of austenite in the appropriate temperature range.

The extremely sharp tantalum resonance has been used to see relaxation effects due to the motion of hydrogen and deuterium (Heidemann et al., 1976, 1978). In these systems, the main effect of the hydrogen is to produce an isomer shift; resolved quadrupole splitting has not been seen. The experimental results shown in Fig. 14 are consistent with theory (cf. Fig. 9).

Another observation of relaxation effects associated with the diffusive motion of point defects was reported by Lindley and Debrunner (1966).

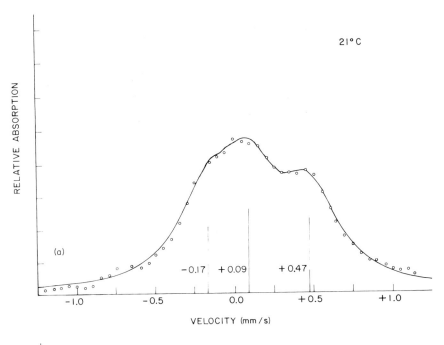

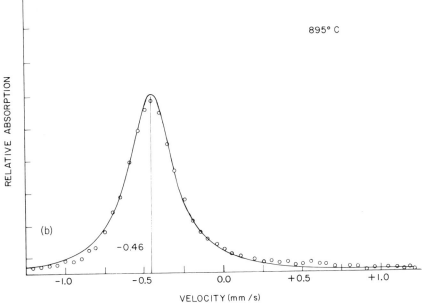

Fig. 13. (a) Mössbauer spectrum for iron–carbon austenite at room temperature (21°C), showing triplet structure as observed by Lewis and Flinn (1968). (b) Mössbauer spectrum for the same material at 895°C, showing the collapse to a singlet (compare Fig. 8).

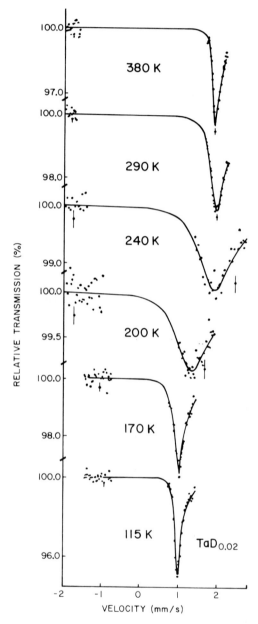

Fig. 14. Changes in the Mössbauer spectrum of Ta $D_{0.02}$ with temperature due to the diffusion of deuterium, as observed by Heidemann *et al.* (1978) (compare Fig. 9).

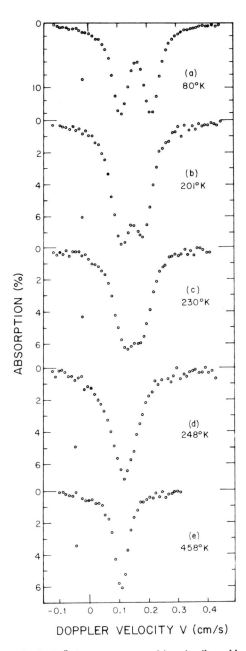

Fig. 15. Changes in the Mössbauer spectrum of iron in silver chloride due to vacancy motion about the iron atom as observed by Lindley and Debrunner (1966) (compare Fig. 10).

They studied ferrous ions present as substitutional impurities in silver chloride. Charge compensation requires a positive ion vacancy for each divalent iron ion, and the binding energy between the ferrous ion and the vacancy is large enough to keep the vacancy in one of the neighboring (100) positions. At low temperatures, the vacancies are relatively immobile and produce static electric field gradients at the iron sites, so that a normal quadrupole splitting is observed. As the temperature is increased, the vacancies jump at an increasing rate among the six (100) sites neighboring each iron position. In the temperature range 200–250 K, the quadrupole splitting collapses as shown in Fig. 15. When the diffusion rate of the vacancies is large relative to the Larmor frequency, a single unsplit line is observed. Again, the experimental results are accounted for by the general theory (cf. Fig. 10).

V. Experimental Results for Liquids

The first systematic studies of the Mössbauer effect in inorganic liquids were outgrowths of attempts to understand the frozen aqueous solution anomaly: the rather general phenomenon that on heating frozen aqueous solutions containing a Mössbauer active isotope, the spectra disappeared at some temperature well below the eutectic and reappeared after further warming, but with different characteristics. The anomaly was explained by Ruby and Pelah (1971) and Ruby (1972, 1973) who showed that the rapidly frozen solution is normally a glass, that the spectrum disappears as a consequence of the glass to liquid transition, and that the new spectrum appears as a consequence of crystallization. It also became clear that the spectrum does not abruptly disappear at the glass transition, but the rapid decrease in resonant fraction and the diffusive broadening make observations difficult. Many systems have been investigated, but in most cases only a narrow temperature range above the glass transition could be studied since crystallization of the supercooled liquid prevented the lengthy measurements needed to observe the weak and broad spectra.

The system phosphoric acid–water proved to be unusually well suited for this type of study since it is highly resistant to crystallization; carefully prepared samples can be cycled repeatedly through the glass–liquid transition and kept for months in the supercooled liquid region without crystallization. The eutectic temperature is low enough so that measurements can be made in the equilibrium as well as in the supercooled liquid. Some spectra are shown in Fig. 16. The general features are in accord with the theory given earlier: below the glass transition (independently determined

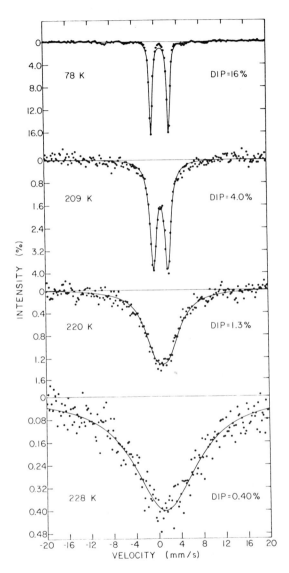

Fig. 16.   Mössbauer spectrum for ^{57}Fe dissolved in $H_3PO_4 + H_2O$ at various temperatures as observed by Flinn *et al.* (1976). The glass-transition temperature of this composition is 155 K. Note the rapid increase in width and decrease in area of the absorption.

by specific volume measurements), the spectrum shows the same temperature dependence as that of a crystalline solid; above the transition, a rapid drop in resonant fraction and a rapid increase in linewidth are observed. A comparison of bulk diffusion measurements and the Mössbauer line broadening results showed that diffusion occurs by extremely small steps, rather than the large jumps of a quasi-crystalline model for liquids (Ruby et al., 1975), since the observed broadening was given by $\kappa^2 D$ [cf. Eq. (29)] rather than by $6D/\langle x^2 \rangle$.

The characteristics of Mössbauer spectra from organic liquids are similar to those from aqueous solutions. Again, many systems have been investigated, but only in a few cases has it been possible to carry out the measurements over an extended temperature range in the liquid. Enriched ferrocene has proved to be a particularly useful solute: it dissolves readily in a variety of solvents, some of which show relatively slow diffusion and correspondingly little diffusive broadening, thus extending the temperature range over which experiments are feasible (Ruby et al., 1976). Ferrocene in o-terphenyl represents an extreme case where no significant diffusive broadening is observable, even at a temperature 16 degrees above the glass transition (Vasquez and Flinn, 1980).

The field of liquid crystals has been extensively investigated by Mössbauer techniques for many years. It provides the intriguing possibility of studying anisotropic liquids, with the orientation of the system controllable by the application of an external magnetic field. Unfortunately, there are also serious difficulties: the systems are complex, the glass transition may be sluggish, and crystallization can occur with inconvenient rapidity. Some early work was later shown to be erroneous: effects attributed to liquid crystals were, in fact, due to suspensions of small crystals.

In spite of the difficulties, some extremely interesting results have been obtained. For example, La Price and Uhrich (1979) have observed effects on both the resonant fraction and diffusive broadening, which are in accord with the general picture for liquids. They investigated the system of 1, 1'-diacetylferrocene (DAF) in n-(p-methoxybenzylidene)-p-butylaniline (MBBA). This system exhibits a glass transition, as determined by calorimetric means, at about 200 K. The transition is not very sharp: it extends over about 10 K. Both liquid and glass are nematic; that is, the molecules have their principal axes parallel, but are otherwise disordered. The solutions were oriented by cooling in a 9 kGs magnetic field. Below about 180 K, the temperature dependence of the resonant fraction showed not only the normal solidlike behavior, but also an orientation dependence: the resonant fraction was larger when the long axis of the molecules was aligned with the gamma-ray direction. Above about 180 K, the rapid drop in resonant fraction characteristics of liquids was observed. The fact that this drop-off begins 20 K below the glass-transition temperature, as determined by calorimetric means, is attributed by the authors to structural complexities of the system; however, the discrepancy could also be due to sluggishness in the transformation. The Mössbauer measurement, being quasi-static, can be expected to give the "true" T_G, while a dynamic method, such as DTA, can easily give a high result for a sluggish transformation.

No line broadening is observed in the lower part of the liquid range (180–220 K), indicating that translational diffusion is too slow to observe. Above 220 K, a rapid increase in linewidth and a rapid decrease in quadrupole splitting are observed. These effects are attributed to rotational diffusion and are in accord with the theory of Dattagupta in the limiting case where the translational diffusion is negligible.

VI. Summary

The theoretical framework necessary for quantitative interpretation of Mössbauer spectra from materials in which diffusive motion is occurring is now reasonably complete. Detailed analyses exist for the broadening due to diffusive motion in solids and liquids of the atoms containing the resonant nuclei, and also for relaxation effects associated with these motions and with motions of neighboring atoms. Mössbauer spectroscopy can now be regarded as a well-developed addition to the set of techniques available for the measurement of diffusive effects. It is, unfortunately, somewhat limited, both in range in D values which are accessible, to the upper range that is found in solids and the very lowest range found in liquids, and in materials suitable for study since only a few isotopes can be used. Obviously, it is unlikely to supplant other methods; its main applications will probably be in conjunction with other methods, such as tracer, NMR, or perturbed angular correlation for the elucidation of mechanism in complicated or ill-understood situations. One or more of its special features: it is a steady-state, nondestructive measurement, does not require single crystals, can be used to study the effects of thermal or other environmental history, and is sensitive to anisotropic phenomena; these are likely to be valuable in a variety of cases.

Mössbauer investigation of diffusive effects will continue to present a formidable challenge to experimenters. In addition to the obvious difficulties of making accurate measurements, including line shape, of weak, greatly broadened resonances, and the problems of working at elevated temperatures in an awkward geometry, with the requirement of accurate temperature measurement, there are additional concerns. For work with polycrystalline materials, it is essential that the crystallographic texture of the samples be accurately determined, so that the proper average over orientation can be used. Since crystal growth can occur during the experiment, x-ray study of the sample after, as well as before, the Mössbauer measurements may be needed. Work on single crystals offers a particular challenge: sufficiently thin crystals of controlled orientation are quite difficult to prepare and handle; in addition, the sharp angular dependence of the anisotropy necessitates the use of solid angles of observation much smaller than those customary in Mössbauer work. The consequent loss of counting rate is hardly welcome in an investigation of a weak and broad resonance. Nonetheless, the extra detail available from single crystal work should make the effort well worthwhile.

Since in many cases the full utility of the Mössbauer work is obtained only when it is combined with the results of other techniques, such as tracer measurements of bulk diffusion, cooperation between investigators versed in various techniques can be quite valuable. Uncertainties in the interpretation of combined results associated with such problems as possible differences in impurity content, ambient atmosphere, temperature calibration, and so on can be greatly reduced by appropriate cooperative effort. By such efforts, we may anticipate a much better understanding of diffusion in solids and liquids than has been possible without the use of Mössbauer spectroscopy.

References

Anand, H. R., and Mullen, J. G. (1973). *Phys. Rev. B* **8**, 3112.
Bender, O., and Schroeder, K. (1979). *Phys. Rev. B* **19**, 3399.
Blume, M. (1968). *Phys. Rev.* **174**, 351.
Boyle, A. J. F., and Hall, H. E. (1962). *Rep. Prog. Phys.* **25**, 441.
Bunbury, D. St. P., Elliot, J. A., Hall, H. E., and Williams, J. M. (1963). *Phys. Lett.* **6**, 34.
Carpenter, J. S., and Cathey, W. N. (1977). *Phys. Lett.* **64A**, 313.
Chudley, C. T., and Elliot, R. J. (1961). *Proc. Phys. Soc. London* **77**, 353.
Dattagupta, S. (1975). *Phys. Rev. B* **12**, 47.
Dattagupta, S. (1976a). *Phil. Mag.* **33**, 59.
Dattagupta, S. (1976b). *Phys. Rev. B* **14**, 1329.
Dibar Ure, M. C., and Flinn, P. A. (1973). *Appl. Phys. Lett.* **23**, 587.
Dibar Ure, M. C., and Flinn, P. A. (1977). *Phys. Rev. B* **15**, 1261.
Egelstaff, P. A. (1967). "An Introduction to the Liquid State," Chapter 11. Academic Press, New York.
Flinn, P. A. (1977). *In Proc. Nassau Mössbauer Conf.* (C. I. Wynter and R. H. Herber, eds.), p. 21. Nassau Community College Press, Long Island.
Flinn, P. A., Zabransky, B. J., and Ruby, S. L. (1976). *J. Phys. (Paris) Colloq.* **C6 37**, 739.

Greenwood, N. N., and Howe, A. T. (1972) *J. Chem. Soc. Dalton* 122.

Heidemann, A., Kaindl, G., Salomon, D., Wipf, H., and Wortmann, G. (1976). *Phys. Rev. Lett.* **26**, 213.

Heidemann, A., Wipf, H., and Wortmann, G. (1978). *Hyperfine Interact.* **4**, 844.

Howe, A. T., and Morgan, G. J. (1976). *J. Phys. C* **9**, 4463.

Howe, A. T., and Tsuji, T. (1977). *J. Solid State Chem.* **21**, 91.

Howe, A. T., Coffin, P., and Fender, B. E. F. (1976). *J. Phys. C* **9**, L61.

Karyagin, S. V. (1974). *Phys. Lett.* **49A**, 183.

Karyagin, S. V. (1975). *Sov. Phys. Solid State* **17**, 1220.

Kidson, G. V. (1963). *Can. J. Phys.* **41**, 1563.

Knauer, R. C., and Mullen, J. G. (1968a). *Phys. Rev.* **174**, 711.

Knauer, R. C., and Mullen, J. G. (1968b). *Appl. Phys. Lett.* **13**, 150.

Krivoglaz, M. A., and Repetskii, S. P. (1966). *Fiz. Tverd. Tela* **8**, 2908 [*English transl.*: *Sov. Phys.–Solid State* **8**, 2325 (1967)].

Krivoglaz, M. A., and Repetskii, S. P. (1971). *Phys. Metall. Metallogr.* **32**, 1.

La Price, W. J., and Uhrich, D. L. (1979). *J. Chem. Phys.* **71**, 1498.

Lewis, S. J., and Flinn, P. A. (1968). *Phys. Status Solidi.* **26**, K51.

Lewis, S. J., and Flinn, P. A. (1969). *Appl. Phys. Lett.* **15**, 331.

Lewis, S. J., and Flinn, P. A. (1972). *Phil. Mag.* **26**, 977.

Lindley, D. H., and Debrunner, P. G. (1966). *Phys. Rev.* **146**, 199.

Lindsey, R. (1976). *Phys. Status Solidi (b)* **75**, 583.

Rahman, A. (1971). *In* "Interatomic Potentials and Simulation of Lattice Defects" (P. C. Gehlen *et al.*, eds.), p. 233. Plenum Press, New York.

Ruby, S. L. (1972). *J. Non-Cryst. Solids* **8**, 78.

Ruby, S. L. (1973). *In* "Perspectives in Mössbauer Spectroscopy" (S. G. Cohen and M. Pasternak, eds.), p. 181. Plenum Press, New York.

Ruby, S. L., and Pelah, I. (1971). *In* "Mössbauer Effect Methodology" (I. J. Gruverman, ed.), Vol. 6, p. 21. Plenum Press, New York.

Ruby, S. L., Love, J. C., Flinn, P. A., and Zabransky, B. J. (1975). *Appl. Phys. Lett.* **27**, 320.

Ruby, S. L., Zabransky, B. J., and Flinn, P. A. (1976). *J. Phys. (Paris) Colloq.* C6 **37**, 745.

Singwi, K. S., and Sjölander, A. (1960). *Phys. Rev.* **120**, 1093.

Song, C., and Mullen, J. G. (1975). *Solid State Commun.* **17**, 549.

Sorensen, K., and Trumpy, G. (1973). *Phys. Rev. B* **7**, 1791.

Tjon, J. A., and Blume, M. (1968). *Phys. Rev.* **165**, 456.

Tsuji, T., Howe, A. T., and Greenwood, N. N. (1977). *J. Solid State Chem.* **20**, 287.

Valov, P. M., Vasil'ev, Y. V., Veriovkin, G. V., and Kaplin, D. F. (1970). *J. Solid State Chem.* **1**, 215.

Vasquez, A., and Flinn, P. A. (1980). *J. Chem. Phys.* **72**, 1958.

Wolf, D. (1977). *Appl. Phys. Lett.* **30**, 617.

Index

This index was compiled
using the UNIXTm
operating system.
UNIX is a trademark of
Bell Laboratories.